MICROMANUFACTURING

MICROMANUFACTURING

International Research and Development

by

KORNEL F. EHMANN
Northwestern University, Evanston, IL,U.S.A..

DAVID BOURELL
University of Texas, Austin, TX, U.S.A.

MARTIN L. CULPEPPER
Massachusetts Institute of Technology, Camridge, MA, U.S.A.

THOM J. HODGSON
North Carolina State University, Raleigh, NC, U.S.A

THOMAS R. KURFESS
Clemson University, Clemson, SC, U.S.A.

MARC MADOU
University of California, Irvine, CA, U.S.A.

KAMLAKAR RAJURKAR
University of Nebraska, Lincoln, NE, U.S.A.

and

RICHARD DeVOR
University of Illinois, Urbana, IL, U.S.A.

 Springer

A C.I.P. Catalogue record for this book is available from the Library of Congress.

ISBN-10 1-4020-5948-5 (HB)
ISBN-13 978-1-4020-5948-3 (HB)
ISBN-10 1-4020-5940-3 (e-book)
ISBN-13 978-1-4020-5949-0 (e-book)

Published by Springer,
P.O. Box 17, 3300 AA Dordrecht, The Netherlands.

www.springer.com

This document was sponsored by the National Science Foundation (NSF) and other
agencies of the U.S. Government under an award from NSF (ENG-0423742) to the
World Technology Evaluation Center, Inc. The Government has certain rights in this
material. Any opinions, findings, and conclusions or recommendations expressed in
this material are those of the authors and do not necessarily reflect the views of the
United States Government, the authors' parent institutions, or WTEC, Inc.

Copyright to electronic versions by WTEC, Inc. and Springer except as noted. WTEC,
Inc. retains rights to distribute its reports electronically. The U.S. Government retains a
nonexclusive and nontransferable license to exercise all exclusive rights provided by
copyright. All WTEC final reports are distributed on the Internet at
http://www.wtec.org. Some WTEC reports are distributed on paper by the National
Technical Information Service (NTIS) of the U.S. Department of Commerce.

Printed on acid-free paper

WTEC Panel on Micromanufacturing

Kornel F. Ehmann (chair)
Northwestern University, Evanston, IL
David Bourell
University of Texas, Austin, TX
Martin L. Culpepper
Massachusetts Institute of Technology, Cambridge, MA
Thom J. Hodgson
North Carolina State University, Raleigh, NC
Thomas R. Kurfess
Clemson University, Clemson, SC
Marc Madou
University of California, Irvine, CA
Kamlakar Rajurkar
University of Nebraska, Lincoln, NE
Richard E. DeVor
University of Illinois, Urbana, IL

Abstract

This international technology assessment study has focused on the emerging global trend toward the miniaturization of manufacturing processes, equipment and systems for microscale components and products, i.e., *small equipment for small parts*. It encompasses the creation of miniaturized units or hybrid processes integrated with metrology, material handling, and assembly to create microfactories capable of producing microprecision products in a fully automated manner at low cost. The study has investigated both the state-of-the-art as well as emerging technologies from the scientific, technological, and commercialization perspectives across key industrial sectors in the United States, Asia, and Europe including medical, electronics, aerospace, and consumer products. This study does *not* include the lithographic-based processes common to the microelectromechanical systems (MEMS) community. The United States receives satisfactory marks for nanotechnology R&D, but its micromanufacturing R&D is lagging behind the rest of the world, particularly in technology transfer and ongoing development. This will undoubtedly have serious long-term implications since it is well-recognized that micromanufacturing will be a critical enabling technology in bridging the gap between nanoscience and technology developments and their realization in useful products and processes. While examples do exist where U.S. government programs are focused on industry-university-government collaboration, the scale of efforts both in Asia and Europe is significantly larger. On this latter point, Europe appears to be very strong, particularly as these partnerships work to refine and fine-tune developments for industry adaptation and commercialization.

WTEC MISSION

WTEC provides comparative assessments of international research and development in selected technologies under awards from the National Science Foundation, the Office of Naval Research, and other agencies. Formerly part of Loyola College, WTEC is now a separate nonprofit research institute. Michael Reischman, Deputy Assistant Director for Engineering, is NSF Program Director for WTEC. Sponsors interested in international technology assessments and related studies can provide support for the program through NSF, or directly through separate grants to WTEC.

WTEC's mission is to inform U.S. scientists, engineers, and policymakers of global trends in science and technology. WTEC assessments cover basic research, advanced development, and applications. Panels of typically six technical experts conduct WTEC assessments. Panelists are leading authorities in their field, technically active, and knowledgeable about U.S. and foreign research programs. As part of the assessment process, panels visit and carry out extensive discussions with foreign scientists and engineers in their labs.

The WTEC staff helps select topics, recruits expert panelists, arranges study visits to foreign laboratories, organizes workshop presentations, and finally, edits and disseminates the final reports.

Executive Editor: Marvin Cassman

Series Editor: R. D. Shelton

WORLD TECHNOLOGY EVALUATION CENTER, INC. (WTEC)

R. D. Shelton, President
Michael J. DeHaemer, Executive Vice President
Y. T. Chien, Vice President for Research
Geoffrey M. Holdridge, Vice President for Government Services

Hassan Ali, Director of International Study Operations
Roan Horning, Director of Information Technology
Maria DeCastro, Director of Publications
Scott Thomason, Editor
Advance work by Erika Feulner and Gerald Hane

TABLE OF CONTENTS

D. Site Reports—Europe

FOREWORD

> We have come to know that our ability to survive and grow
> as a nation to a very large degree depends upon our scien-
> tific progress. Moreover, it is not enough simply to keep
> abreast of the rest of the world in scientific matters. We
> must maintain our leadership.[1]

President Harry Truman spoke those words in 1950, in the aftermath of
World War II and in the midst of the Cold War. Indeed, the scientific and
engineering leadership of the United States and its allies in the twentieth
century played key roles in the successful outcomes of both World War II
and the Cold War, sparing the world the twin horrors of fascism and totali-
tarian communism, and fueling the economic prosperity that followed. Today, as the United States and its allies once again find themselves at war,
President Truman's words ring as true as they did a half-century ago. The
goal set out in the Truman Administration of maintaining leadership in science has remained the policy of the U.S. government to this day. Dr. John
Marburger, the Director of the Office of Science and Technology (OSTP)
in the Executive Office of the President, made remarks to that effect during
his confirmation hearings in October 2001.[2]

The United States needs metrics for measuring its success in meeting
this goal of maintaining leadership in science and technology. That is one
of the reasons that the National Science Foundation (NSF) and many other
agencies of the U.S. government have supported the World Technology
Evaluation Center (WTEC) and its predecessor programs for the past 20
years. While other programs have attempted to measure the international
competitiveness of U.S. research by comparing funding amounts, publication statistics, or patent activity, WTEC has been the most significant public domain effort in the U.S. government to use peer review to evaluate the
status of U.S. efforts in comparison to those abroad. Since 1983, WTEC
has conducted over 60 such assessments in a wide variety of fields, from
advanced computing, to nanoscience and technology, to biotechnology.

The results have been extremely useful to NSF and other agencies in
evaluating ongoing research programs and in setting objectives for the future. WTEC studies also have been important in establishing new lines of

---

[1] Remarks by the President on May 10, 1950, on the occasion of the signing of the law that
created the National Science Foundation. *Public Papers of the Presidents* 120: p. 338.

[2] http://www.ostp.gov/html/01_1012.html.

xi

communication and identifying opportunities for cooperation between U.S. researchers and their colleagues abroad, thus helping to accelerate the progress of science and technology within the international community. WTEC is an excellent example of cooperation among the many agencies of the U.S. government that are involved in funding research and development; almost every WTEC study has been supported by a coalition of agencies with interests related to the particular subject at hand.

As President Truman said over 50 years ago, our very survival depends upon continued leadership in science and technology. WTEC plays a key role in determining whether the United States is meeting that challenge, and in promoting that leadership.

Michael Reischman
Deputy Assistant Director for Engineering
National Science Foundation

PREFACE

Over the history of technological invention, we have seen many successful applications of the notion *larger is better*. The Boeing jumbo-jet 747, the SUV, large-screen TV and Nimitz-class super-carriers are good examples of this concept. But, with recent advances in basic sciences, the concept that *smaller is better* has been receiving increasing attention from researchers and practitioners. The entire balloon of nanotechnology is the direct outcome of this notion. However, it is interesting to note what is in the middle of these size ranges—the larger range on the order of 10 to 10^3, and the smaller range on the order of 10^{-8} to 10^{-9} meters. We refer to the size range of centimeters to micrometers as the meso/micro scale. This is about the size of the tiny robots in Steven Spielburg's science fiction film, *Minority Report*, in which tiny devices individually and collectively, in a distributed manner, monitored an entire human society.

Inspired by how ants work, the concept of *small equipment for small parts or even large parts* grew out of Japan in mid-1990s in which four discrete manufacturing processes were fit into a portable, less than 0.12 m^3 space, 34 kg suitcase. Imagine the efficiency and the effectiveness of a similar machine located in a hospital operating room where doctors could create a customized implant just right for your needs during the operation. Imagine such modular, but highly reconfigurable, equipment located in your neighborhood where a customized design could be realized from art-to-part with required precision and functionality. Imagine how such equipment could empower individual creativity, the underlying reason why the United States has dominated and grown in this very competitive global market. Perhaps it is as hard to imagine this much like it was hard to imagine the future of computers back in the 1970s.

Recognizing the enabling nature of small, highly flexible and reconfigurable equipment for the United States in the competitive global economy, the National Science Foundation (NSF) in partnership with the Office of Naval Research (ONR), Department of Energy (DOE), and the National Institute of Standards and Technology (NIST), commissioned a worldwide study on the status of micromanufacturing with particular emphasis on the United States, Europe and Asia to be conducted by the World Technology Evaluation Center (WTEC). In the context of this study, micro and meso-scale manufacturing refers to manufacturing processes and systems capable of fabricating parts with three-dimensional micro-scale features and high relative accuracy (10^{-3} to 10^{-5} meters) from a wide range of engineering materials, including stainless steel, titanium, brass, aluminum, platinum, iridium, plastics, ceramics and composites. This size range is a critical link between the nano and macro worlds.

During this project, I had the privilege to work with many dedicated program directors/managers at various government agencies who care deeply about how the U.S. is moving in terms of technological advances and how to best support our advancement in manufacturing and develop our young talent. Many thanks go to the following individuals for their continued support and participation in this WTEC study: Drs. Khershed Cooper and Ralph Wachter of ONR, Dr. Phyllis Yoshida of DOE, Dr. Amit Bagchi of NIST/ATP, Dr. Michael Reischman of NSF Directorate for Engineering, Drs. George Hazelrigg, Delcie Durham, Warren DeVries and Kevin Lyons of NSF/DMI (Division of Manufacturing and Innovations), Drs. Yip-Wah Chung, Masayoshi Tomizuka and Mario Rotea of NSF/CMS (Civil Mechanical Systems), Dr. Alfonso Ortega of NSF/CTS (Chemical and Transport Systems), Dr. Lynn Preston of NSF/EEC (Engineering Education & Centers), Dr. Bruce Hamilton of NSF/BES (Bioengineering & Environmental Systems) and Dr. Sreeramamurthy (Rama) Ankem of NSF/DMR (Division of Materials Research).

Any success attributed to this study and this book could not have been achieved without the intellectual contributions and the great devotion of panel members during this year-long study. As a government participant representing the National Science Foundation, I accompanied the group on site visits and participated with them in report writing and planning of events and can attest to the panel's dedication. At this moment, it is particularly important for me to thank Prof. Ehmann for his extraordinary leadership, Prof. DeVor for his unmatchable wisdom, Prof. Bourell for bridging materials to processes, Prof. Culpepper for his now famous quote on design at the micro/meso scale, "That which was not seen tells the story better than that which was observed," Prof. Hodgson for his expertise in operational systems, Prof. Kurfess for his in-depth knowledge of machine design and metrology, Prof. Madou for his breadth in many different fields, and Prof. Rajurkar for his broad insights on the landscape of manufacturing processes. Last, but not least, I would like to thank WTEC personnel and consultants for their excellence in managing this study under the leadership of Dr. Shelton, particularly, Roan Horning, Gerald Hane and Hassan Ali. All these people were so knowledgeable and so efficient that it has made the work extremely enjoyable.

Looking back, it has been quite an exciting journey from the encouragement of Dr. Delcie Durham—"Do something that you feel passion about and that you can enjoy along the way"—on a quiet day in early December at her NSF office, to the first one-hour meeting in the afternoon of Dec. 19, 2003 with Dr. Robert D. Shelton, President of WTEC, and Dr. Y. T. Chien, VP of Research at WTEC, to the interagency exploratory meeting on March 11, 2004, which led to the workshops and site visits in 2004 and 2005, and finally to the publication of this book now in 2006. During and after the WTEC study trips, the interest from the international community on this particular topic has been overwhelming. A striking example is Takashima Sangyo Co. Ltd.'s Desktop Manufacturing (DM) plant which consists of about 120 desktop-sized machines operating in a mere 300 m^2 factory developed since our December 2004 visit to Asia. Concurrently, numerous Japanese companies are starting to offer specialized products ranging from assembly, joining, metrology, to processing and other equipment, along with supporting component technology products such as sensors, actuators, and controllers that support the Desktop Manufacturing Paradigm. Foreign government activities include, but are not limited to: a) renewed funding by the Japanese government for AIST's efforts with a consortium of companies to develop the Desk Top Factory (DTF) concept and lines of supporting processes and equipment; b) the large national Microfactory Program in Korea, headed by the Korea Institute of Metals and Machinery (KIMM); c) the large scale project funded by the European Union on Evolvable Ultra-Precision Assembly Systems (EUPASS) focusing on micro-assembly headed by Philips Applied Technologies with participation from key European universities and leading companies; d) another large EU project, MasMicro, focused on developing various desktop machines for mass production of microparts. The funding level for each of these activities is in the tens of millions of dollars over a three to five year span.

Globalization has pushed manufacturing into an economically lean mode. This is the time to invest and to find the competitive edge with which the U.S. can lead and can excel. It is not time to abdicate our leadership position to others. A large nation like the United States should continuously contribute to the manufacturing science base and the technologies that transfer raw materials and energy into products that people can use to enrich their lives and that of the environment that surrounds us.

Science fiction frequently stimulates imagination long before the realization of physical inventions. In the famous Chinese traditional novel, "Journey to the West," written by Wu Chen-En circa 1590, Monkey King, in a single leap, could travel as much as 33,000 miles (more than enough to go around the globe), had eyesight like an X-ray, and many other superpowers that modern technologies are still trying to realize. In the fictional Star Trek universe, a *replicator* can convert energy into matter—any inanimate matter as long as the desired molecular structure is on file. Likewise, I hope that you will enjoy the rest of this book and allow your imagination to run unbridled as the future of manufacturing unfolds.

Jian Cao, Ph.D.
Department of Mechanical Engineering
Northwestern University
Evanston, IL, U.S.A.

August 2006

LIST OF FIGURES

LIST OF TABLES

Executive Summary

Kornel Ehmann and Richard DeVor

PURPOSE AND SCOPE OF STUDY

In an effort to better understand the current status and emerging directions of R&D efforts in micromanufacturing worldwide, the National Science Foundation (NSF), the Office of Naval Research (ONR), the Department of Energy (DOE), and the National Institute of Standards and Technology (NIST) have commissioned a study by a team of U.S. experts. The team first organized a workshop in August, 2004 to survey U.S. activities in the field. The team then visited selected government, industry, and university sites in both Asia (Japan, Korea, and Taiwan) and Europe (Austria, Germany, Netherlands and Switzerland), conducted under the auspices of the World Technology Evaluation Center (WTEC). Detailed site visit reports can be found in Appendices C and D. The sponsors of this study selected a panel of experts to make the site visits and prepare the report. The expertise of the panelists spans the range of issues to be examined including design, materials, processing, metrology, applications, and business and economics. Detailed biographical information on the panelists can be found in Appendix A.

This international technology assessment study has focused on the emerging global trend toward the miniaturization of manufacturing equipment and systems for microscale components and products, i.e., *Small Equipment for Small Parts*. This trend is referred to with increasing frequency as the Microfactory Manufacturing Paradigm or Desktop Manufacturing Paradigm. It encompasses the creation of miniaturized unit or hybrid processes integrated with metrology, material handling and assembly to create microfactories capable of producing microprecision products in a fully-automated manner at low cost. The study has investigated both the state-of-the-art, as well as emerging technologies from the scientific, technological, and commercialization perspectives across key industrial sectors in Asia and Europe including medical, electronics, aerospace, and consumer products. This study does NOT include the lithographic-based processes common to the microelectromechanical systems (MEMS) community. This related topic was previously the subject of a similar WTEC study, "Microsystems Research in Japan," published by WTEC in September of 2003.

In planning for the study, and in order to bring into focus the issues for which the panel sought answers, a series of questions were crafted by the panel and sent in advance to the hosts at each site to be visited. These questions can be found in Appendix B. The guiding principles driving the creation of this set of questions can be broadly classified into three categories that relate to miniaturization and for which answers have been sought. These are:

Scientific

- Impact of scaling laws on manufacturing processes/equipment
- State-of-the-science; gaps, deficiencies and needs in fundamental process knowledge
- Understanding of multi-disciplinary science-based requirements

Technological

- Driving forces for miniaturization needs
- State-of-the-art; gaps, deficiencies and needs for miniaturization of manufacturing
- Bridging between scales; nano to micro to macro
- Results from proof-of-concept testbeds

Commercialization

- Understanding principal current and future applications
- Economics of microscale manufacturing
- Societal benefits and broad-based impact of miniaturization
- Possibility of creating a disruptive manufacturing technology
- Results from proof-of-concept testbeds

SUMMARY OBSERVATIONS AND INSIGHTS

What follows is a synopsis of the findings of the panel given in terms of the state of worldwide R&D initiatives, specific technology trends and observations, and the interactions among universities, government institutes and labs, and private industry in the development of micromanufacturing technologies.

Worldwide Micromanufacturing Initiatives

1. Emerging miniaturization technologies are clearly driving developments in microscale processes, machines, metrology to meet needs re-

lated to part size, feature definition, accuracy and precision, and materials development. The study has revealed that there is a lot of activity in both Asia and Europe in this regard.

2. In both Asia and Europe, starting with MEMS, the approach tends to be more mechanics-centric rather than electronics-centric in nature. While 2D lithography-based technology development is present, it is not dominant, i.e., there is a more balanced approach. R&D tends to be very product-oriented with patient and sustained efforts aimed at refining the technology, i.e., more emphasis on *down-to-earth* as opposed to *blue-sky* initiatives.

3. The trend toward miniaturization of machines is evident in both Asia and Europe, with commercialization of desktop machine tools, assembly systems, and measurement systems well underway. The Microfactory Paradigm is more evident in Asia than Europe with several concept systems developed as far back as ten years ago. For example, in Japan, at the National Institute for Advanced Industrial Science and Technology (AIST), strong efforts have been directed toward developing small-scale micromanufacturing machine tools and microfactories. In Europe, the focus is more at the machine/process level.

4. Both in Asia and in Europe, many of the issues under study and technical challenges embraced, were those identified at the U.S. workshop in August of 2004, viz., the need for smaller-scale machines, the need for multi-functional hybrid machines, attention to part handling and fixturing, the integration of metrology and processing, and the need for microscale process development.

Technology-based Trends and Observations

Summaries are provided in this section related specifically to the technology trends observed during the Asian and European site visits.

Design

1. The results of the assessment indicate the state-of-the-art in technologies that support the design of/for non-lithography-based micro/mesoscale parts is far from ready to provide adequate support for designers, due in large part to the nascent state of the technology and the fact that design researchers have yet to become aware of the design challenges in this field.

2. Both Asia and Europe show evidence of well-funded and focused efforts that are aimed at developing nanoscale and microscale design knowledge. There are a few efforts that are targeted at understanding how to simultaneously model and simulate multi-scale and multi-physics in engineering applications.

3. It was clear from the site visits in both Asia and Europe that standards for micro/mesoscale parts are currently in an early stage and require better definition. The standards for measurement and evaluation of part characteristics are of particular importance. Without these standards, it will be difficult for designers to talk with others (e.g. vendors, customers, etc.) about the specifications that drive the design and fabrication of their parts.

4. At present, the gap between the existing and the ideal in the application of stochastic techniques in micro/mesoscale design appears to be a *practice* gap. Designers simply are not using these powerful techniques. This was quite apparent from the visits to both Asia and Europe. The most important recommendation to be made here is to increase awareness of the benefits of stochastic methods and how they should be used during the design process.

5. In the area of design modeling tools, it was evident that European designers were focused upon generating accurate and robust modeling tools. The most successful designers were observed to have utilized a hybrid approach in which they augmented existing software. This indicates a technology gap that is probably best addressed by software vendors.

Materials

1. The visits to Asia and Europe revealed that typically, materials used for micromanufacturing are the same as those used in macromanufacturing. They encompass the full range of metals, polymers and ceramics/glasses. However, a feature unique to micromanufactured materials is the need for clean, inclusion-free materials.

2. Many efforts were found to be focused on improving the understanding of material behavior at smaller-size scales and how this would affect fabrication processes. In particular, grain size effects were found to be heavily researched, including effects of grain size on machinability, surface finish, and materials properties.

3. A major materials issue that was identified particular to micromanufacturing is the lack of an economic driving force for materials development, primarily due to small quantity needs.

Processes

1. The visits to industries, universities and research organizations in Asia and Europe revealed a broad spectrum of processing methods and equipment now in use for microscale manufacturing. It was observed that many microscale components/products are being manufactured using existing macroscale or reduced-size precision manufacturing proc-

esses and equipment. This approach is, however, exposing the difficulties related to the smallest unit of amount of material removal, addition and forming per cycle and achievable precision. In particular, issues such as material properties, generation and delivery of small amounts of energy, the effects of scaling on the process mechanism, process-material interactions, and related heat transfer issues are being revealed.

2. In both Asia and Europe the visits revealed a tremendous amount of activity on the miniaturization of processing equipment. Numerous examples could be found of both R&D activity and commercialization efforts in developing reduced-size and desktop/tabletop-size processing equipment and systems, with particular emphasis on multi-functional machines, e.g., processing, assembly, and metrology. This includes considerable evidence of interest and activity in the microfactory paradigm.

3. At the same time, the site visits revealed a great deal of activity in new process development to address specific needs and issues in microscale manufacturing. This was particularly evident in Japan, Taiwan, and Korea where such development frequently crosses the boundaries of mechanical, electrical, and chemical methods and encompasses technologies developed for both MEMS- and non-MEMS-related applications.

4. While the emphasis on new process development was found to be strong, activity directed at achieving a fundamental understanding of the mechanisms and performance characteristics of these new processes based on first principles and modeling efforts was proportionally less evident when compared to device development and experimentally based performance evaluation.

5. The Asian and Europe site visits revealed that issues related to process modeling and simulation, process-material interactions, monitoring and control, process capabilities, tool and equipment design, metrology, economics, and application, are yet to be fully addressed.

6. The visits demonstrated a belief that processes performed in a desktop factory environment could have a dramatic impact on society. Sankyo Seiki, for example, believes that its desktop factories (DTFs) might revive manufacturing in Japan and Korea. The Japanese Government has just started a new desktop factory project.

7. Although there is a great deal of past and continuing research on the directed assembly of small-scale parts (much of it driven by fiber optics and microphotonics), it was evident from the Asian and European

visits that there is a need for improved assembly and integration technology at the microscale.

Metrology, Sensors and Control

1. It was observed that there is a variety of metrology systems available for microcomponent inspection. However, few of these systems are three-dimensional in nature. Furthermore, all of the systems are relatively slow and expensive, making them reasonable choices for research and development but less than desirable choices for production lines from both a robustness perspective, as well as an inspection speed perspective.

2. Many of the standards that are applicable to macroscale metrology are not available for microscale metrology. Tools such as interferometers, ball bars and even gage blocks and gage balls are not available at the microlevel for testing and calibrating micrometrology systems. Thus, calibrating these systems and determining their capability is limited.

3. A variety of sensors for microcomponent metrology were observed during the visits. Many of them circumvent calibration issues, which are difficult to address, by either incorporating their own standards, or by employing procedures that generate data that can be interpreted without the need for precise calibration.

4. While machine control has progressed quite well, process control has not. This is primarily due to a lack of models, process understanding, and experience. Thus, significant efforts are needed in developing micromanufacturing process models and the controllers and control algorithms to utilize these models to improve the overall process and, ultimately, the product.

5. Controllers are also becoming more flexible to address the variety of processes that are being used in the microfactory. While they are becoming more reconfigurable, e.g., controlling a lathe and a mill, etc., major control system manufacturers in Asia are not looking toward open architecture controllers. These companies control the majority of the computer numerical control (CNC) market, and their customers are not requesting open architecture controllers. Thus, there is no great incentive to move in this direction.

Government Strategies and Funding Patterns

1. Research-to-technology refinement-to-commercialization appears more organized at the national government levels in Asia and Europe than in the U.S. in terms of both direction and government financial assistance for the long-term, resulting in more sustained efforts to refine and fine-tune new developments.

2. Both Japan and Korea support large, multi-year country-wide programs in micromanufacturing and microfactories, although in Korea this has been a very recent phenomenon. In Japan, the 10-Year Micromachine Program (1991-2001) constituted a major government investment that jump-started a number of initiatives with industry that continues today. Major successes include micromanufacturing and assembly systems at Olympus, Seiko, Hitachi, Fanuc, and Mitsubishi. In Korea, the Korean Institute of Machinery and Metals (KIMM) was awarded a major government contract for microfactory development.

3. In Japan, both the National Institute for Advanced Industrial Science and Technology and RIKEN (The Institute of Physical and Chemical Research) have missions heavily oriented toward R&D for industrial application, and both had major efforts directed toward micromanufacturing with very impressive results. In both labs, the R&D programs are producing very sophisticated, complex, and highly innovative processing methods.

4. In Taiwan, there is some institutional government investment, but it is mostly through large corporations with strong product focus, typical of Japan's *branding* strategy. The Industrial Technology Research Institute (ITRI) is the major government-supported laboratory conducting research in support of Taiwan's high-technology industries with a large segment being devoted to micromanufacturing research and development. Another government facility, the Metal Industries Research Institute (MIRI), is initiating a program in micro/mesoscale manufacturing methods (M^4).

5. In Europe, there has been much government investment in institutions at both federal and state levels. The emphasis seems to be on creating an enabling infrastructure to support the conversion of research results into technologies to the point that they are attractive to companies for application and commercialization. The major success story seems to be the Fraunhofer Institutes in Germany that are spread throughout the country. Each is focused on a particular technology, is co-located with major universities engaging students as staff members, and works closely with companies.

6. In Germany, the *Fraunhofer System* is a major driver of micromanufacturing research, technology development, and commercialization. With strong ties to the university system and industry, the Fraunhofers unite the three partners and the results are impressive. Efforts tend to be long-term, sustained, and lead to commercialization. State-based institutes are also common in Germany, again usually co-located with a local university.

Corporate Strategies and Observations

1. In terms of overarching corporate strategies, several points are worth noting:

    *   *Large R&D Budgets:* In general, R&D budgets abroad can be larger than in the U.S. for both large and small companies. For example, in 2001 at Samsung, R&D investment was 4% of sales (about the same as the average U.S. company's R&D investment), but climbed to 8.5% in 2004, where it will remain for the future. At Kugler in Germany, a small precision machine tool manufacturer, the R&D budget is about 20% of total company expenditures.
    *   *Sustained Efforts:* In both Asia and Europe, companies tend to be able to develop and sustain R&D projects over the long term. For example, FANUC's ROBOnano (multi-purpose micromachine that sells for $1 million) was the realization of a 17-year effort involving dozens of researchers and engineers.
    *   *Strong Government Partnerships:* The new Carl Zeiss microcoordinate measuring machine (CMM) was a joint effort with the German government, which funded 30% of the project.
    *   *Close Institute/University Ties:* German companies provide significant investment and have staff located at Fraunhofer Institute and university locations to jointly develop technologies.

2. In Japan, the companies that have been strong over the past two to three decades in manufacturing leadership, e.g., FANUC (controls), Matsushita Electric (consumer products), Mitsubishi Electric (electronic products, devices) and Olympus (optics), seem to have invested heavily in micromanufacturing technologies continuously over the last fifteen or so years. Emphasis on robotics and mechatronics has driven this investment, which focuses on automated assembly of small devices and systems, and, application-driven new process development to meet specific part needs in terms of requirements related to geometric features, surface finish, relative accuracy, and materials properties. Notable examples are the microfactory concept developed by Olympus, primarily as a microlevel, automated assembly system and the ROBOnano machine tool developed by FANUC.

3. In Japan, it is interesting to note that the majority of the micromanufacturing equipment developed could be classified as somewhat exotic in nature, directed toward sophisticated, low-volume, high-precision needs of specific products and devices, and requiring a significant investment, with costs in the several $100K to $1M range. On the other

hand, there was little evidence found to support the notion that Japan might be considering the development of lower-cost, higher-volume commodity micromanufacturing equipment at this time.

4. In contrast, in Germany, there was abundant evidence of the desire to commercialize smaller micromanufacturing machine tools and accessories on a commodity basis, examples including Kugler's Flycutter and MicroTURN machines, the Carl Zeiss F25 small-scale CMM, and the Klocke Nanotechnik microscale robotic systems.

Institute/University Strategies and Observations

1. In both Asia and Europe, university research tends to be more device-development oriented with longer-term projects aimed at developing devices and associated integrated systems. Activity in the areas of process fundamentals, particularly, modeling and simulation, was less evident.

2. In Japan, university research programs tend to be more fundamental and professor-centric than in the United States. University/industry collaborations were less evident during site visits. Both Korea and Taiwan follow a similar pattern. In Japan, universities and government laboratories appear less connected.

3. In Germany, virtually all universities visited were associated with a Fraunhofer Institute and heavily engaged in industry-based research and development projects. Similar tendencies were seen in Switzerland and the Netherlands (e.g., ETH Zurich, Eindhoven). In all cases, laboratory facilities were excellent, based primarily on government funding, and faculty appeared to have more time to focus on research and less time on funding issues than in the United States.

4. In Japan, desktop manufacturing via micromachine tool and microfactory efforts were found at several university locations, including multi-functional machine and robotic devices. Perhaps the most important message gleaned from the university visits is that micromanufacturing in the context of this study will continue to see strong growth, and demand continuing research on a broad range of related topics, e.g., new materials, process understanding, new concepts for micromanufacturing equipment.

5. Regarding the relationship between the universities and companies in Japan, it was observed on several occasions that companies expect the universities to teach fundamental principles and provide broad scientific education, while the companies generally provide focused and application-oriented special training during the early years of employment. It was noted that government policy related to intellectual property seems to provide a favorable situation for industry regarding

university-based innovations and inventorship under government funding. Companies can purchase licenses from the government, which owns all such funded intellectual property (IP), to commercialize university-based inventions. The faculty involved can be required to work with the companies free-of-charge. However, the universities in Japan are in the process of adopting the U.S. model of funding research.

6. In Germany, the Fraunhofer Institutes are co-located with major universities, some of whose departments seem to mirror the institute structure. The focus tends to be on specific technologies at each location, e.g., laser, production methods, machine tools, etc. There is a mix of government and private/industry funding, and projects tend to be longer-term. Emphasis is on refining and fine-tuning technologies to make them commercially attractive and easily adapted. A wide range of services are available, including consulting, feasibility studies, basic research, technology transfer, systems integrations, and quality assurance/quality control (QA/QC). Links with universities seem to be very important to success. The Fraunhofer institutes are extremely well equipped with state-of-the-art facilities. It is noteworthy that a high percentage of staff, approximately 10%-15%, later start companies.

MICROMANUFACTURING R&D: A U.S., ASIA, AND EUROPE COMPARISON

Table ES.1 indicates that while the United States receives adequate marks for nanotechnology R&D, emphasis in the on micromanufacturing R&D is lagging far behind the rest of the world. This will undoubtedly have serious long-term implications, since it is well-recognized that micromanufacturing will be a critical enabling technology in bridging the gap between nanoscience and technology developments and their realization in useful products and processes. The Uniteed States gets particularly low marks for government funding of micromanufacturing R&D and the development and nurturing of industry, government, and university interactions and collaborations. On this latter point, Europe appears to be very strong, particularly as these partnerships work to refine and fine-tune developments for industry adaptation and commercialization.

Table ES.1
Summary of the Relative Status of International Micromanufacturing Technology Development

Activities	Japan	Taiwan	Korea	Europe	U.S.
Government funding in micromanufacturing	****	****	***	*****	*
State of the micromanu-facturing technology	*****	****	***	*****	**
Industry/University/ Gov't partnership	***	***	****	*****	*
State of nanotechnology*	****	**	*	***	****

A PERSPECTIVE ON FORWARD PLANNING FOR THE U.S.

One overarching conclusion to be reached in reflecting on the observations made during this study is that MEMS and nanoelectromechanical systems (NEMS) advances are highly oversold in in the United States. While it is true that the U.S., over the last twenty years, has emphasized lithography-based MEMS with outstanding research results and a dominant market position, many MEMS products have become commodity products, and therefore, the Asian countries stand to reap more benefits in the near future from them. Perhaps there is an important lesson to be learned here.

Although less advertised, non-lithography micromanufacturing, practiced mostly in highly competitive, private companies such as Sankyo Seiki, Samsung, Olympus, etc., is more likely to continue to lead to more practical products faster. These products include lenses for cameras in telephones, flat panel displays, a myriad of automotive parts, microfuel cells, microbatteries, micromotors and, of course, desktop factories. Based on the state-of-the-art and current investment levels, Europe, especially Germany and Switzerland, and Asia, particularly Japan and Korea, will gain the most from developments in non-lithography-based micromanufacturing as they have a long tradition in it and have invested more heavily in this field.

We believe that to succeed in non-lithography-based manufacturing, a stronger-than-usual link between industrial partners and academia is required since micromanufacturing is very applied and product-driven. In this regard we are now behind in the United States, although it was here that the trend of academia/industry collaborations started. The links between industry and academia are now better in both Europe and in Asia. It

---

* Reflects assumptions about relative level of funding for nanotechnology net qualitative assessment. This panel did not evaluate quality of nano research.

is a commonly held belief that technology transfer offices in U.S. academia have become so unwieldy that they prevent smoother and better collaboration with industry. This will need to change.

In some showrooms of the Asian visit, it became apparent that none of the products on display were manufactured in the United States anymore. New product demands are stimulating the invention of new materials and processes. The loss of manufacturing goes well beyond the loss of one class of products. If a technical community is dissociated from the product needs of the day, say those involved in making larger flat-panel displays or the latest mobile phones, such communities cannot invent, and eventually, can no longer teach effectively. Yet, a more sobering realization is that we might invent new technologies, say in nanofabrication, but not be able to manufacture the products that incorporate them. It would be naïve to say that we will still design those new products in the United States. For a good design, one needs to know the latest manufacturing processes and newest materials. We are quickly losing ground in developing new manufacturing processes and materials, and we must reverse this trend quickly.

To stem the hollowing out of their manufacturing base, the governments of many developed countries have made huge investments in the miniaturization of new products (MEMS and NEMS) and the miniaturization of manufacturing tools, for example DTFs. These efforts are intended to regain a manufacturing edge. To illustrate this point, Olympus' Haruo Ogawa, leader of that company's MEMS team, says that MEMS may help rebuild Japan's power as a manufacturing nation. Sankyo Seiki representatives believe that its DTFs might revive manufacturing in Japan. In Korea, the government just started a new DTF project. Finally, in some quarters in the United States, nanotechnology is seen as a means for the United States to remain a high-technology innovator.

An approach for the United States would be to launch a concentrated effort in advanced manufacturing techniques and re-introduce the societal merits and value of actually making things. With the information technology sector depressed, and high-paying jobs still scarce, this is a good time to launch such an effort. The current WTEC study could be a first attempt toward this goal. Hybrid manufacturing approaches, incorporating top-down and bottom-up machining approaches, could be key in attracting a new generation of motivated engineers and scientists into the science and engineering of manufacturing.

CHAPTER 1

INTRODUCTION

Richard E. DeVor and Kornel F. Ehmann

BACKGROUND AND SCOPE

For the purpose of this study, the term micromanufacturing refers to the creation of high-precision three-dimensional (3D) products using a variety of materials and possessing features with sizes ranging from tens of micrometers to a few millimeters (See Figure 1.1). While microscale technologies are well established in the semiconductor and microelectronics fields, the same cannot be said for manufacturing products involving complex 3D geometry and high accuracies in a range of non-silicon materials. At the same time, the trends in industrial and military products that demand miniaturization, design flexibility, reduced energy consumption, and high accuracy continue to accelerate—especially in the medical, biotechnology, telecommunications, and energy fields. By and large, countries with traditional strengths in manufacturing, such as Japan and Germany, have continued to invest heavily in recent years in micromanufacturing R&D for several reasons. First, the demand from the global market for ever-smaller parts and systems at reasonable cost and superior performance is strong. This demand tends to drive the high-end research. Second, the prospects of multidisciplinary research are causing companies increasingly to blend material science, biology, chemistry, physics, and engineering to speed up technology innovation and thereby new applications based on microtechnology. Third, strong government actions in the form of national R&D initiatives have resulted in more effective R&D infrastructures that are conducive to advanced research and education.

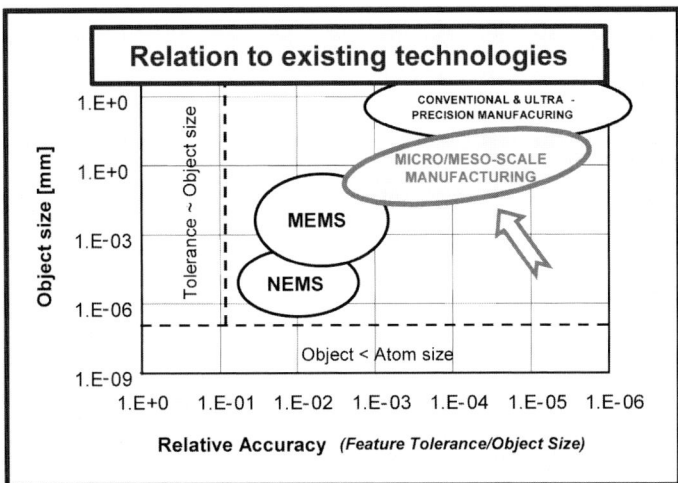

Figure 1.1. Micromanufacturing size/precision domain.

The explosion of microscale product development in consumer markets including healthcare, communications, and electronics, to name a few, has the potential to significantly improve quality of life and personal well-being. In the healthcare field alone, advances in cardiovascular system remediation and non-invasive surgery have been tremendous. In the communications industry, microminiaturization advances are providing new possibilities for functionality but will require new and low-cost manufacturing capabilities to be realized. Below is a list of some of the specific applications that today are driving the need for micromanufacturing research and development:

- Cardiovascular and *in vitro* diagnostic devices
- Cardiovascular clot removal catheters
- Medical implants
- Portable electronic devices
- Cameras
- Wireless devices
- Microscale fuel cells
- Fluidic microchemical reactors
- Microscale pumps, valves, mixing devices
- Microfluidic systems
- Fiber optic components
- Micronozzles for high-temperature jets

- Micromolds
- Deep X-ray lithography masks
- Optical lens assemblies

Micromanufacturing is an important new technology because:
- It is an *enabling* technology for the widespread exploitation of nanoscience and nanotechnolgy developments—it bridges the gap between the nano- and macroworlds;
- It is a *disruptive* technology that will completely change our thinking as to how, when, and where products will be manufactured—e.g., on-site, on-demand in the hospital operating room or on-board a warship;
- It is a *transforming* technology that will redistribute manufacturing capability from the hands of a few to the hands of many—micromanufacturing becomes a cottage industry;
- It is a *strategic* technology that will enhance the competitive advantage of the U.S.—reduced capital investment, reduced space and energy costs, increased portability, increased productivity.

The motivation for this international technology assessment study comes from the global attention being paid to product miniaturization, from the related trends toward the miniaturization of manufacturing equipment and the development of systems for microscale components and products, and from the emerging microfactory manufacturing paradigm. This paradigm encompasses the creation of miniaturized unit or hybrid processes integrated with metrology and material handling and assembly to create "microfactories" that are capable of producing microprecision 3D products in a fully-automated manner and at low cost. In this study the paradigm is specifically applied to processes and applications that may contain 3D (free-form) surfaces, employ a wide range of engineering materials, and have relative accuracies (tolerance-to-feature size) in the 10^{-3} to 10^{-5} range, as distinct from MEMS, which is primarily based on exploiting silicon planar lithography as the core technology (Figure 1.1). Hence, MEMS processes and applications are 2 - 2 ½ D and are limited in the engineering materials employed and involve relative accuracies in the 10^{-1} to 10^{-2} range. While interests in this study range from the nano level to the microlevel and even beyond, processes and applications that are more commonly associated with the semi-conductor industry will assume less emphasis.

The study was aimed at investigating both the state-of-the-art and future directions of micromanufacturing from the perspectives of science, technology, and commercialization. It was designed to provide a first-hand look at the latest R&D developments overseas and compare them to those in the United States. In particular, the study encompasses three important scientific and technological areas:

- Microscale process fundamentals and modeling
    o Grain size and orientation effects
    o Influence of interfacial phenomena
    o Machine-process dynamics interactions
- Microscale process innovations and device development
    o Miniaturized processing equipment
    o Sensors and actuators
    o Microspindles, monolithic motion devices
- Integration of microscale devices into autonomous systems
    o System architecture/topology
    o Communications and control
    o "Art-to-part" support

The specific charge for the panel was:

- To assess the emerging global trend toward the miniaturization of manufacturing systems, equipment and processes for the manufacture of micro/mesoscale components and products (Microfactory manufacturing paradigm; processes and equipment distinct from those for semiconductors and MEMS);
- To identify scientific, technological, commercialization and other barriers;
- To assess potential economic and environmental impact of these technologies;
- To identify trends and critical R&D needs and directions; and
- To assess micromanufacturing R&D at leading sites abroad and evaluate the competitiveness of the U.S.

METHODOLOGY

The basic framework of the study emerged from an initial meeting held at the National Science Foundation (NSF) in Arlington, VA on March 11, 2004. The meeting participants included representatives from potential sponsoring agencies, national labs, industry, academia, and WTEC. Presentations addressed various aspects of micromanufacturing to begin to define the scope of the study. At subsequent meetings, the committed sponsors, WTEC, and the panel chair established the scope, time table, panelists, and logistics of the study. The panel members are listed in Table 1.1.

The expert panelists were selected to cover a broad range of background and expertise relevant to micromanufacturing technologies, including:

- Material science and technologies

- Manufacturing processes, automation, and robotics

- Mechanical engineering and engineering design

- Micromanufacturing processes

- Manufacturing systems and metrology

- Economics and education

The study commenced with a two-day U.S. kick-off workshop, held at NSF on August 12 and 13, 2004. The technical part of the workshop consisted of 12 presentations by representatives from eight U.S. companies, three national labs and one university with active programs in micromanufacturing technology applications, development, and advanced research. The primary objective was to provide baseline information on U.S. activities as a starting point for the study, to solicit suggestions for potential sites to visit abroad, and to provide feedback on the scope of the study. Presentations were followed by breakout sessions in which all workshop attendees participated. On the second day of the kick-off workshop, the panel members, sponsors, and WTEC representatives met to rank the potential sites, firm up the schedule of the study, reach decisions on the technical content and format of the final report.

During the closing session, several key recurring themes were summarized and became important benchmarks for the study:

- The difficulty in achieving *feature size and accuracy* with current macroscale manufacturing equipment;

- The need for *multi-functional machines*, e.g., machines that combine processes such as micromachining and measurement;

- Problems associated with *part handling, fixturing, and repositioning,* driven by part size and tolerance requirements;

- *Assembly* as a critical process, perhaps providing more challenges than processing;

- *Metrology* as a critical process, and the need for *in situ* measurement and inspection methods;

- *Equipment cost reductions*, driven by the capital-intensive nature of micromanufacturing;

- The need to better understand the *influence of scaling on process performance;*

- *Materials issues,* in particular, microstructure effects, consistency of properties, and multi-material integration;

- *Intellectual property* (IP) issues that inhibit cooperation and free exchange of information; particularly between suppliers and their customers; and
- The need for *"out-of-the-box" thinking*.

Following the workshop, the presenters were given the opportunity to revise and extend their presentations and remarks. The collection of the final viewgraphs from their presentations is available at WTEC's website at http://wtec.org/micromfg/us_workshop/. The members of the panel also compiled a final report that summarized the key findings of the workshop. Copies of the report, along with the complete viewgraphs of the presentations, were provided to hosts prior to the site visits.

Table 1.1
Panel Members

#	Panelist	Affiliation
1	Kornel F. Ehmann	Northwestern University (Panel Chair)
2	Beth Allen	University of Minnesota
3	David Bourell	University of Texas at Austin
4	Martin L. Culpepper	Massachusetts Institute of Technology
5	Richard E. DeVor	University of Illinois at Urbana-Champaign
6	Thom J. Hodgson	North Carolina State University
7	Thomas R. Kurfess	Georgia Institute of Technology
8	Marc Madou	University of California at Irvine
9	Kamlakar Rajurkar	University of Nebraska at Lincoln

Following the U.S. kick-off workshop, the panel traveled to sites in Asia and to Europe. Site visits in Japan (19 sites), Korea (six sites), and Taiwan (eight sites) took place between December 3 and 16, 2004. Site visits in Austria (one site), Germany (10 sites), the Netherlands (two sites), and Switzerland (four sites) took place between April 4 and 9, 2005. Table 1.2 summarizes the sites visited.

Table 1.2
Sites Visited in Asia and Europe

#	Country	Site	#	Country	Site
1	Japan	National Institute of Advanced Industrial Science and Technology (AIST)	25	Korea	Korean Advanced Institute of Science and Technology (KAIST)
2	Japan	FANUC, FA & Robot	26	Korea	Korean Institute of Machinery and Materials (KIMM)
3	Japan	Hitachi Chemical R&D Center in Tsukuba	27	Korea	LG Chemical
4	Japan	Matsuura Machinery Corporation	28	Korea	Samsung Electro Mechanics Corporation, R&D Center

Table 1.2
Sites Visited in Asia and Europe

#	Country	Site	#	Country	Site
5	Japan	Matsushita Electric Industrial Company	29	Korea	Seoul National University
6	Japan	Mitsubishi Electric Corporation	30	Korea	Yonsei University
7	Japan	RIKEN (Institute of Physical & Chemical Research)	31	Austria	Zumtobel Staff Werkzeugmaschienen
8	Japan	Center for Cooperative Research in Advanced Science and Technology	32	Germany	Institut für Mikrotechnik Mainz GmbH
9	Japan	Kyoto University	33	Germany	Laser Zentrum Hannover e.V.
10	Japan	University of Electro Communications	34	Germany	Karlsruhe Research Center (Forschungszentrum Karlsruhe)
11	Japan	Seiko Instruments Incorporated (SII)	35	Germany	Kugler GmbH
12	Japan	Department of Microsystem Engineering, Nagoya University	36	Germany	Carl Zeiss Industrielle Messtechnik, GmbH
13	Japan	Olympus Corporate R&D Center	37	Germany	Fraunhofer IPK/Institute for Machine Tools and Factory Management Technical University
14	Japan	University of Tokyo	38	Germany	Robert Bosch, GmbH
15	Japan	Laboratory of Structure and Morphology Control, Nagoya University	39	Germany	Manufacturing Engineering and Automation (IPA), Fraunhofer Institute
16	Taiwan	Industrial Technology Research Institute (ITRI)	40	Germany	Institute of Reliability and Microintegartion, Fraunhofer Institute Berlin (IZM)
17	Taiwan	ITRl Mechanical Industry Research Laboratories	41	Germany	Fraunhofer Institutes for Production and Laser Technology (IPT)/(ILT)
18	Taiwan	ITRI Nano Technology Research Center (NTRC)	42	Germany	Klocke Nanotechnik
19	Taiwan	National Taiwan University (NTU)	43	Netherlands	Philips CFT - Philips Applied Technologies

Table 1.2
Sites Visited in Asia and Europe

#	Country	Site	#	Country	Site
20	Taiwan	Metal Industries Research and Development Center (MIRDC)	44	Netherlands	Technical University of Eindhoven (TU/e)
21	Taiwan	National Cheng Kung University	45	Switzerland	Baselworld
22	Taiwan	National Science Council	46	Switzerland	Swiss Federal Institute of Technology (ETHZ) and Institute of Robotics and Intelligent Systems (IRIS)
23	Taiwan	Asia Pacific Microsystems, Incorporated (APM)	47	Switzerland	École Polytechnique Fédérale de Lausanne (EPFL) and Institut de Production & Robotique (IPR)
24	Taiwan	Precision Instrument Development Center (PICD)			

Panel members, sponsor representatives Jian Cao (NSF), Kershed Cooper (ONR) and Gerald Hane, and WTEC personnel Roan Horning and Hassan Ali, were divided between two groups that each visited, on average, two sites per day. As a result, no panel member had first-hand contact with all of the visited sites. In most cases, pre-designated members of the panel completed site reports immediately following each visit, in consultation with the other attendees. Site reports were then sent to the hosts for concurrence and corrections. After revisions to account for their comments, they were posted on WTEC's website and are now included, after final editing, as Appendices to this report.

Following the site visits and the composition of the site-visit reports, the panel held a final Workshop on April 21 and 22, 2005 at the Arlington Hilton and Towers, Arlington, VA, to disclose the findings of the study. The first day of the workshop was devoted to a presentation of findings to the sponsor organizations and the second day to the public. Approximately 65 people attended the public workshop. The collection of viewgraphs from the presentations by the panelists and the sponsors are posted at http://wtec.org/micromfg/workshop/proceedings.html. The present report constitutes the final step in this study.

OVERVIEW OF THE REPORT

The sponsors and panelists relied on an iterative process to determine the structure of the report. An attempt was made to address key areas relevant to micromanufacturing that span the spectrum from the conception of products and processes to their ultimate societal impact. The sponsors and panelists selected the following technical areas to be presented in the form of analytical chapters:

Table 1.3
Chapter Designations

Chapter Name	Authors
Design	(Culpepper, Kurfess)
Materials	(Bourell, Rajurkar)
Processes	(Rajurkar, Madou)
Metrology, Sensors and Control	(Kurfess, Hodgson)
Applications	(Madou, Rajurkar)
Business, Education and the Environment	(Hodgson, Culpepper)

The chapters were constructed by extracting the relevant information from the site reports.

FORMAT FOR THIS REPORT

This report is organized in three distinct parts. First, this Introduction provides an overview of the purpose, goals, and scope of the study and includes a summary of some of the key overarching observations and conclusions of the study team. The next section is a series of brief analytical chapters that organize the findings of the study according to a number of technology-based perspectives: design, materials, processing, metrology, applications, and business and economics. Appendices include biographical sketches of the members of the panel, the list of questions asked by the panel, and the site visit reports.

ACKNOWLEDGEMENTS

The development of this report would not have been possible without the assistance of the representatives of the sponsoring agencies (National Science Foundation, The Office of Naval Research, The Department of Energy and the Advanced Technology Program at the National Institute of Standards and Technology) who traveled with the panel and contributed significantly to the planning and execution of the study. This report would also not have been possible without the presentations of the participants of the WTEC kick-off workshop or without the support, assistance and hospitality of the hosts and colleagues in the Asian and European countries visited by the panel. The panel gratefully acknowledges the participation of all of these persons.

CHAPTER 2

DESIGN

Martin L. Culpepper and Thomas R. Kurfess

ABSTRACT

The purpose of this chapter is to review the existing capabilities that may be used to design parts that will be micromanufactured without lithography-based processes. In the past five to 10 years, non-lithography-based meso- and microscale (NLBMM) parts have seen increased use in medical applications, consumer products, defense applications, and several other areas. These technologies promise to have an impact on the economy, improve health and safety, raise our standard of living, and form a middle-scale stepping stone by which the benefits of nanotechnology may be accessed. The technologies used to design NLBMM parts and the processes and equipment used to fabricate them are in a nascent stage. These technologies have been, for the most part, borrowed from the design practices of macroscale engineering and very large-scale integration (VLSI). At present, most designers have difficulty ascertaining the appropriate time to use pre-existing design knowledge, theory and tools. Designers must be able to assess the suitability of pre-existing technology for the design of NLBMM parts. Otherwise, design processes will be long and iterative, with the result that the products' benefits will be either delayed or lost. As designers, we must understand the nature of this new technology and work hard to generate the design knowledge, theories and tools that will enable the widespread and rapid advance of NLBMM technology.

Given the nascent state of NLBMM technology, this chapter focuses on the technologies that "should be." Some of these technologies may be borrowed or adapted from the macroscale and VLSI design domains, whereas others will have to be fundamentally different. This chapter aims to explain the aspects of NLBMM parts design that are fundamentally unique and the circumstances in which these unique differences call for new design knowledge, theories and tools. Toward this end, this chapter contains discussions on the following topics: (1) the reasons why unique requirements exist for the design of NLBMM parts, (2) the existing knowledge and practices that may be borrowed to design NLBMM parts, and (3) the gaps between existing technologies and the requirements for NLBMM part design. The topics are arranged so that members of disparate design communities (e.g., decision theorists, hardware and mechanical designers, design theorists, industrial designers, and process design specialists) may make use of the knowledge gained during this study.

THE APPROACH TAKEN TO ASSESS THE STATE-OF-THE-ART IN NLBMM DESIGN

To assess the state-of-the-art in NLBMM design, the first step is to understand the fundamental differences between traditional design technologies and design technologies that are needed to support the design both "of" and "for" NLBMM components. This distinction is made here to accommodate the disparate and polarized design camps that support either:

1. The "design of," meaning the design of specific machines, equipment, software, and other devices; or

2. The "design for," meaning general technologies that support design theory and process.

The needs of NLBMM design span the interests of many design communities. The assessment conducted in this chapter takes a balanced approach, concentrating on design knowledge that is considered important by most design communities. Great effort has been expended to construct the chapter so that (1) it has value to members of many design communities and (2) the members of the different communities will understand the need for collaboration among the different communities. Although the focus of this chapter is on technical challenges, it is important to note that the low level of collaboration between the different design communities is a primary barrier to solving the general problem of NLBMM design.

UNIQUE REQUIREMENTS FOR THE DESIGN OF NLBMM PARTS, PROCESSES AND EQUIPMENT

Designers need to know why NLBMM design is different from macro-scale design and they need to understand the fundamental reasons for these differences. Figure 2.1 shows a generic representation of a multi-scale system in which distinct parts (e.g., rectangles) are grouped according to characteristic size scale. The reason for examining a multi-scale system will become apparent in a moment. The different-scale parts are coupled to other parts via cross-scale links. In general, these links may be characterized as (1) function/performance links; (2) form/geometric links such as physical interfaces, (3) flows of mass, momentum, or energy; (4) physical phenomena that govern mono- and multi-scale behaviors; and (5) fabrication and integration constraints. Some of these links are material in nature (e.g., physical interfaces) and others are non-material (e.g., the compatibility of parts fabricated by different processes).

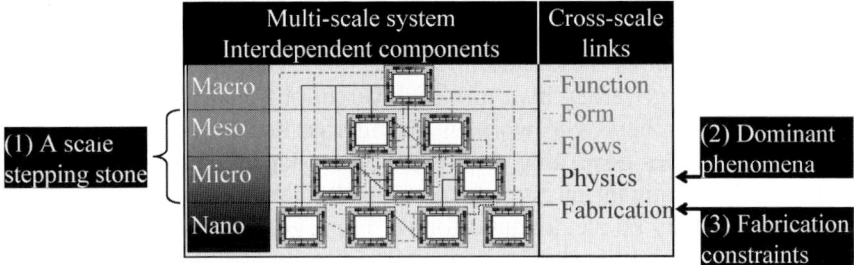

Figure 2.1. Multi-scale system with cross-scale links between components.

The uniqueness of NLBMM parts may be traced to the elements of Figure 2.1 that are highlighted by keyed arrows:

Stepping Stone for Characteristic Size and Time Scales

The microscale is perfectly situated as a "stepping stone" between familiar large-scale technologies and emerging nanoscale technologies. In many cases, the benefits of the nanoscale cannot be realized at the larger scales without the help of microscale components that link the larger and smaller scales. Microscale components therefore occupy a unique position within the scale hierarchy and this position is a critical link for the future utilization of nanoscale technologies.

Meso- and microscale parts are often integrated with parts of different scale, thereby coupling the design of the meso- and microscale parts to the requirements and constraints of parts from different scales. The coupling is particularly important when size and time scales of the meso- and microscale parts differ significantly from the scales of the macro- and nanoscale parts with which they will interact. To some extent, packaging may address the interfacing of different-size parts, but this is a half-measure and not a general solution. Packaging works well for VLSI devices and microsensors, but this approach is not particularly well suited for devices that must share moving mechanical interfaces with parts and devices of other scales. Other size scale issues exist; for instance, situations arise wherein the ratio of a part's minimum feature size to the grain size approaches unity. This issue significantly affects performance that is not accurately captured by traditional, macroscale modeling and design techniques. Temporal scales are also an issue. For example, the orders of magnitude mismatch in time constants between different scales. These and many other issues point to the need for a new design perspective in which length and temporal scales are critical for determining the approach to designing small-scale parts and multi-scale assemblies.

Dominant Physical Phenomena

The nature of the physical phenomena that dominate the behavior of components generally changes between the meso-and microscales. As such, the behavior of meso- and microscale components within multi-scale systems may be governed by a mix of different physical phenomena. Engineers and researchers who work in the microscales have learned that the physical principles governing part behavior are dependent on part size. For instance, the dominant physics that govern the behavior of a meso- or microscale part may be electrostatic forces. For a geometrically similar macroscale part, the dominant principles that govern behavior may be linked to body forces. This has several implications for the design of NLBMM parts and systems. First, the general design of the parts may require design tools capable of multi-physics modeling. Second, the need to interface and/or integrate microscale parts with parts of a different scale may require multi-scale modeling tools to predict system-level behavior. Both implications point to a need for a departure from traditional macroscale design models and simulation tools.

Fabrication

Fabrication processes are often scale dependent. For instance, traditional macroscale techniques such as milling are not generally applicable at the nanoscale. Likewise, the nanoscale processes used by nature to build biological systems are not generally applicable to the fabrication of some large-scale parts. The range of utility for a specific fabrication process generally terminates within the meso- or microscale. This is an important point for designers to realize; designers must "design for" compatibility of microscale parts with parts that were fabricated using a macro- or nanoscale fabrication processes. Microelectromechanical systems (MEMS) and very large scale integration (VLSI) designers encounter this issue in the form of packaging challenges.

The link between design and manufacturing has led to design for X (DFX) methods (Boothroyd et al., 1994) and concurrent design practices (Syan and Menon, 1994) that are used to help designers select appropriate design-fabrication process combinations. Although the general idea of DFX and concurrent engineering may be considered scale-independent (the design of parts to be made with fabrication processes that are cost and/or time appropriate), the implementation of these practices depends upon the fabrication processes that are to be used. The small-size scale of NLBMM parts makes it necessary to use new or adapted versions of existing manufacturing technology. As a result, new DFX rules will be needed to help designers make design-process choices that ensure scale-specific manufacturability and cross-scale compatibility.

THE DESIGN PROCESS AND THE IMPORTANT ELEMENTS OF DESIGN OF/FOR NLBMM PRODUCTS

At the most basic level, "design" is a process that is used to generate, evaluate and select a solution for a given problem. There are several approaches to engineering design; creative concept design followed by deterministic modeling (Slocum, 1992), decision making in the presence of risk and uncertainty (Hazelrigg, 1998, Hazelrigg, 1999), robust design (Taguchi and Subir-Chowdhury, 2004), six-sigma methods (Creveling et al., 2002, Yang and El-Haik, 2003), axiomatic design (Suh, 1990, Suh, 2001) and complexity theory (Suh, 2005), customer-centered product and industrial design (Ulrich and Eppinger, 2003), and systematic design approaches (Pahl and Wolfgang, 1999). For the purposes of this assessment, the details of a given approach are less important than an examination of the technologies and processes that are needed to support the practice of design. This section is devoted to (1) developing a design framework that most design communities can understand and agree upon for the sake of discussion, (2) relating this framework to the specific challenges in the design of NLBMM parts, and (3) acting as a segue to the subsequent section, which covers the state-of-the-art (SOA) and the gaps in design knowledge, theory and tools.

Most design approaches utilize the combination of steps shown in Figure 2.2.

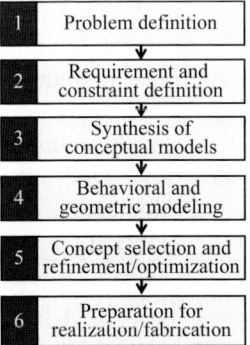

Figure 2.2. Primary steps that are generally followed during design.

The knowledge and approaches taken in steps 1, 2 and 5 have traditionally been independent of specific problems. At the end of this assessment, the author has yet to find a reason that this should change for meso- and microscale parts. The unique characteristics of NLBMM parts (e.g., scale, physics and fabrication processes) are clearly linked to steps 3, 4, and 6; and therefore the technology gaps are associated with these steps. To make this link clearer, it is important to develop a shared understanding of the elements that are important to NLBMM design. These elements are independent of the design "school of thought" held by the different design communities. Figure 2.3 shows a triumvirate of elements that are necessary for most design processes. The triumvirate is divided into the following categories:

1. *Multi-scale/physics design knowledge*—The knowledge of physical phenomena that describes the behavior of elements within particular concepts or designs. The knowledge described here extends beyond the realm of the scientific. Design knowledge consists of basic and applied knowledge that may be used in combination to enable one to create a useful and realizable design. This knowledge forms the starting point for generating the concepts (step 3) and forming the models (step 4) that link customer and design requirements with design performance.

2. *Simulation and modeling tools*—Design tools that help designers link knowledge with the requirements of specific concepts/designs. Modeling and simulations tools are used to generate and assess first-order concept models (step 3) and to optimize designs. These tools provide the information that is used to make design decisions.

3. *Design theory and design process*—The approach taken to guide the application of the acquired knowledge and the modeling tools in the formation and optimization of design concepts. This category covers a range of issues that are important for ensuring that parts have been properly designed. The approach includes the selection of overall system/part architecture (e.g., coupling and complexity), consideration of stochastic issues (e.g., risk, and how this applies to engineering decision making), design rules (e.g., DFX and best practices) and a vernacular that designers, manufacturers, and others may use to "converse" with each other (e.g., standards).

Figure 2.4 shows how the design process, driven by design requirements, draws upon an existing base of design knowledge and then uses modeling and simulation tools to process this knowledge and generate designs. Designers must understand all aspects of the triumvirate in order to manage the transformation of design knowledge into a good design.

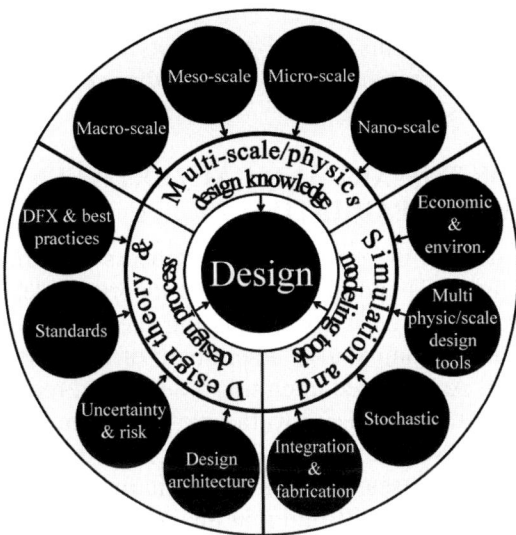

Figure 2.3. Design triumvirate and example elements within each triumvirate.

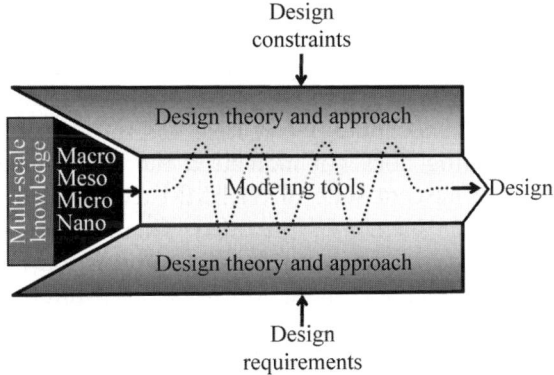

Figure 2.4. The knowledge, theory, modeling, design transformation process.

While the process shown in Figure 2.4 is well developed for most macroscale design domains, it is fairly undeveloped for NLBMM design. The undeveloped state of NLBMM design is due to a mismatch between the nascent state of the technology and the length of time required for design researchers to learn of, and then contribute, to this new field. At present, most designer researchers have yet to learn of this new field. As such, the elements required for the design of/for NLBMM parts are far from fully developed. During the foreign site visits and two workshops, the panelists noted little evidence of substantial efforts that are aiming to address the unique design requirements of NLBMM parts. Most approaches are borrowed from VLSI or macroscale design and then modified to enable engineers to "make a design work."

THE STATE-OF-THE-ART AND GAPS BETWEEN EXISTING AND REQUIRED CAPABILITIES

This section utilizes the generic design process (Figure 2.2), the design triumvirate (Figure 2.3) and the knowledge-theory-modeling-design transformation process as a basis for discussing the design technology gaps. Herein is provided a comparative assessment of the state-of-the-art in the U.S., Europe, and Asia. Based upon this assessment, recommendations for improving the SOA are tendered to the global design community. Due to the nascent state of NLBMM technology, most of the areas shown in Figure 2.3 have yet to be researched with specific focus on NLBMM applications. As such, the reader will find that the recommendations of this chapter map with high correlation to the elements shown in Figure 2.3.

U.S.-Europe-Asia Comparative Assessment

Figure 2.5 shows a comparative assessment of the SOA in the design of NLBMM parts. Grading is based upon a five-star system in which five stars represents technology that is well-suited to most design applications. For instance, achieving five stars in "multi-scale/physics design knowledge" indicates that the technology is mature and suitable for most applications. A five-star rating does not indicate a full understanding has been achieved. Given the nascent state of NLBMM technology, ratings with a quantitative foundation cannot be provided. This assessment reflects the opinions of the primary author formed using information obtained via the site visits, literature search and the U.S. workshops. More information was available to assess the SOA in Europe and this may have led to the slightly higher grades for that region.

Countries / Activities	Europe	Asia	US
Multi-scale/physics design knowledge	**	**	**
Design theory and process	**	*	*
Modeling and simulation tools	**	*	*

Figure 2.5. Comparative assessment of worldwide efforts in the design of/for NLBMM products.

Regardless of the limitations, the results of the assessment indicate that the SOA in technologies that support the design of/for NLBMM parts is far from ready to provide adequate support for designers. This is primarily due to the nascent state of the technology and the fact that many design researchers have yet to become aware of, and address the design challenges in this field.

SOA and Gaps in Multi-scale, Multi-physics Design Knowledge

There are several well-funded and focused efforts aimed at developing nanoscale and microscale design knowledge. Few efforts are targeted at understanding how to simultaneously model and simulate multi-scale, multi-physics design problems. Many efforts have been focused upon improving the understanding of material behavior at smaller-size scales and understanding how this behavior would affect fabrication processes. Two notable areas in which design knowledge was being generated are:

1. Grain size—The Technical University of Eindhoven, where a focused research effort is underway to (a) better understand, model, and experimentally verify the effects of length scales upon the physical properties of materials, and (b) to better understand how this behavior may change with respect to specific design parameters; for instance, quantifying how the coupling of grain size and a part's minimum feature size affects the performance of structural parts. These relationships are easily plotted so that designers may develop "rules of thumb" and use these rules during the synthesis of concept designs. The experimental and analytic techniques used to produce these rules may also be used to enable rigorous modeling of behavior, thereby providing information that is suitable for concept selection and design optimization. There are complimentary analytic and experimental efforts under way in the Advanced Materials Processing Laboratory at Northwestern University. Specifically, there are efforts aimed at (1) understanding how grain size and material history affect the quality of formed microscale components, (2) creating simulations/design rules that capture the behavior, and (3) linking fundamental physics to process/part design. More efforts of this flavor are required to enable designers to model and design parts with a high degree of certainty.

2. Process design—At the Fraunhofer Institute of Industrial Laser Technology (Aachen) there is a focused effort to understand the fundamental interactions between laser radiation and materials. The aim is to develop the DFX rules and fabrication processes that result in designs with a high probability of success. This approach enables the design of microscale welded joints in metals and polymers and the fabrication of precision micromolding tools by laser ablation.

The strength of the European efforts in developing design knowledge is due in large part to the strong university-industrial collaborations that take place via mechanisms such as the Fraunhofer Institutes. These mechanisms serve as a "meeting place" wherein scientific knowledge and applied knowledge meet to form a knowledge base that is readily used to design NLBMM parts.

SOA and Gaps in Design Theory and Process

As mentioned before, NLBMM technology is in a nascent stage and most design researchers have yet to discover this field. As such, little work has been done toward developing general theories and approaches that could be used to design NLBMM parts. Efforts in Europe, specifically in Germany, have been driven by university-company collaborations via the Fraunhofer Institutes and similar entities. As such, they are often initially focused upon a specific technology or product. Once the technology or product is proven successful, the knowledge gained during the design and engineering process is extended to create similar products via many low-risk, but economically significant advances. These methods have proven to be successful in developing a large body of DFX rules that may be used to design a number of robust and commercially viable products. General design theory and process work has yet to be addressed. Surprisingly, the use of stochastic methods within the design process was not seen or highlighted during the site visits or at the first U.S. workshop.

DFX and Best Practices

The fabrication processes for a given part must be designed in parallel—that is, concurrently engineered—with the design of corresponding parts. Fortunately, designers may use process-specific metrics (cost of a process, rate of manufacture, and the quality of a part) and design/DFX rules to help generate designs that are compatible with the available fabrication technologies. Investment in research will determine the new rules that will in turn enable designers to select appropriate design-fabrication process combinations.

At the macroscale, designers wish to reduce part count for several reasons: the cost and effort associated with the fabrication of interfaces, the validation of an interface's alignment geometry, and the overhead associated with interfaces. Compared to lithography-based processes, NLBMM processes are more prone to multi-piece designs that require directed assembly. Although a great deal of past and continuing research is focused upon the directed assembly of small-scale parts—much of it driven by fiber optics and microphotonics—assembly and integration technology needs to be improved. New processes are also needed to produce monolithic parts without lithography-based processes. For instance, processes such as microcasting, microstamping, and microembossing are capable of creating monolithic geometries and more homogeneous systems. These new methods and processes will require new DFX rules.

Standards

Standards for NLBMM parts are currently in a nascent stage and require better definition. Of particular importance are the standards for measurement and evaluation of part characteristics. Without these standards, designers will have difficulty talking with vendors, customers, and others about the specifications that drive the design and fabrication of their parts. These standards should cover geometry, material properties, optical properties, electromagnetic properties, and other related characteristics. Designers need to understand how to design NLBMM parts so that they are compatible with standard measurement methods. The existing techniques for measuring the characteristics of NLBMM parts are far from ideal. Many of these techniques are slow, provide only 2D information, and require destructive evaluation. Clearly, improved measurement tools are needed, as will be explained in a later chapter of this report. A set of rules that govern the design of NLBMM parts "for measurement" does not exist. As a result, most designers use intuition and advice from metrology vendors to design parts that may be measured in a way that does not affect the rate of manufacturing and does not lead to long time constants for manufacturing processes control. Metrology technology is traditionally slow to develop. As such, early designs must be compatible with available technology. "Design for Standard Measurement" will be important to the rapid adoption of NLBMM parts.

Uncertainty and Risk

Uncertainty and risk are unavoidable when designing in the absence of full knowledge. Uncertainty stems from the lack of knowledge required to model performance. The risk that is associated with a particular design may be ascertained if one is able to use models to define a probability of success. In short, it is better to be in a position to ascertain risk than to be in a position of uncertainty. Throughout this assessment, little evidence has been found to suggest that stochastic methods have been employed during the design of NLBMM parts. Given the nascent state of the NLBMM technology, one finds variation in fabrication processes, metrology tools, material properties, and other aspects which must be considered during the design process. The ability to model these variations and use such models to assess risk is important for two reasons:

1. Knowing the risks may help to prevent the stifling of design concepts in the face of erroneously perceived high risks.

2. A good understanding of risks may be used to guide the distribution of resources toward designs that have a higher probability of success.

At present, the gap between the "existing" and "ideal" in the application of stochastic techniques appears to be a "practice" gap. Designers are not using these techniques, even when the supporting tools/simulators are made available. It is not clear whether new stochastic-based design theories need to be developed for NLBMM design. It is most important then to increase awareness of the benefits of stochastic methods, how they can help in the design of NLBMM parts, and how they should be used during the design process.

Design Concept Architecture

The architecture of a design concept sets limitations on performance, the ease of manufacture, engineering resource requirements, time to market, and the cost of the final product. The ability to generate optimal design concepts and ascertain the best way to integrate meso- and microscale components should be a core competence for a designer. Designers need to understand and apply the concepts of functional decoupling, sensitivity, complexity and signal to noise ratios. There are several general methods that have been developed to guide designers in setting the architecture of parts and systems, for instance Axiomatic design (Suh, 1990, Suh, 2001), six sigma methods (Creveling et al., 2002, Yang and El-Haik, 2003) and robust design (Taguchi and Subir-Chowdhury, 2004). As discussed previously, physics and fabrication fundamentals change at the meso- and microscales. As a result, traditional perspectives and approaches to the generation of architectures must account for this. Recently, efforts have been made to integrate the unique physics and characteristics of the small-scale (e.g., functional) periodicity into the design process (Suh, 2005). More work is needed to better understand how to design concept architectures so that the fundamental issues associated with the physics and fabrication of meso- and microscale components have been addressed.

Given the difficulty of assembling small-scale parts, an important question to consider when constructing a concept's architecture is: "Should this product be designed as a heterogeneous or homogeneous system?" At present, the most appropriate use of the fundamental approaches that dominate macroscale designs (heterogeneous architectures and packaging), VLSI designs (homogeneous), or a mix of the two is not always readily apparent. Given the link between product architecture, cost, engineering resources, and time to market, additional research in product architecture for small-scale and multi-scale applications is in order.

The SOA and Gaps in Modeling and Simulation Tools

The SOA in modeling and simulation relies primarily upon the use of legacy tools that are based upon macroscale physics and continuum-based modeling. Several commercially available tools are capable of simultaneously modeling multiple physical phenomena within the limits of continuum models (COMSOL, 2005, ANSYS, 2005, COSMOSM, 2005, ALGOR, 2005, NEiNASTRAN, 2005, MSC Software, 2005, ABAQUS, Inc., 2005, CFDRC, 2005, MEMSCAP, 2005, IntelliSense, 2005). However, these tools often fail to provide accurate results when the type or nature (continuum or non-continuum) of the dominant physical phenomena changes. Four approaches are currently used in practice:

1. Designers use experimentation to develop empirical models of phenomena and experience.

2. Designers apply correction factors to map erroneous, continuum-based simulation results onto experimental results.

3. Designers use the appropriate physics to generate custom computer-based modeling and simulation tools.

4. Designers collaborate with established software vendors to develop custom modeling and simulation tools.

The panel observed the latter approach several times during site visits in Europe. The overall impression is that the European designers were focused upon generating accurate and robust modeling tools. The most successful designers were observed to have utilized a hybrid approach in which designers augmented existing software to meet their needs. This indicates a technology gap that is probably best addressed by software vendors. It would be very useful to have more open source modeling and simulation tools. At the very least, customers should be willing to more readily adapt them to accurately model custom applications. The open source approach to multi-scale modeling exists to some extent in tools such as FEMLAB (COMSOL, 2005), though a fair amount of expertise in programming is required to take advantage of this. Designers should look for the capability to customize commercial modeling tools when they are seeking to purchase modeling and simulation tools for NLBMM part design.

Integration and Fabrication Design Tools

Given the challenges associated with the selection of concept architecture, the fundamentally different nature of NLBMM micromanufacturing processes, and the lack of trained human resources, integration and fabrication issues will likely dominate initial design and engineering efforts. At present, designers often develop their own means of modeling and simulation via experiment, or via a combination of extensive experimentation and custom-designed simulation tools. Clearly, this is not the most cost-, effort-, or time-efficient means of ensuring that products will be well-suited for fabrication and integration. Researchers should be encouraged to generate modeling and simulation tools that reduce the cost and time associated with extensive experimentation.

Stochastic Design Tools

There is a need for modeling and simulation tools that link the performance of components to variations in the characteristics of the part and the variations in the performance of a system to variations in the performance of its parts. In many cases, the preceding may be addressed by custom-made, stand-alone simulations (e.g., Monte Carlo simulations). Unfortunately, many of these tools are difficult to learn and they are not integrated with existing geometric and behavioral modeling tools. This is perhaps the reason that stochastic design tools are not widely used in engineering design. Other modeling tools, for instance computational fluid dynamics (CFD) and finite element analysis (FEA)—both of which are integrated into computer-aided design (CAD) programs—find widespread use in engineering design. Computer-aided application software (CAX) vendors should consider adding this functionality to their programs and designers should look for this functionality when purchasing CAX tools.

Multi-physics, Multi-scale Design Tools

Behavioral modeling—Designers require tools that are capable of multi-scale, multi-physics behavioral modeling. Such modeling includes simulation of part function, assembly function, and the simulation of fabrication processes that are to be used to make the parts. Several commercial simulation packages (COMSOL, 2005, ANSYS, 2005, COSMOSM, 2005, ALGOR, 2005, NEiNASTRAN, 2005, MSC Software, 2005, ABAQUS, Inc., 2005, CFDRC, 2005, MEMSCAP, 2005, IntelliSense, 2005) are capable of multi-physics modeling at the macro-, meso-, and microscales. The development of a theoretical basis for multi-scale design tools is in the early stages. A few annual conferences have been seeded and journals have been started (Multi-scale Modeling and Simulation, http://epubs.siam.org/ sam-bin/dbq/toclist/MMS, International Journal for Multi-scale Computational Engineering, http://www.edata-center.com) to disseminate news of early advances. To date, the theories are application-specific, and are both derived for and applied in fields other than manufacturing. A micromanufacturing focused effort in multi-scale, multi-physics modeling tools is needed.

Geometric modeling—Commercial CAD tools are readily used to model the geometry of small-scale parts. Although these tools may be used to "get parts made," their utility and impact would benefit from specialization for small-scale design. This would be particularly useful when new DFX rules are required to restrict modeled geometries to be compatible with available fabrication processes. For instance, geometric modeling programs for MEMS automatically generate mask patterns and provide both guidance and constraint for DFX (MEMSCAP, 2005, IntelliSense, 2005).

The same capabilities would be helpful when NLBMM parts are fabricated using non-traditional processes.

Tools for Modeling Economic, Environmental, and Other Impacts

The potential impact of NLBMM technology is difficult to ascertain without appropriate modeling tools. For instance, the economic and environmental impacts of this technology have the potential to be large. The economic impact is described in a preceding section. The environmental impact is potentially significant due to the reduction in energy costs, pollution, and the material waste associated with smaller-scale parts and the smaller machines that make them. Given the perceived risks associated with new technologies, it will be important to have tools that model the impact and thereby help to justify the need for NLBMM products.

Gaps in the Processes Used to Develop Human Resources

Synthesis is a process in which knowledge, skill, and experience are used to guide a process wherein several concepts are created and then refined to create practical designs. Different perspectives exist on what synthesis, precisely, is. At one end of the spectrum, synthesis is considered to be a process by which existing knowledge, design rules, and concepts are combined to derive new concepts. In another approach, synthesis is the development of new concepts by generating new design rules; this is often called invention. In yet another view, synthesis is the result of mathematical models that are manipulated to provide alternative designs.

The importance of these different perspectives for our discussion will be explained shortly. For now, the key point here is that synthesis is in part dependent upon a designer's perceptions and experience. Although computer-based programs have shown great promise in performing synthesis via the first and third methods described above, they must be "fed" rules, and they are therefore not yet capable of competing with designers in the second method. Two important observations come from this: designers are needed to codify the rules, and designers are needed to generate new rules. An important problem for the design of NLBMM parts is that most designers are "preloaded" with intuition that enables them to understand and make design rules for macroscale applications; however, macroscale intuition is not always applicable at the meso- or microscale. Without meso- and microscale experience, designers will not be able to assess and mitigate the risks associated with designs. As such, meso- and microscale design education is needed for those who will practice synthesis in this field or for those who design synthesis tools for use in this field. Knowledge development must be facilitated in three ways:

1. *Existing knowledge and logic*: Undergraduate, graduate and professional courses are needed to help disseminate knowledge of the domi-

nant physics for meso- and microscale parts. MEMS/VLSI-specific fundamentals (Campbell, 2001) are commonly taught in MEMS manufacturing courses (Madou, 2002). The existing materials do not fully cover the required knowledge and design rules in sufficient depth for NLBMM design. As a result, either complimentary or new materials must be developed to disseminate this knowledge.

2. *Experience and skill*: Designers must obtain experience in synthesizing, modeling, fabricating, integrating and implementing meso- and microscale parts and systems. The means for designers to gain experience and skill should be integrated into the classroom, research projects and internships. In many cases, the resources required to do this (e.g., fabrication, testing, etc.) are expensive. Therefore, the combination of measurement tools, computer simulations and haptics (Ferreira et al., 2001), may be cost-effective "hands-on" experiences.

3. *New knowledge*: As NLBMM design and manufacturing are nascent fields, the broad and rapid advance of these technologies will best be served by a first generation of designers who are capable of generating knowledge to fill the gaps between existing and required design technology. This talent pool must be developed via training programs in basic and applied research.

   Other topics that are important to add to education are;
   i.   Simulation and experimental methods
   ii. Handling risk and uncertainty
   iii. System architecture/design

SUMMARY AND CONCLUSIONS

The design of/for NLBMM parts is unique due to: (1) the ability of the meso- and microscales to serve as a "handshake" between the macro- and nanoscales, (2) the transition in dominant physics that often takes place over the meso- and microscales, and (3) the change in the fundamental nature of fabrication processes that occurs over the meso- and microscales. These unique qualities give rise to gaps in the knowledge that supports the design of NLBMM parts, the design theory and process that are used to specify the embodiment of a design, and the design modeling and simulation tools that provide information for design decisions. Design researchers and practitioners have not had adequate time to solve the nascent and unique challenges that are associated with design of/for NLBMM technology. Continued research is required, with particular emphasis on encouraging work that helps to streamline the process of transforming knowledge into designs.

Another important issue for the design of NLBMM parts is that most designers are "preloaded" with intuition that enables them to understand and make design rules for macroscale parts and manufacturing processes. As such, meso- and microscale design education (beyond that available in VLSI/MEMS education) is needed for those who will practice design in this field. Design of NLBMM parts will require a new design perspective in which the behavioral dependence upon geometry/time scales is inherent to the design process. Designers therefore require tools that are capable of multi-scale, multi-physics behavioral modeling. This includes simulation of part function, assembly function, and the simulation of fabrication processes that are to be used to manufacture the parts. Existing geometric modeling tools, such as CAD, are sufficient to "get parts made," though their utility and impact would benefit from modular specialization for small-scale design. For instance, modules or "plug-ins" work within the CAD tool to provide feedback on the manufacturability of a given geometry. This will be particularly useful when DFX rules are substantially different from those that are used at the macroscale. This leads us to the need for new DFX rules and guidelines that will help designers select appropriate design-fabrication process choices, and the means to codify the rules so that they may be incorporated into behavioral/geometric modeling tools.

REFERENCES

ABAQUS, Inc. http://www.abaqus.com (Accessed October 20, 2005).

ALGOR. http://www.algor.com (Accessed October 20, 2005).

ANSYS. http://ansys.com/products (Accessed October 20, 2005).

Boothroyd, G., P. Dewhurst, W. Knight. 1994. *Product Design for Manufacture and Assembly*, New York: Dekker.

Campbell, S. A. 2001. *The Science and Engineering of Microelectronic Fabrication*, New York: Oxford University Press.

CFDRC. http://www.cfdrc.com (Accessed October 20, 2005).

Creveling, C. M., J. L. Slutsky, D. Antis. 2002. *Design for Six Sigma in Technology and Product Development*, Upper Saddle River, NJ: Prentice Hall PTR.

Ferreira, A., J. -G. Fontaine, S. Hirai. 2001. Visually servoed force and position feedback for teleoperated microassembly based virtual reality. In *Proceedings of the 32nd International Symposium on Robotics*, April, Seoul, South Korea, 676-681.

Hazelrigg, G. A. 1998. A framework for decision-based engineering design, *ASME Journal of Mechanical Design*, 120: 653-658.

———. 1999. An axiomatic framework for engineering design, *ASME Journal of Mechanical Design*, 121: 342-347.

IntelliSense. http://intellisensesoftware.com (Accessed October 20, 2005).

Madou, M. J. 2002. *Fundamentals of Microfabrication: The Science of Miniaturization*, Boca Raton, FL: CRC Press.

MEMSCAP. http://www.memscap.com (Accessed October 20, 2005).

MSC Software. http://www.mscsoftware.com (Accessed October 20, 2005).

NEiNASTRAN. http://www.nenastran.com (Accessed October 20, 2005).

Pahl, G. and B. Wolfgang. 1999. *Engineering Design: A Systematic Approach*, Second Edition, London: Springer.

Slocum, A. H. 1992. *Precision Machine Design*, Englewod Cliffs, New Jersey: Society of Manufacturing Engineers.

Suh, N. P. 1990. *The Principles of Design*, New York: Oxford University Press.

———. 2001. *Axiomatic Design: Advances and Applications*, New York: Oxford University Press.

———. 2005. *Complexity: Theory and Applications*, New York: Oxford University Press.

Syan, C. S. and U. Menon. 1994. *Concurrent Engineering: Concepts, Implementation and Practice*, Dordrecht, Netherlands: Kluwer Academic Publishers.

Taguchi, G. and Y. W. Subir-Chowdhury. 2004. *Taguchi's Quality Engineering Handbook*, Hoboken, NJ: Wiley-Interscience.

Ulrich, K. and S. Eppinger. 2003. *Product Design and Development*, Third Edition, New York: McGraw-Hill/Irwin.

Yang, K. and B. S. El-Haik. 2003. *Design for Six Sigma: A Roadmap for Product Development*, New York: McGraw-Hill Professional.

CHAPTER 3

MATERIALS

David Bourell and Kamlakar Rajurkar

ABSTRACT

Materials play an important role in manufactured goods. Materials must possess both acceptable properties for their intended applications and a suitable ability to be manufactured. These criteria hold true for micromanufacturing, in which parts have overall dimensions of less than 1 mm. This chapter begins by reviewing materials usage in Asian and European research in micromanufacturing, categorized by manufacturing process. Following that, specific treatment is given to materials factors that are unique to micromanufacturing.

MATERIALS FOR MICROMANUFACTURING

Machining

Dr. Ohmori at RIKEN has developed a number of impressive machines and processes for micromanufacturing, such as the electrolytic in-process dressing (ELID) feature in which tools are continuously dressed during machining. Key to the process is use of diamond particulate tools with a cast iron matrix. Parts include silicon lenses, ceramic parts and complex die steel molds. Machines include grinders and three- to six-axis computer numerical control (CNC) machines. Micromolds with 50 μm features have been made. Ohmori has done considerable work in surface microgrooving. Surface profiles in copper were produced with 1,000 nm, 500 nm and 250 nm grooves. At 100 nm, grain size became an issue in producing reliable features in chip-free fashion. Finally, precision grinding has produced 8-inch diameter silicon wafers with 50 μm thickness.

Kugler makes ultraprecision air-bearing machine tools/flycutters; multi-axis systems for micromachining; mechanical and laser cutters with laser 3D-measuring equipment; reflective optical components for spectra ranging from the infrared (IR) to the visible (VIS) with submicron accuracy; and beam-bending components and systems for high-power CO_2 lasers such as focusing heads, bending units, and lightweight components. One of their major manufactured products is a copper mirror used for beam bending of up to ~40 kilowatt CO_2 lasers.

Philips CFT produces microlenses with a surface finish at 2 nm using ELID-poligrind. During the site visit, they demonstrated a Linear Fresnel Lathe equipped with a diamond tool. A tool grinding apparatus was also developed in this center.

FANUC has developed the ROBOnano ultra high-precision multifunction machine tool. Using a cutting tool made from a single-crystal diamond, ROBOnano can machine hard materials such as tungsten carbide lens cores.

The National Institutes of Advanced Industrial Science and Technology (AIST) has developed a microfactory that includes a microlathe and micromill. The lathe is used to machine 2 mm diameter brass starting stock to 500 μm. The cutting tool was a diamond wedge. The micromill uses a 700 μm carbide end mill cutter to hollow out a cylindrical housing in 900 μm diameter brass. These parts and their assembly are shown in Figure 3.1.

In the University of Tokyo's Department of Engineering Synthesis, soda glass has been micromachined using an oscillating diamond tool. The grooves are 3–4 μm deep and 150–200 μm wide. Laser machining of lithium niobate glass with a 355 nm ultraviolet (UV) laser has been performed, creating 10 μm deep, 300 μm wide grooves for optical waveguide applications.

Also at the University of Tokyo, a computer/strain gage-controlled three-axis table with a 200k RPM tool has been used to create 100–500 μm parts in poly-methyl methacrylate using one or more fine carbide tools mounted on air spindles. In a second nanomanufacturing project, a preparatory controlled atmosphere chamber has been used to fast-atom-beam (FAB) machine thin-layered silicon and polyimide. CHF_3 accelerated by electric ionization, acceleration, and neutralization created the beam. The FAB was projected onto a micromask produced using a proprietary process. The target material was then selectively sputtered. One part production was silicon "microtorii." The shrine shape was machined in ~10–20 μm thick silicon using FAB machining. The long dimension was 100 μm. The part was then moved using 15 μm diameter electrically charged probes. Two square holes were FAB-machined and loaded with solder paste prior to positioning on the microtorii vertically into the holes. The solder was then laser reflowed to generate the final part. Another demonstration of micromachining was creation of a polyimide cabin. The cabin dimensions were 500 μm by 400 μm by 500 μm tall with walls/roof each about 80 μm thick. The walls were machined by laser cutting and were positioned using the electrically charged probes. The roof was also laser cut, ridge-line creased and folded prior to positioning atop the structure. Two scanning electron microscope (SEM) imaging systems were available for visualization of the process.

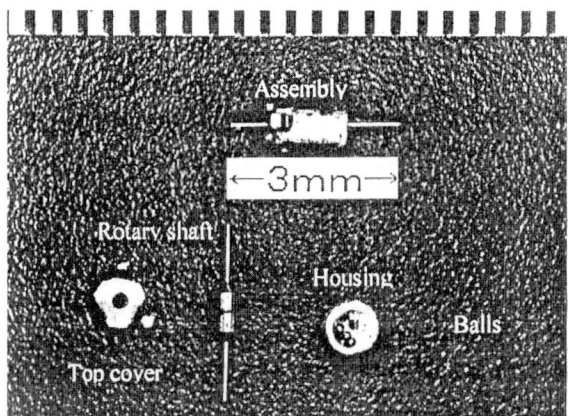

Figure 3.1. A brass spindle assembly created and assembled using the AIST microfactory.

Professor Higuchi at the University of Tokyo developed a four-axis machine commercialized by Toshiba Machine Company. Control to 1 nm is available. It has been used to produce 30 mm curved lenses, mirrors and immersion gratings with microscale surface features. A copper optical lens and micromachined forceps in 303 stainless steel have been produced.

Kyocera designs and manufactures machine vision lens and illumination systems that are extensively used in semiconductor manufacturing, electronic assembly, laser guidance devices, as well as for photography, scanners, microscopes, and charged-couple device (CCD) TVs. Kyocera produces about 2 million lens elements per month. Grinding and polishing are the fundamental processes used for making spherical, cylindrical and aspheric lenses of sizes ranging from 1.8 mm to 300 mm. Kyocera is also a leading Japanese manufacturer of silicon and germanium lenses for infrared applications.

Kyocera produces cutting tools for use in the machining of a variety of materials. Recently the company introduced two new chemical vapour deposition (CVD)-coated grades for cutting ductile cast iron at the International Manufacturing Technology Show (IMTS) in September 2004. Kyocera also offers a line of microsize cutting tools. They make 2 million tools each year including 1.5 million tools which are smaller than 0.020 inch in diameter for applications in resistance welding, injector nozzles, extrusion

dies, medical/dental components, mold-making and semiconductor manufacturing. Kyocera has produced electro-discharge machining (EDM) electrodes of tungsten-copper, chromium-copper, tungsten carbide, graphite
and tantalum of diameter 0.001 inch and aspect ratios of 15 and 70 (depending on the electrode material) with a high-volume consistency in diameters from 0.005 inch to 0.250 inch, and tolerances of ±50 millionths
with excellent surface finish.

National Cheng Kung University has a micromilling capability that includes use of diamond-coated metal tools with 100 μm shanks and 50 nm
cutting edge radii. Confirmed by theoretical modeling, a key difference between macromilling and micromilling is the tendency of the latter to induce plowing during machining due to the large cutting edge radius relative to the depth of cut.

Kyoto University demonstrated a high-speed CMC machine. Using a
vanadium carbide (VC)-coated carbide tool for milling and a cubic boron
nitride tool for grinding and cutting, the CMC machine is capable of machining hardened steel (HRC53) at speeds approaching 5 m/s. Resolution
is in the range of 5–6 μm.

Researchers at Sansyu Finetool Co. Ltd. manually cut tungsten carbide
micromilling cutters with a final diameter of 20 μm using diamond tools.

An ultraprecision two-axis machining center has been developed at the
Korean Institute of Machinery and Materials (KIMM). It uses a single-
crystal diamond cutting tool. Temperature control limits the accuracy to
the sub-micron level.

Mitsubishi Electric Corporation has developed a laser cutting system
capable of achieving accuracies on the order of 2.5 μm. A variety of sheet
parts of various materials have been cut using this technology (see Figure
3.2).

Figure 3.2. Laser-machined parts from Mitsubishi Electric
Corporation.

Focused ion beam (FIB) milling is used at AIST to produce glassy carbon-embossing molds. The low coefficient of thermal expansion makes glassy carbon well suited since part removal is facilitated. The ion source was gallium (Ga) and the FIB-etching depth was linearly proportional to the accumulated beam current dose. Submicron embossing details are possible.

The Institut für Mikrotechnik Mainz GmbH uses X-ray lithography, electroplating and molding (LIGA) manufacturing, and laser micromachining technologies for microstructuring technologies. High-aspect ratio shapes, 500 μm tall with 85 μm wide features, can be built using SU-8 photoresist. Electrode structures have also been created in gold, platinum, aluminum and silicon nitride. Features as small as 1 μm thick and 10 μm wide were obtained. A second application was the creation of platinum resistors spanning microbridge structures. Microfluidic structures may be produced using advanced silicon etching (ASE™), as shown in Figure 3.3. Channel widths as narrow as 40 μm and 300 μm deep may be produced in silicon as well as in silver and copper. Silicon deep etching of membranes and cantilevers may be produced using ASE™ with 0.3–50 μm width and 0.1–10 μm height. A variety of materials are amenable to the process including silicon, silicon nitride, tungsten silicide, nickel and copper.

Two-micron diameter hole arrays have been produced at the Institut für Mikrotechnik Mainz GmbH for isopore filter membranes using UV-lithography. LIGA techniques have been used to produce molds for plastic injection molding a variety of shapes. Molds from nickel, steel and polyoxymethylene (POM) have been produced; one application has been in a three-stage planetary gear that is ~50 μm in diameter.

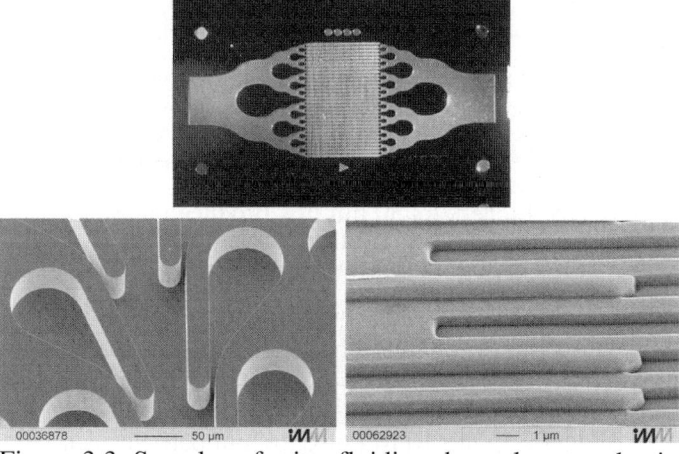

Figure 3.3. Samples of microfluidics channels created using Advanced Silicon Etching™, from the Institut für Mikrotechnik Mainz GmbH.

The Institut für Mikrotechnik Mainz GmbH has used the lithographic approach to create Interdigital Electrode Structures (IDS). A mold is formed in silicon and is used to produce small polymer and nickel parts. Using oblique sputtering across grooved surfaces, it is possible to generate discrete sputter zones for sensors. Probes for brain insertion are prepared using a proprietary process. Wires are end-finished to a 30–35 μm probe tip with a linear or helical sensor array further away from the tip. Ultraprecision machining includes use of diamond cutters for turning, milling and grinding of nonferrous metals and polymers. A surface roughness of less than 10 nm and a shape error of less than 0.1 μm/100 mm are obtained. Used for optical applications such as hot embossing tools for polycarbonate, lenses of ~315 μm in diameter can be produced. A pyramidal surface feature in aluminum and brass is created using a rotational table.

A major category of research at Yonsei University is the application of laser and ion beam technologies for microfabrication. Projects in this area include work in nanomold fabrication using FIB machining, fabrication of patterned media using immersion holography lithography, and laser micromachining and monitoring. Several types of holographic patterns have been produced with typical pattern geometric parameters, including pitch sizes of 200 nm, line widths of 100 nm and aspect ratios of 1.18.

Seiko Instruments Inc. presented to the panel a new nanofabrication technology for manufacture of 100 nm apertures. First, 20–50 nm tips were prepared by etching. These were used to puncture thin aluminum film to create the apertures. The tips and film holes are shown in Figure 3.4. Similar microprobe arrays have been produced at Yonsei University.

Tip and Aperture

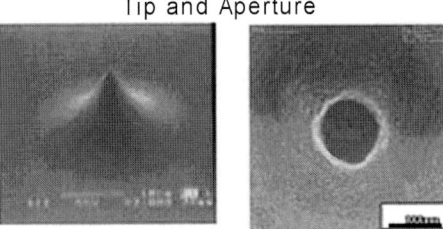

Tip Radius: 20~50 nm

Figure 3.4. Etched 20–50 nm tips produced at Seiko Instruments Inc. and their use in producing apertures in aluminum film.

Hole drilling is improved using femtosecond pulsed lasers at the Laser Zentrum Hannover. The quality of the hole includes considerations of cracking avoidance and clean, sharp hole features with no localized melting, burring or material build-up, Figure 3.5. It is possible to deposit gold nanobumps on glass. Using 10 nJ/pulse, 60 nm particles can be deposited. For patterning, 100 nm gold particles are arrayed to create photographic images microns in size.

Figure 3.5. Clean hole drilling in steel using a femtosecond laser. Approximately 200 μm in diameter. From the Laser Zentrum Hannover.

The "Shalom" ultra-high-speed milling machine at AIST uses a 30-μm carbide cutting tool rotating at 300,000 RPM to create parts with features as small as 30 μm in aluminum and hardened steel. The depth of cut is 20 μm to preserve tool life, and jerk is high, up to 2.1 g acceleration in 30 seconds.

A novel microlens process has been demonstrated at the National Cheng Kung University. Laser masks are created and used to selectively laser machine polycarbonate. A 248 μm KrF laser is used to create parts with accuracy on the order of 1 μm and roughness of 10 nm. The lenses are 600 μm in diameter and 60–100 μm thick. Of particular interest was the mask design, which included rotated triangular openings and planetary-rotated four-blade fan shapes. A model that predicts machining depth and profile backed the shapes.

A single point diamond machine has been developed at the Industrial Technology Research Institute (ITRI) Mechanical Industry Research Laboratories. Equipment was modified in-house to enable nano- and microscale feature generation for applications such as microlens arrays and diffraction gratings. Ductile mode machining is routinely used for brittle materials.

Electrical Discharge Machining

Research using both wire micro EDM and plunge/sink EDM units was being conducted at the Matsushita Electric Industrial Company. Wire EDM was used exclusively to manufacture carbide EDM tools. Carbide wire 100 μm in diameter was cut and wire EDM-ed to reduce the diameter and form the profile shape. For plunge micro EDM, electrodes were made from carbide or tungsten. Micro-EDM workpieces include 304 stainless steel, tungsten carbide and silicon. Micro EDM has also been used to create punch and die sets for stamping gears less than 20 μm in size.

The Institut für Mikrotechnik Mainz (IMM) GmbH is making an effort in precision machining and micro EDM. Parts are used in watches, injection molding, stamping dies and the medical field. Structural features as small as 0.5 μm are obtainable. Equipment includes two computer numerical control machines with high-speed rotational spindles and with 0.1 μm resolution, a Fehlman Picomax 54 2D CNC and a Fehlman Picomax 60 3D CNC. They have several Swiss-made EDM machines, an AGIE HSS150F 0.025–0.25 mm diameter wire EDM and an AGIE Vertex EDM with 0.02-0.2 mm diameter wire. Researchers at the IMM have used this approach to create ~80 μm diameter shafts with axial holes and 100 μm features. EDM machining time is about 20–30 hours depending on the shape produced.

In collaboration with the University of Limerick, researchers at the Institut für Mikrotechnik Mainz GmbH created a 6 mm by 6 mm fan housing. Rotating parts have a 25 μm clearance, and the fan blade is 15 μm at the thinnest section. The three-step process is to cut out a rotor blank 50 μm in size using wire EDM. Second, the part is precision-machined using sink EDM. Last, a second sink EDM process is used to finish the blades.

Slit arrays 80 μm wide on about 200 μm centers have been produced in steel at the Institut für Mikrotechnik Mainz GmbH. LIGA microextrusion electrodes have been machined to create cross or hexagonal extrusion die openings. End mill tools 30 μm in diameter have been made using wire EDM in titanium boride, a conducting ceramic. EDM turning of tungsten carbide is used to point the ends of wire. A nominally 100 μm reduced section flares into a 150 μm sphere followed by a tip about 80 μm long and 10 μm in diameter at the end. Sink EDM tools with 80-300 μm features have been produced in tungsten carbide or copper. These tools are then used as sink EDM electrodes inserted in holes and run around the hole inner diameter to create internal non-circular cross section holes. EDM drilling of 50 μm holes in platinum and 25 μm holes in stainless steel foil has been demonstrated. Applications of micro EDM include superfocal mixer components, oval gear wheels and rib electrodes. Tolerance on the order of ±1 μm is achievable in wire EDM of stainless steel.

Sink EDM has been used at the Mitsubishi Electric Company to create meso- and microscale features and components. Examples are lens dies 500 μm in diameter with 0.5 μm surface finish and tungsten carbide inserts with 3 μm surface finish. Wire EDM has been used to create fine slits 100 μm wide and 125 μm holes in 800-μm carbide.

Researchers at the National Cheng Kung University are developing micro-ECM and -EDM processes for production of micromilling tools.

At National Taiwan University, micro EDM is used for production of parts using high-aspect ratio 8 μm tools. They also have a wire EDM unit that uses 20 μm brass wire for vertical, horizontal and slanted cutting. An impressive list of parts includes outer gears as small as 600 μm without tooth taper, 324 μm micropinions, and even a Chinese pagoda just over 1 mm in size, as shown in Figure 3.6.

Figure 3.6. A Chinese pagoda built using wire EDM at the National Taiwan University. The part is just over 1 mm in cross section and 1.75 mm tall.

Professor Masuzawa at the University of Tokyo Center for International Research on Micro Mechatronics has extensive experience in micro-machining, particularly micro EDM. He has done pioneering work in these areas that has led to a number of commercial successes. One of his most notable accomplishments is the development of the micro-EDM process and machine that incorporates a wire electro-discharge grinding (WEDG) unit for the *in situ* preparation of the electrodes. This development has allowed for a dramatic increase in accuracy and micro-EDM capabilities. Panasonic has commercialized this concept. He has also conceived the concept of uniform electrode wear in micro EDM and the micro-EDM lathe that allows for the creation of intricate external and internal features, Figure 3.7.

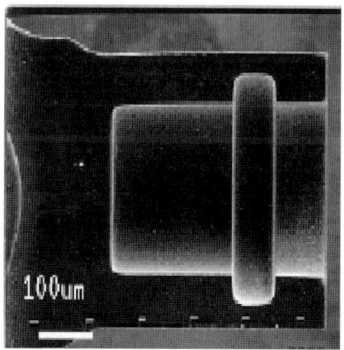

Figure 3.7. Microhole with internal groove machined by EDM lathe at the University of Tokyo.

Sansyu Finetool Co. Ltd. uses a wire EDM machine with 30 μm wire for mold manufacturing.

Deformation Processes

A Micro/Mesomechanical Manufacturing R&D Department has been organized at the Metal Industries R&D Centre in Kaosiung, Taiwan. Included are facilities for microforming, microstamping and microextrusion. Development was in partnership with the University of Erlangen/Nuremberg in Germany. They have successfully extruded thick bar stock aluminum hydrostatically to produce 50 μm diameter wire.

Researchers at AIST have created a microfactory that includes a micropress. Brass foil 120 μm thick is punched into a top cover for spindle assembly (refer to Figure 3.1) in the six-station micropress. The assembly is completed by pinching the three top-cover prongs onto the housing using microrobot tweezers.

Fundamental research at Nagoya University centers on deformation processing of aluminum and its alloys to refine the grain morphology. Using compressive torsion from room temperature to about 300°C above room temperature, they have demonstrated a tenfold reduction in grain size by several minutes of processing.

Researchers at Matsushita Electric Industrial Company have stamped 18 μm copper foil to create demonstration gears.

Additive Manufacturing

Dr. Ikuta in the Department of Micro-system Engineering at Nagoya University has worked for many years with microstereolithography (μSLA). The process, termed the "IH Process," consists of numerous improvements to traditional SLA designed to facilitate formation of microparts. Also included are: cross-linking slightly below a surface glass slide to facilitate unagitated thin layering as small as 200 nm; short focal length to produce a small diameter laser spot, a high-frequency (350 nm) laser; and a high-viscosity polymer to facilitate overhang formation without a support structure. Parts have been produced including a gear of 15 μm diameter in about four to five minutes, moving microturbines (15 μm diameter) rotating gears (15 μm diameter), grippers and a three degree-of-freedom (DOF) manipulator for cells that is 10 μm in size and 250 nm thick.

Dr. Chichkov at the Laser Zentrum Hannover has developed two-photon polymerization (2PP), a freeform technique for creating freeform objects in a photopolymeric liquid like SU-8 photoresist. The laser interacts with the liquid only near the focal point where the photon density is high, creating radicals that initiate polymerization of the resin. Exquisite parts, such as miniature replicas of statues, demonstrate the versatility of the technique (see Figure 3.8). Part dimensions are on the order of 20–100 μm. The polymer resin is an inorganic/organic hybrid polymer (ORMOCERS). Other parts include nano-sized three dimensional scaffold mesh structures with 220 nm lines on 450 nm centers. Lateral resolution of the 2PP process is 120 nm. Other application examples include spiders with 1 μm legs that are ~50 μm across and ~30 μm tall, free-standing LEGO hollow columns approximately 50 μm in diameter and 300–400 μm tall, three-dimensional tapered waveguides 2–10 μm in width, and nanoembossing.

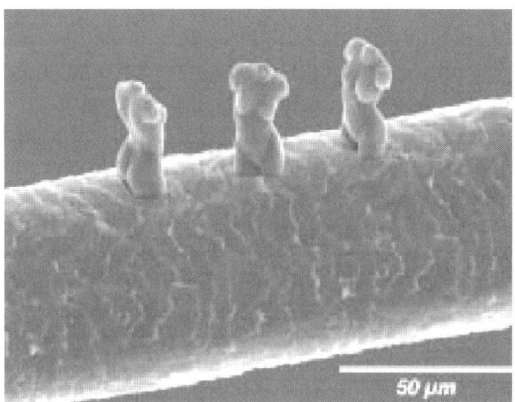

Figure 3.8. Images of a microscaled Venus statue built on a human hair using two-photon polymerization. From B. Chichkov, Laser Zentrum Hannover.

Dr. Regenfuβ at the Laserinstitut Mittelsachsen e.V. has developed a laser microsintering device. Powder is spread over a work surface under vacuum and sintered using a scanning laser beam. Structures with feature size on the order of 100 μm have been produced in a variety of metals including tungsten, Figure 3.9. Ceramics laser microsintering is under development at this time.

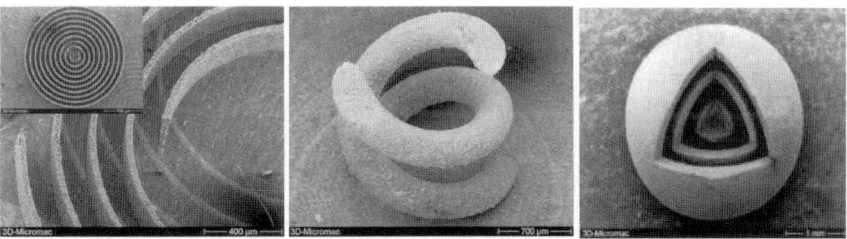

Figure 3.9. Tungsten Parts created using laser microsintering. The central spring is 1.6 mm across with a "wire" diameter of approximately 350 μm. From P. Regenfuβ at the Laserinstitut Mittelsachsen e.V.

Joining

The Micro-Joining Lab at the Metal Industries R&D Centre has developed plasma-welding, laser-welding and resistance-welding capabilities. For example, Fe-Ni-Co caps 50 μm thick are microwelded onto the iron headers of diodes. Another example was production of 1 mm wide plasma welds in 200 μm thick 304 stainless steel to create 50 μm thin-walled tubes.

Coatings

Researchers at AIST have produced coatings using lead zirconate titanate (PZT) and alumina. Submicron particles are formed into an aerosol of dried air or helium. Particles are accelerated to about 200 m/s and are selectively deposited onto silica, soda glass or copper substrates through use of a laser machined or photoresist mask. Deposit width and thickness values are several hundred microns and 10-20 microns, respectively.

A project on "Micro Spark Coating" is underway at the Mitsubishi Electric Corporation. The process forms a hard surface of ceramic or metal cladding about 1 mm thick on metallic substrates. An electrode of semi-sintered powder moves 100–200 μm over the surface of the substrate. Discharge energy melts the electrode and substrate and deposits the former onto the latter. The process has been demonstrated on aircraft engine components.

The microprecision surface treatment lab at the Metal Industries R&D Centre has facilities for ion implanting, surface coating and physical vapor deposition (PVD). DVD molds have been coated with diamond-like carbon for wear resistance.

Researchers at the National Cheng Kung University are developing a laser nanoimprinting process. The goal is to selectively imprint nanocrystalline nickel onto the surface of stainless steel. The nickel is coated on a glass slide, and a laser is transmitted through the slide to transfer the nickel onto the steel surface.

Molding and Solidification Processes

The Sansyu Finetool Co., Ltd. has extensive polymer microinjection molding facilities. Working with materials including polybutylene terephtalate (PBT), pentabromobenzyl acrylate (PMMA), acrylonitrile butadiene styrene (ABS), polycarbonate (PC), POM and thermo plastic elastomer (TPE), they specialize in over-molding of dissimilar materials and in the use of multi-slide core molding of very small components with tight tolerance. Sample parts, shown in Figure 3.10 and Figure 3.11, include feature sizes in the 7–100 μm range.

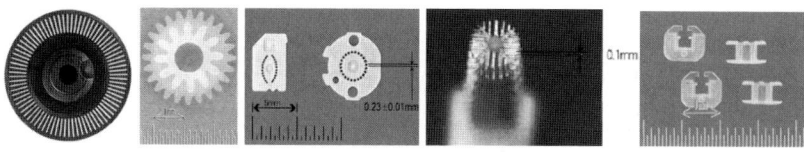

Figure 3.10. Polymer injection-molded parts from Sansyu Finetool Co. Ltd. (a) Encoder disk, (b) Miniature gear, (c) Miniature holes, (d) Fins, (e) Wire coil winding bobbin.

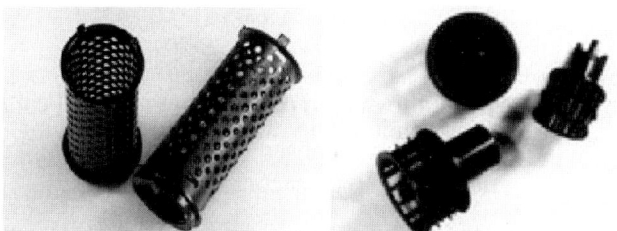

Figure 3.11. Components requiring molds with moving cores.

A project is underway at AIST to injection-mold acrylic resin optic lenses that will be 14 mm in diameter. The primary challenge is thermal control of the molten resin prior to injection. The material transforms into the solid state prior to injection through the nozzle to form the part.

A miniaturized, tabletop injection molding machine is under development with an industrial partner at ITRI Mechanical Industry Research Laboratories.

Samsung Electro Mechanics Corporation R&D Center has developed a micromotor for auto-folding cell phones. The motors are embedded in 20 mm long zinc die-casting.

Professor Higuchi at the University of Tokyo has demonstrated the "pulling" of paraffin domed cylinders on 15 μm centers. A laser locally melts the material where the cylinder is to be built. Forces that were undisclosed to the panel visiting the site are then applied to the melt under the laser beam which result in rising of the melt along the laser beam axis. The cylinder aspect ratio was approximately 5:10.

Embossing

Dr. Takahashi at AIST has led the development of a hot embossing unit. Pyrex and quartz are heated above their softening point and are pressed into previously prepared glassy carbon molds. Maximum press force and temperature are 10 kN and 1,400°C, respectively. Features in pyrex are on the order of 3–10 μm long and are submicron in width (see Figure 3.12).

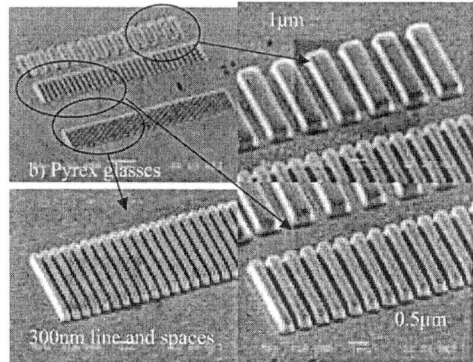

Figure 3.12. Hot embossed pyrex glass at AIST. Features are submicron in width. The embossing parameters are 600°C, 0.22 MPa and molding time equal to 60 seconds.

Polycarbonate lens patterns have been prepared by laser machining and coated with 1–2 μm nickel by electroforming at the National Cheng Kung University. The polymer is removed and the nickel is backed to create a microembossing tool for polycarbonate sheet. Lens size is on the order of 60–100 μm.

Research at Yonsei University includes molding parts using a nickel stamper with a self-assembled mono-layer. Figure 3.13 shows microlenses and a microlens array produced as part of this research.

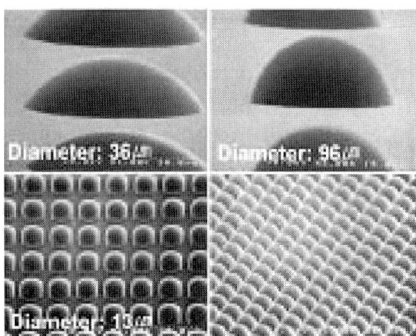

Figure 3.13. Microlenses and a microlens array produced by nickel stamping at Yonsei University.

Dr. Chichkov at the Laser Zentrum Hannover used two-photon polymerization to create molds with negative features that, upon being embossed onto polymers, produced raised structures for optics applications.

Philips CFT uses a machine to make indents on surfaces for microlenses. The machine indents a form in the surface in the desired lens shape.

Composites

Dr. Kanno at Kyoto University has developed a method for sputtering a variety of materials, particularly PZT, using powder-processed targets. The substrate was magnesia and $SrTiO_3$ to get the right holing of the PZT. Applications of the materials include microfluidic devices and micropumps.

Using a process termed "μ-ARTS," or micro-arranged tube technology, Hitachi Chemical R&D Center has created polyimide polymer films 500 μm thick with 250 μm diameter linear composite fibers "sewn in." The fibers may be polyetheretherketone (PEEK), perfluoroalkoxy (PFA), optical fibers, or copper wire. The wires are laid on the surface following application of an adhesive using ultrasonic vibration. A second layer of polyimide is applied to sandwich the fiber assembly. Applications are heat exchangers for microelectronics devices, motherboards for microreactors, and bio-applications.

Hitachi Chemical R&D Center has also developed a large number of polymer-based systems, including ANISOM™, a non-conducting polymer matrix containing metal-coated glass spheres (5 μm diameter). This anisotropic-conducting film can be made conductive in the through thickness direction by compressing the film. The plane of the film remains an insulator.

Dr. C-K. Chung at the National Cheng Kung University has produced several case studies of micro-nano technology. The first case study was monolithic microelectromechanical systems (MEMS)-based lithographic microstructures for thermal inkjet (HP) and bubble-jet (Canon) printhead applications. The active material was PZT. The second project was creation of an ionic polymer-metal composite (IPMC) actuator. The actuator finds potential application in waveguides, biomimetic sensors and artificial muscle. Gold nanoparticles are embedded in Nafion by dissolution and casting. The Nafion is embossed before being electroless plated with gold and electroforming with nickel. The resulting IPMC will bend significantly under applied voltage. For example, a 3 V driving voltage can result in 6.8 mm displacement, 0.222 grams force and more than a 90° bend. The third case study was lithographic nanostructured TaSiN films. Two-hundred eighty nm thick SiO was generated on a silicon substrate followed by about 900 nm of TaSiN. The hardness was about 15 GPa and the reported grain size was about 1 nm.

Dr. Kobashi at Nagoya University has been developing a pressureless infiltration technique that uses an exothermic reaction in the powder phase to enable pressureless infiltration. He has demonstrated this approach to produce titanium carbide infiltrated with either magnesium or aluminum. Work is also underway to use combustion synthesis to produce porous titanium with controlled microstructure.

MATERIALS ISSUES IN MICROMANUFACTURING

Products of micro- and nanofabrication have size scales spanning from about one thousand micrometers down to tens of nanometers. Independent of the specific manufacturing process, these size scales impact material structure and concomitant mechanical and physical properties.

During the site visits, panelists identified several materials features that are unique to micromanufacturing. One of the issues discussed by representatives of U.S. industry at the Review Workshop in Washington, D.C. on August 12, 2004, was the economic driving force for the development of new materials specifically suited for micromanufacturing. For macro-manufacturing, material quantities of significant size can drive development of new materials. This is nonexistent in micromanufacturing, where annual volume production of parts is often measured only in cubic centimeters. This fact was confirmed and reiterated in discussions with Mr. A. Kamiya, Chairman of Sansyu Finetool Co. Ltd. and W-S. Hwang, president of the Metal Industries Research and Development Centre.

Another feature unique to micromanufactured materials is the need for clean, inclusion-free materials, as noted by Professor Masuzawa of the University of Tokyo Center for International Research on Micro Mechatronics. For example, manganese sulfide stringers in steel may be over 100 μm long, on the order of some part dimensions.

According to Dr. Ohmori of RIKEN, a grain size effect occurs on surface finish as the result of the groove machining of copper. When the groove dimension falls below about 100 nm, it becomes increasingly difficult to control surface finish.

For polymers, the ability to predict the dimensional stability and accuracy of injection-molded products requires correct modeling of the specific volume. The effect of thermo-mechanical history on the resulting specific volume of semi-crystalline polymers is needed (the influence of high cooling rates and mechanical deformation). A new dilatometer is being designed and built at the Technical University of Eindhoven for experimental validation of current pVT constitutive equations.

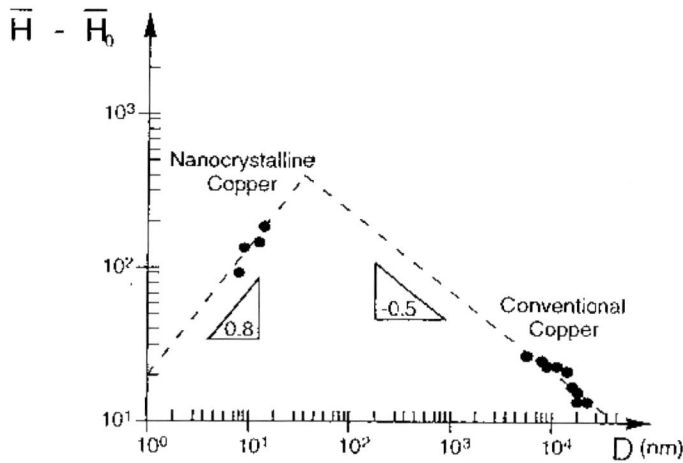

Figure 3.14. Plot showing the influence of grain size on part strength. (Chokshi et al. 1989, reprinted with permission).

A principal feature of materials is the presence of grains of size d. The boundaries between grains in polycrystalline materials are disordered crystallographically and serve as high-mobility pathways for atomic transport. Typical grain size varies between a few micrometers to approximately 100 µm. As the overall part size decreases below this size range, grains must either be removed or refined. If removed, the resulting part becomes a single crystal. Most physical and mechanical properties of single crystals are anisotropic, leading to directional variations in properties (Nye, 1985). For the latter, reduction in grain size results in significant property changes. The strength of conventional polycrystalline materials obeys the well-known Hall-Petch relationship, in which strength is proportional to the reciprocal square root of grain size (Chokshi et al., 1989). Extension of this relationship to micro- and nanoscale grains results in sizeable strength increases. Departure of material behavior from this Hall-Petch relationship may occur at the nanoscale, with reports of an inverse linear relationship and even grain-size weakening (Chang et al., 1995), such as that shown in Figure 3.14.

Dr. M-C. Tsai and co-workers at the Metal Industries Research and Development Centre in Kaosiung, Taiwan have found evidence that grain-size effects are diminished when the test sample size decreases below about 1 mm, an observation that has import for micromanufacturing (Tsai and Chen, 2004, Tsai et al., 2004, Tsai, 2004). Shown in Figure 3.15 are stress-strain plots for compression testing of 5 mm diameter copper (UNS12000) cylinders. The grain size d varied between 18 µm and 200 µm. According to conventional literature (Bardes, 1979), the strength difference based on Hall-Petch should be 100-200 MPa. Tension results are more telling. Figure 3.16 shows tension results for UNS12000 copper samples with grain size d between 23–113 μm and gage thickness T varying between 100–1,000 μm. The strength is reduced with an increase in d/T, not the Hall-Petch grain size d. Furthermore, the tensile strength may apparently be described by a rule of mixtures between surface grains with a grain-size independent strength and interior grains that obey Hall-Petch.

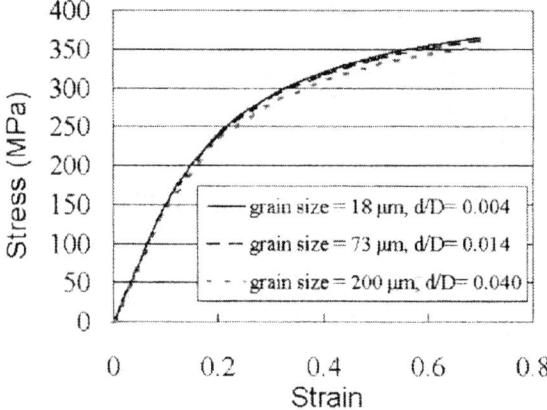

Figure 3.15. Compression-tested UNS12000 copper cylinders where d is the grain size and D is the sample diameter (5 mm). A lack of grain-size effect on strength is evident (Courtesy Tsai et al. 2004).

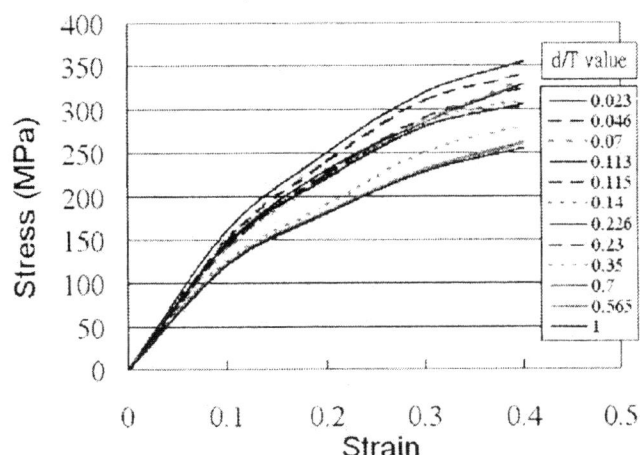

Figure 3.16. Tension-tested UNS12000 copper. d is the grain size and T is the sample gage thickness. The strength appears to correlate to d/T, not d (Tsai et al. 2004).

Similar thickness effects have been observed in pure aluminum at the Technical University of Eindhoven in the Netherlands. As shown in Figure

3.17, the tensile strength decreases with decreasing part thickness in the range of 0.1–0.5 mm.

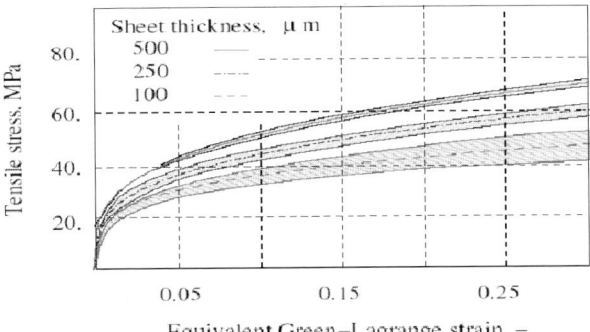

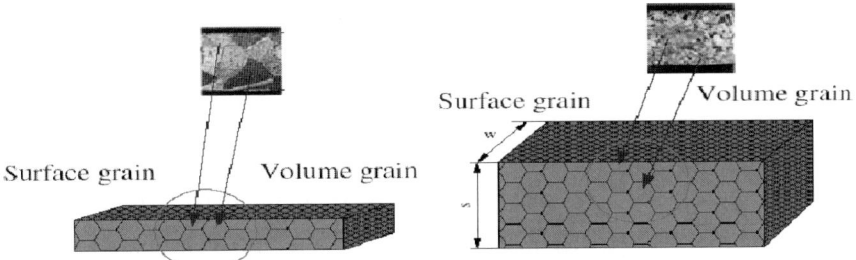

Figure 3.17. Tensile tests of aluminum sheet of constant grain size and varying overall thickness. From M. Geers, Technical University of Eindhoven, Netherlands.

The rule-of-mixtures approach to strength mentioned previously may be expressed mathematically. For a tension sample that is T thick and βT wide, the number of surface grains N_s may be expressed as a fraction of the total grains N_g in the gage section:

$$N_S = \frac{2(\beta + 1)Td}{\beta T} N_g$$

For sheet specimens, $\beta \rightarrow \infty$, and the fraction becomes simply 2d/T. It is interesting to note that in Figure 3.16, the strength becomes d/T independent when d/T>0.5, associated with 100% of the grains possessing a free surface. Applying the rule of mixtures to the N_s surface grains and N_i (=N_g-N_s) interior grains, one obtains

$$\sigma_y = \frac{N_s\sigma_{ys} + N_i\sigma_{yi}}{N_g} = \frac{2(\beta+1)}{\beta}\frac{d}{T}\left(\sigma_{ys} - \sigma_{yi}\right) + \sigma_{yi}$$

where σ_{ys} is the grain-size independent strength of surface grains and σ_{yi} is the Hall-Petch strength of interior grains. Figure 3.18 is a plot of the calculated flow stress from this equation compared to the experimental flow stress at $\varepsilon = 0.4$ from Figure 3.16.

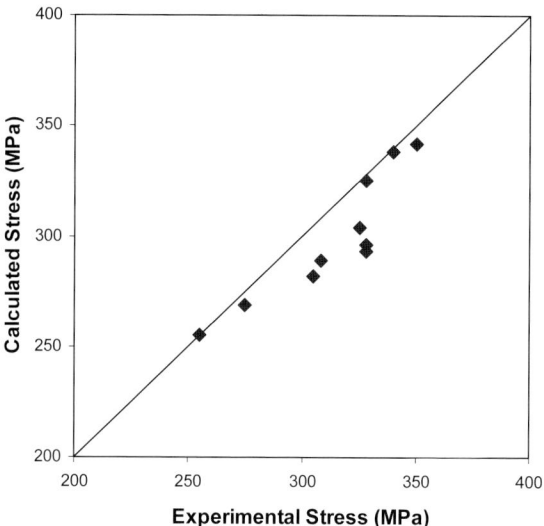

Figure 3.18. Calculated flow stress compared to experimental values for a strain equal to 0.4 (Figure 3.16), based on a rule of mixtures between surface and interior grains.

For nanoscale-grained materials, as many as 15-50% of atoms are present in the disordered grain boundary regions (Padmanabhan and Hahn, 1996). This gives rise to grain boundary instability. For example, nanocrystalline palladium has shown grain growth at $0.16T_m$ (T_m is the absolute melting point), whereas typical materials do not exhibit significant grain growth until temperatures reach approximately $0.5T_m$ (Weissmüller et al., 1995). Ultrafine-grained materials have been shown to obey the traditional power-law coarsening relationship

$$d^n - d_0^n = kt$$

where d is the grain size, d_o is the starting grain size, n is a constant equal to approximately 2, k is a constant, and t is coarsening time (Malow and Koch, 1996). As the starting grain size is reduced, coarsening time drives grain growth. Effectively, reductions in grain size are ineffective in delaying grain coarsening. Several traditional methods have been used to retard grain growth in ultrafine-grained materials, including second phase particles, solute effects and even fine porosity (Weissmüller, 1996). All serve as drag forces on moving grain boundaries.

Grain boundary-size reduction affects chemical diffusivity. The large volume fraction of atoms that are associated with the disordered grain boundary regions produce atomic mobility orders of magnitude higher than traditionally observed (Schuhmacher et al., 1989). One effect of this is increased solubility, particularly for interstitial atoms such as hydrogen (Mütschele and Kirchheim, 1987, Eastman et al., 1993, Stuhr et al., 1995). Another effect is a reduction in the temperature associated with high-temperature plasticity. This latter effect has engendered a class of deformable ceramics with grain size less than 300 nm at temperatures on the order of 1,200°C (Wakai and Kato, 1988, Nieh and Wadsworth, 1989).

For most metals, mechanical properties are dominated by the presence and mobility of dislocations. As grain size is reduced, the maximum spacing between a dislocation and a grain boundary is reduced. Since grain boundaries may act as a sink for dislocations, at grain sizes less than 20–30 nm, dislocations may literally vanish from the microstructure or may be present only in non-mobile, sessile orientations (Padmanabhan and Hahn, 1996). The inability to move dislocations results in high strength and usually vanishingly small ductility.

Reduced grain size also impacts the magnetic response of certain materials. As the grain size is reduced to below the domain size (on the order of 10–100 nm) in ferromagnetic materials, the transition from multidomain behavior to single domain behavior occurs and easy magnetization disappears (Weissmüller, 1996). As grain size is further reduced, the coercivity decreases until a superparamagnetic state is reached. These effects allow materials designers to produce a wide variety of materials with outstanding hard and soft magnetic properties (Yoshizawa et al. 1988, Herzer, 1991, Koon and Das, 1981, Croat, 1982).

Several researchers have noted that size reduction can produce non-equilibrium crystallographic phases in certain materials, often associated with high-pressure phases. For example, in nano-sized zirconia the equilibrium monoclinic phase is partially converted to the high-pressure tetragonal phase (Skandan et al., 1994). Refractory metals including Cr, Mo and W have been reported in nanocrystalline form in the A15 structure, rather than the usual body-centered cubic form (Kimoto and Nishida, 1967, Ganqvist et al., 1975, Saito et al., 1980).

Ceramics and other notch-sensitive materials are strengthened as their volume decreases. This characteristic is helpful in micromanufacturing and results in resistance to fracture. For example, at Hitachi Chemical R&D Center, fiber optics and glass fibers were curved through reasonably tight radii in the micro-arranged tube technology. Another example is shown in Figure 3.19 for silicon nitride tested in tension at various sample sizes along with a small four-point bend sample (Soma et al., 1986). The phenomenon is described macroscopically using Weibull statistics. Specifically, the cumulative probability of failure F is given as

$$F = 1 - \exp\left[-kV\left(\frac{\sigma_{max}}{\sigma_0}\right)^m\right]$$

where k is a dimensionless load or structure factor, V is the effective volume of the part experiencing tension loading, σ_{max} is the maximum stress at failure, σ_0 is a normalizing parameter, and m is the Weibull modulus which typically varies between 10 and 20.

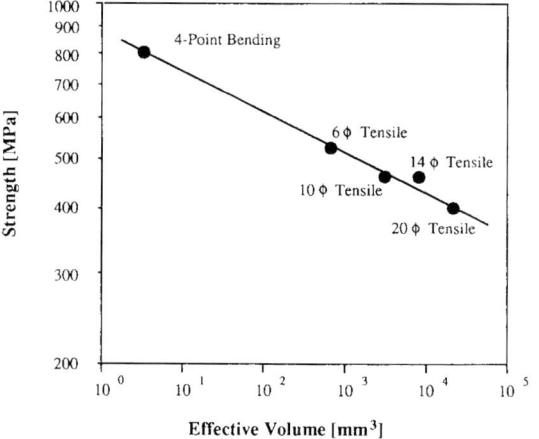

Figure 3.19. Weibull plot of strength of silicon nitride of varying sample sizes in tensile-testing modes. (Soma et al. 1986, reprinted with permission.)

SUMMARY AND CONCLUSIONS

This chapter provided an overview of materials usage in micromanufacturing at sites visited by the panel throughout Asia and Europe. Typically, materials used for micromanufacturing are the same as those used in macromanufacturing. They encompass the full range of metals, polymers and ceramics/glasses.

Materials-related issues that are particular to micromanufacturing include the lack of an economic driving force for materials development; grain size-effects that affect machinability; an increase in the significance of impurities and inclusions as the part volume decreases; and, for brittle materials, improvement in the resistance to failure that accompanies decreasing part volume. Grain size is a significant feature of material microstructures. A significant and controversial finding from the site visits is that material sensitivity to grain-size strengthening decreases as the sample volume also decreases.

REFERENCES

Bardes, B. P. ed. Metals Handbook: Ninth Edition. 1979. Volume 2, 285 in *Properties and Selection: Nonferrous Alloys and Pure Metals*, (Bruce P. Bardes, Ed.), American Society of Metals, 1979.

Chang, H., C. J. Altstetter, R. S. Averbach. 1995. Nanophase Metals—Processing and Properties. In *Advanced Materials and Processing*, vol. 3, K.S. Shin, et al., eds, Kyongju, Korea: Korean Institute of Metals and Materials, 2107.

Chokshi, A. H., A. Rosen, J. Karch, H. Gleiter. 1989. On the Validity of the Hall-Petch Relationship in Nanocrystalline Materials. *Scripta Metall* 23, 1679-1684.

Croat, J. J. 1982. Permanent Magnet Properties of Rapidly Quenched Rare Earth-Iron Alloys. *IEEE Trans Magn* 18, 1442-1447.

Eastman, J. A., L. J. Thompson, B. J. Kestel. 1993. Narrowing of the Palladium-Hydrogen Miscibility Gap in Nanocrystalline Palladium. *Phys Rev B* 48, 84-92.

Ganqvist, C. G., G. J. Milanowski, R.A. Buhrman. 1975. A15-Type Structure of Chromium Films and Particles. *Physics Letters* 54A, 245-246.

Herzer, G. 1991. *Materials Science and Engineering A* 133, 1-5.

Koon, N. C. and B. N. Das. 1981. Magnetic-Properties of Amorphous and Crystallized $(Fe_{0.82}B_{0.18})_{0.9}Tb_{0.05}LaO_{0.05}$. *Applied Physics Letters* 39, 840-842.

Kimoto, K. and I. Nishida. 1967. An Electron Microscope and Electron Diffraction Study of Fine Smoke Particles Prepared by Evaporation in Argon Gas at Low Pressures II. *Applied Physics* 6, 1047-1059.

Malow, T. R. and C. C. Koch. 1996. Grain Growth of Nanocrystalline Materials—A Review, 33-46. In *Synthesis and Processing of Nanocrystalline Powder*, D. L. Bourell, ed., Warrendale PA: TMS.

Mütschele, T., R. Kirchheim. 1987. Hydrogen as a Probe for the Average Thickness of a Grain Boundary. *Scripta Metallurgica* 21, 1101-1104.

Nieh, T. G. and J. Wadsworth. 1989. Characterization of Superplastic Yttria-Stabilized Tetragonal Zirconia by a Hot Indentation technique. *Scripta Metallurgica* 23, 1261-1264.

Nye, J.F. 1985. *Physical Properties of Crystals*, Oxford: Clarendon Press.

Padmanabhan, K. A. and H. Hahn. 1996. Microstructures, Mechanical Properties and Possible Applications of Nanostructured Materials, 21-32. In *Synthesis and Processing of Nanocrystalline Powder*, D.L. Bourell, ed., Warrendale, PA: TMS.

Saito, Y., K. Mihama, R. Uyeda. 1980. Formation of Ultrafine Metal Particles by Gas-Evaporation .6. BCC Metals Fe, V, Nb, Ta, Cr, Mo and W. *J Applied Physics* 19, 1603-1610.

Schumacher, S., R. Birringer, R. Strauss, H. Gleiter. 1989. Diffusion of Silver in Nanocrystalline Copper Between 303 and 373K. *Acta Metallurgica* 37, 2485-2488.

Skandan, G., H. Hahn, M. Roddy, W. Cannon. 1994. Ultrafine-Grained Dense Monoclinic and Tetragonal Zirconia. *J American Ceramics Society* 77, 1706-1710.

Soma, T., M. Matsui, I. Oda. 1986. Tensile strength of a sintered silicon nitride, 361-374. In *Non-Oxide Technical and Engineering Ceramics,* S. Hampshire, ed., New York: Elsevier.

Stuhr, U. H. Wipf, T. J. Udovic, J. Weissmüller, H. Gleiter. 1995. Inelastic Neutron Scattering Study of Hydrogen in Nanocrystalline Pd. *NanoStructured Materials* 6, 555-558.

Tsai, M. -C. and Y. -A. Chen. 2004. Size effects in micro-metal forming. In *MIRDC Government Annual Report*, 18-27 (2004) [in Chinese].

Tsai, M. -C., Y. -A. Chen, C. -F. Wu, F. -K. Chen. 2004. Size effect in micro-metal forming of copper and brass. *Forging*, 13:2, 41-46 [in Chinese with English abstract].

Tsai, M. -C. 2004. Cu15Zn microstructure effects on micro-metal forming. *MIRDC Government Annual Report* [in Chinese].

Wakai, F. and H. Kato. 1988. *Adv Ceramic Materials* 3, 71.

Weissmüller, J. 1996. Nanocrystalline Materials—An Overview, 3-20. In *Synthesis and Processing of Nanocrystalline Powder*, D.L. Bourell, ed., Warrendale PA: TMS.

Weissmüller, J., J. Löffler, M. Kleber. 1995. *Nanostructured Materials* 6, 105-114.

Yoshizawa, Y., S. Oguma, K. Yamauchi. 1988. New Fe-Based Soft Magnetic-Alloys Composed of Ultrafine Grain-Structure. *J Applied Physics* 64, 6044-6046.

CHAPTER 4

PROCESSES

Kamlakar Rajurkar and Marc Madou

ABSTRACT

Manufacturing processes convert raw material into desired parts to make usable and saleable products. All manufacturing processes are evaluated and then selected for specific applications based on the type and amount of energy involved, the process mechanism and its capability (including accuracy and repeatability), environmental effects, and economy. In addition to these measures, micromanufacturing processes also need to be evaluated on the quality of the removal (or plastic deformation or addition) of the smallest amount of material in one cycle, as well as the achievable precision of the related micromanufacturing equipment. This chapter begins by describing the status of the micromanufacturing processes observed during the WTEC visits to Asia and Europe. The state-of-the-art of micromanufacturing processes in the U.S. is also included in this chapter. The sites visited in Asia and Europe include industry, universities and research organizations. Specific issues of process mechanism, modeling and simulation, surface integrity, and scaling effects are summarized in the second part of this chapter.

MICROMANUFACTURING PROCESSES AND EQUIPMENT

Many different types of micromanufacturing processes were observed during the visits. To provide a broad view of the micromanufacturing, these processes are categorized into three main types—subtractive, near-net-shape, and additive. Some of these processes are downscaled versions of existing traditional macromanufacturing technologies, and others are innovative methods using various physical and chemical effects.

Subtractive Processes

The material subtractive processes include mechanical micromachining such as turning, drilling, milling, and grinding; electro-physical and chemical processes such as electro-discharge machining (EDM) and electro-chemical machining (ECM); and energy-beam machining such as laser, electron, and focus ion beam.

Mechanical Micromachining

Mechanical micromachining processes are naturally downscaled versions of the existing macro-level processes. In these processes, the tools are usually in direct mechanical contact with the workpieces and therefore, a good geometric correlation between the tool path and the machined surface can be obtained. Compared to microelectronic fabrication methods, they have higher material removal rate and the ability to machine complex 2D and 3D microshapes in a variety of engineering materials.

FANUC in Japan has commercialized an ultra-high precision micromachine named ROBOnano Ui (Figure 4.1a). It can function as a five-axis mill, a lathe, a five-axis grinder, a five-axis shaping machine, and a high-speed shaper (Figure 4.1b, c, d). Cutting is nominally accomplished by a single crystal diamond tool. For milling, an air turbine spindle is employed for rotational speed up to 70,000 rpm. Shaping is done using a high speed shuttle unit capable of producing three grooves per second, and a 3 kHz fast tool servo using a PZT actuator (FANUC ROBOnano Ui, 2004).

A FANUC series 15i controller implements numerical control of ROBOnano. The resolution of the linear axes is 1 nm and 0.00001° for the rotational axes. The actual trace rate to NC command is ±2nm at straight line and ±5/100,000 degrees at indexing. This excellent servo trace capability is ascribed to a "totally friction-free nano servo" technology (all linear slides and screws are static air bearings). Furthermore, the system is enclosed in an air shower temperature-controlled unit, whose temperature is controlled to 0.01°C to eliminate thermal growth errors.

Turning spindle   Grinding tool
Milling spindle

(a) ROBOnano Ui.

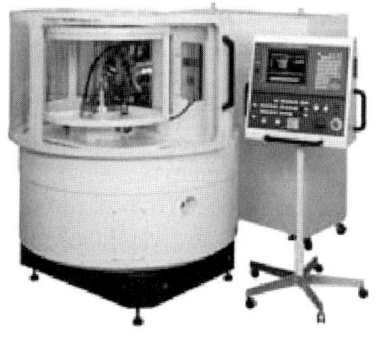

(b) Grinding/Milling.

(c) Turning.   (d) Shaping.

Figure 4.1. Ultra high-precision micromachine FANUC
ROBOnano Ui. (Source: FANUC ROBOnano
alpha-0iB machining samples October 2005;
http://www.fanuc.co.jp/en/product/ROBOnano/s
ample/index.html)

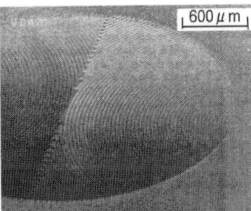
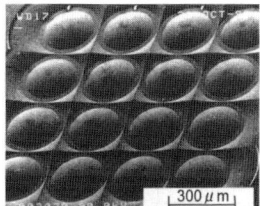

(a) Trapezoid   (b) Double-focused   (c) Microlens array.
grooves.   lens mold.

Figure 4.2. Microfeatures machined by ROBOnano.
(Source: FANUC ROBOnano alpha-0iB machining samples October 2005;
http://www.fanuc.co.jp/en/product/ROBOnano/s
ample/ index.html)

ROBOnano is used to machine mirrors, filters, gratings, liquid crystal
display (LCD) panels, small lenses, micromoldings, and other small ultra-
precise parts. Most of the parts require approximately five minutes to be
machined. Figure 4.2a shows a piece of brass with trapezoid grooves with
a high-aspect ratio, opening angles of 3 degrees, and groove pitches of 35
μm. Microgrooves with high-aspect ratio are popular in semiconductor
processing. However, the roughness of sidewall and the sharpness of the
bottom edge of ROBOnano machined grooves are much better than many
other methods in use for semiconductor processing. Figure 4.2b is a mold
for double-focused lens consisting of continuous curved V-grooves of 35
μm in width and with changeable V-angle and depth. In addition, 3D free-
curved surfaces such as a 4 × 4 convex lens array (Figure 4.2c) can also be
machined by ROBOnano. The material can be brass or silicon, the lens

pitch 290 μm, the diameter 236 μm, the height 16 μm, and R 448 μm (FANUC ROBOnano alpha-0iB machining samples, 2005).

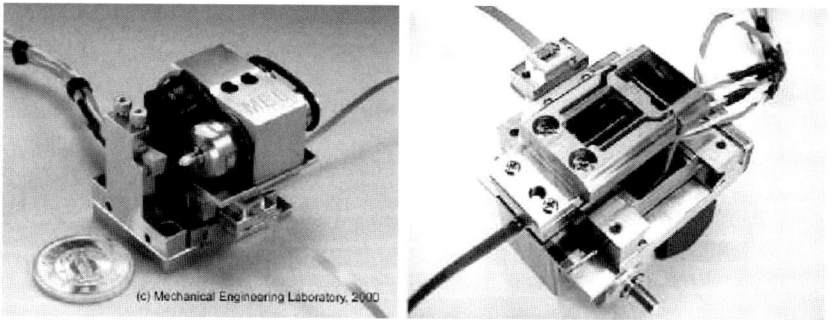

Figure 4.3. Microlathe.

The National Institute of Advanced Industrial Science and Technology (AIST) in Japan has developed a microlathe that is 32 mm long, 25 mm wide, 30.5 mm high, and weighs only 100 grams (see Figure 4.3). It comprises a friction-driven inchworm system, a main shaft device driven by a micromotor, and a toolrest. Modified inchworm type microsliders are employed, in which slides are guided and held with friction. It has positioning resolution of 25 nm. By optimizing the drive pattern, smooth feed of 400 μm/s can be achieved despite its stepwise feed action. Although the main spindle motor has only 1.5 watt rated power, the rotating speed can reach at 10,000 rpm (Okazaki and Kitahara, 2000, Okazaki, 2000).

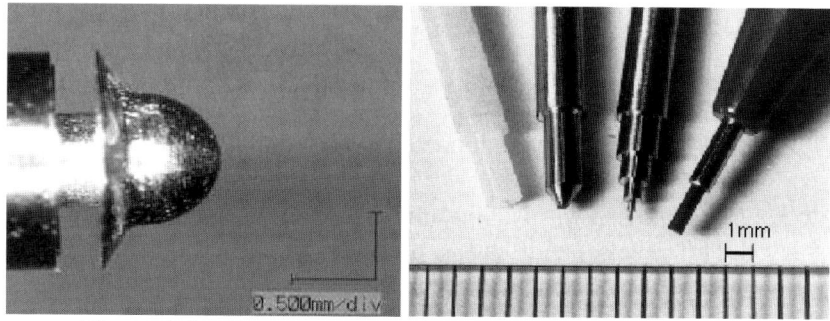

Figure 4.4. Microcomponents machined by microlathe.

The microlathe can cut brass with a 1.5 μm roughness in the feed direction. Figure 4.4 shows the microparts machined by the microlathe. The minimum diameter of the workpiece machined in the experiment is 60 μm.

Figure 4.5a shows a micromilling machine developed in AIST. The machine consists of an X-Y stage and Z-axis drive unit holding a spindle unit. The Z-axis drive unit uses a hollow drive screw, and is driven by a direct drive alternating current (AC) motor. The spindle and the drive screw are allocated in a line. The spindle with diameter of 27 mm is rated 80 watt and offers 300,000 rpm of maximum rotation speed. The spindle end holds a milling tool with a direct-machined collet chuck of 1 mm. High-spindle rotation, high-feed rate, and small depth-of-cut contribute to the reduction of machining force. Motion control is achieved by a full closed-loop digital servo system, obtaining a feedback signal from a linear scale with a resolution of 50 nm.

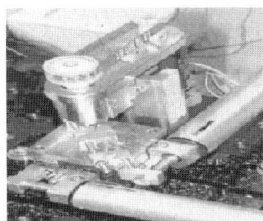  

(a) Micromilling ma-   (b) Rouge figure pock-   (c) Honeycomb 4
    chine.   eting.   mm pitch and 2 mm
                                             deep.

Figure 4.5. Micromilling machine and generated features. (Okazaki, K. and T. Kitahara 2000, reprinted with permission)

With small diameter end-milling tools, the micromilling machine has been used to machine hard aluminum alloys and pre-hardened steel. As shown in Figure 4.5b and c, 3D microfeatures have been generated. The sample shown in Figure 4.5b was completed in just 130 seconds (Okazaki and Kitahara, 2000, Okazaki, 2000, Kawahara et al., 1997, Kitahara et al., 1998).

Nanoimprint, with advantages of simple processing, low cost and high throughput, can overcome the limitations of conventional lithography technology. It has great potential in nanoelectronics, data storage, opto-electronics and biomedical applications. Industrial Technology Research Institute (ITRI) in Taiwan has developed a nanoimprint equipment system with advanced technologies in parallelism, pressure uniformity and rapid heating (see Figure 4.6). The parallelism and pressure variation is within 5% and the heating rate is expected to exceed 100°C per second. The objective of nanoimprint patterning is to fabricate a nanoscaled structure (<100 nm) using a nanomold made by electron beam lithography and dry etching (ITRI Annual Report, 2004, Ehrfeld et al., 1996).

Figure 4.6. Nanoimprinter MIRL 30B. (Ehrfeld, W. et al.
1996, reprinted with permission)

A focus of research at the Center for Micro Systems Technology
(ZEMI, Berlin Adlershof), jointed by Berlin University of Technology's
Institute for Machine Tools and Factory Management (IWF) and Fraunhofer Gesellschaft Institute for Production Systems and Design Technology (IPK), is on miniaturizing equipment to machine single parts and small
batch economically. Tools made of tungsten carbide with geometrically
defined cutting edges (Figure 4.7a, b) are used for microcutting. Researchers at ZEMI also work on the optimization of tool geometry and process
parameters for turning and milling to machine optical structures and surfaces. Suitable correction algorithms have been developed and lead to
high-precision features (Figure 4.7c) (IPK and IWF handout, 2004).

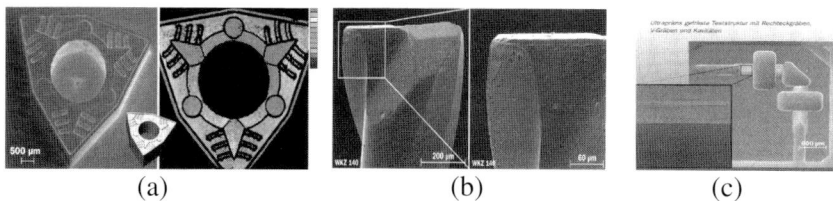

      (a)   (b)   (c)

Figure 4.7. Micromilling tools and machined features.

At ZEMI, finite element methods (FEM) modeling (Figure 4.8a) and
mechanical analysis (Figure 4.8b) have been used as tools for process optimization, such as manufacturing strategies and technological parameters.
It is obvious that ever-improving modeling and simulation techniques
largely reduce the need for expensive and time-consuming experiments
(IPK and IWF handout, 2004).

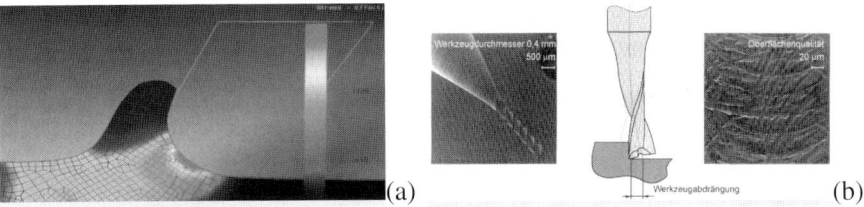

Figure 4.8. Simulation and modeling at ZEMI.

As a technical highlight, Institute for Production Technology (IPT, in RWTH-Aachen University, Germany) is capable of simultaneously measuring the hole in drilling, boring and cutting. Based on the measurement feedbacks, the cut position is corrected dynamically during the process. An actuator captures and compensates the deformations of the spindle. This eliminates subsequent finishing steps. Such an intelligent boring tool named "iBo" is capable of manufacturing holes with cylindricity of less than 5 µm (Fraunhofer IPT, 2005). The micromilling at IPT utilizes a traditional macroscale precision machine to drive microscale cutting tools. An effort is to utilize hard coatings to reduce tool wear, and the major challenge is to control the coating thickness and maintain the geometry of the microtool. However, the coating has demonstrated positive results on wear resistance as shown in Figure 4.9a. Highly defined microscale features are milled at IPT as shown in Figure 4.9b.

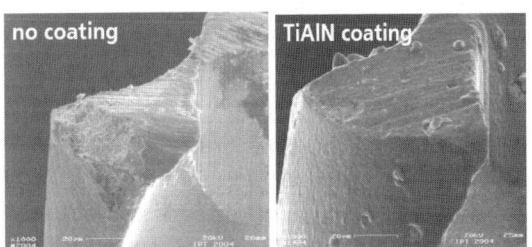
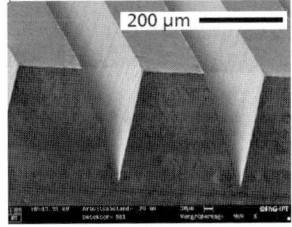

(a) Comparison of the wear on micro end mills without (left) and with (right) hard coatings.

(b) Micromilled features on microscale objects.

Figure 4.9. Micromilling at IPT.

As part of a research project at the University of Twente (Netherlands) named "Fabrication of Microstructures by Powder Blasting," a particle jet is directed towards a target for mechanical material removal (Figure 4.10a). The nozzle is scanned over the target several times to achieve a uniform etched surface. It is a fast, inexpensive and accurate directional etch technique for brittle materials such as glass, silicon and ceramics. The time to etch through a 500 mm thick Pyrex wafer with one nozzle is approximately 20 minutes. Figure 4.10b shows a typical microfeature obtained by this technique (Micro-Blast, 2005).

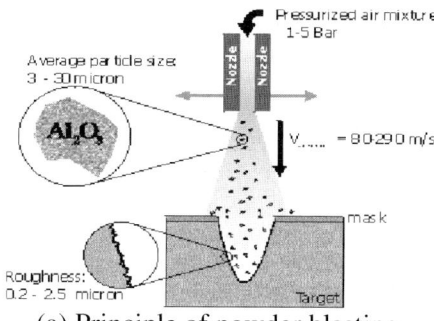 

(a) Principle of powder blasting.   (b) 400 μm deep-blasted glass structure.

Figure 4.10. Micropowder blasting. (Lammerink, T.S.J and M.C. Elwenspoek, 2005, reprinted with permission)

The mechanics of mechanical micromachining are the subject of intensive experimental and modeling studies at the University of Illinois at Urbana-Champaign (UIUC) and Northwestern University; set-ups are shown in Figure 4.11a and b. Experimental work includes the size effect and minimum chip thickness effect, elastic-plastic deformation, and microstructure effects in micromachining. Modeling studies include molecular dynamics methods, FEM, mechanistic modeling, and multiscale modeling. Researchers at UIUC are studying surface generation mechanisms in the micro end milling of both single-phase and multiphase workpiece materials. FEM simulations for the ferrite and pearlite materials were conducted to find the minimum chip thickness. A cutting force model for the micro end-milling process was also proposed to predict the effects of the cutter edge radius on the cutting forces (Vogler, DeVor and Kapoor, 2004).

(a) Microturning machine   (b) Three-axis horizontal   (c) Machined
(NWU).   milling machine (UIUC).   components.

Figure 4.11. Mechanical micromachining in NWU and UIUC.

Moore Nanotechnology Systems LLC, a U.S. company, is dedicated to the development and supply of ultraprecision CNC machine systems. Single-point diamond turning and deterministic microgrinding technologies are used for the production of plano, spherical, aspheric, conformal, and freeform optics. The provided systems (Nanotech 220UPL, 350UPL, 350FG and 500G) are available in two, three, four, or five axes for on-axis turning of spherical, aspheric, and toroidal surfaces. Capabilities also include slow-slide-servo machining (rotary ruling) of freeform surfaces and raster fly-cutting of freeforms, linear diffractive, and prismatic optical structures. The Nanotech 220UPL (Figure 4.12a) is considered to be the most accurate diamond-turning machine in the world, because of its groove-compensated air-bearing work spindle that provides motion accuracy of less than 50 nm through its 1,500 rpm speed range (IPK and IWF handout, 2004).

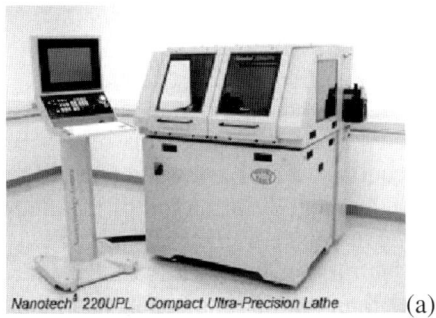 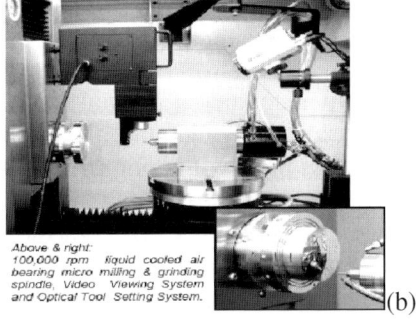

Nanotech 220UPL.   Micromilling machine.

Figure 4.12. Moore nanotechnology systems. Electro-
physical and chemical processes.

With the introduction of the newest family of automatic micro-EDM instruments, Panasonic Factory Automation in Japan has significantly pushed forward its technology and dramatically expanded the potential uses of micro EDM. Their micro-EDM systems have position accuracy of 1 μm and can drill various holes with diameters from 5 μm to 1 mm. The typical holes drilled are 50 μm in stainless steel with tolerance of ±1 μm. Micro-EDM equipment is increasingly applied to machine the components (for accelerometers, force-balanced transducers, fiber optic light-detector fixturing, microshafts, gears, springs, micromolds, and dies) other than only drill precision microholes. Results such as precision optics and magnetic heads for digital video camera obtained at Panasonic were remarkable. A handheld EDM machine was also shown to the panel representatives during the site visit.

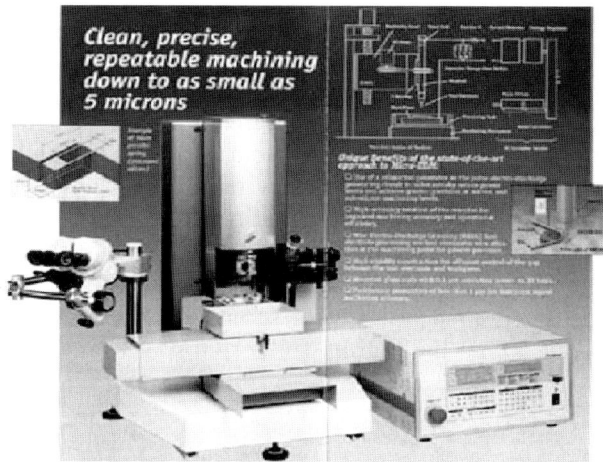

Figure 4.13. Panasonic Model 82 Micro EDM.

As it machines a coaxial microhole in a small spindle, a mechanical lathe can hardly maintain the concentricity of the tool tip. This results in generated parts with poor dimensional accuracy. EDM is a good method to solve the problem since it is free from mechanical deformation of the electrode (tool in EDM). An implosion that takes place at the end of an EDM pulse generates a force that helps eject the molten material from the interelectrode gap. The force is much weaker than that is required to mechanically remove a chip from a solid material. Additionally, the shape of an EDM electrode can be much simpler than that of the tool used for mechanical boring inside a hole because electro-erosion takes place on all surfaces of the electrode and removes the material from the workpiece efficiently.

(a) Microhole with inter-   (b) Stepped cavity.
nal groove.
Figure 4.14. Machined structures by micro-EDM lathe
(Courtesy Masuzawa, T. 2001).

Such a micro-EDM lathe has been developed at Professor Masuzawa's laboratory at the University of Tokyo. The typical cavities machined are shown in Figure 4.14 (Masuzawa, 2001). Almost all of the companies visited by the panel in Japan (including AIST, RIKEN, and Panasonic) prided themselves on having made their own micro-EDM machines.

A multi-functional four-axis high-precision micro-CNC machining center (Figure 4.15a) has been developed at the Mechanical Engineering Department in National Taiwan University (NTU).

(a) Multi-functional four axis.   (b) Pagoda machined by microwire EDM.

Figure 4.15. Micro-computer numerical control (CNC) machining center.

The machining accuracy of the system is 1 µm. It is capable of performing four functions: 1) Low-speed (0~3,000 rpm), middle-speed (1,000~30,000 rpm), and high-speed (2,000~80,000 rpm) milling operations using a micromilling cutter of 0.2 mm diameter. The depth of the cut varies from 0.1 mm to 0.3 mm with a feed rate of 20 mm/min. 2) for micro EDM, microelectrodes as small as 8 µm with high-aspect ratio using a wire electro-discharge grinding (WEDG) unit. 3) Microwire EDM using 0.02 mm brass wire. Microwire tension control and wire vibration suppression strategies have been developed and implemented, and a novel mechanism designed by the research team permits vertical, horizontal and slanted wire cutting. 4) Online measuring of the flatness, circularity, concentricity, dimensions, and geometric tolerances of the workpiece using a specially designed module. In general, the machine is capable of performing standard machining operations (milling and EDM) to make micro- and mesosize 3D complex molds and dies. It has successfully fabricated complex features and parts, especially the impressive Chinese pagoda (1.25 mm × 1.75 mm) shown in Figure 4.15b.

A micro-EDM research program conducted at Seoul National University in Korea uses kerosene as a dielectric medium for smaller dimensional machining. This program is used to make the masks for organic thin film transistors (OTFTs). A combination of EDM and ECM was used to make a mask with an aspect ratio of 20. When using a faster pulse rate in ECM, better precision and shape conformity have been obtained. In addition, simulation of the ECM process has been conducted for predicting the shape of the tool in order to get the desired final workpiece shape precisely. Thermal effects, which are detrimental to the work piece final shape, are considered in the process modeling and simulation.

In Europe, the objectives of micro-EDM research activities at ZEMI are to develop microadjusted technologies that define design rules for micro-tool electrodes and introduce wear compensation strategies and flushing techniques. Current work on microwire EDM (Figure 4.16) at ZEMI focuses on the development of new machine tool components and clamping systems to meet the special requirements of micromanufacturing (IPK and IWF handout, 2004).

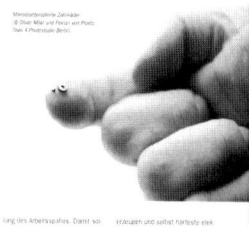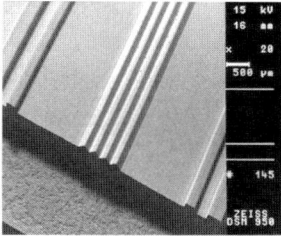

Figure 4.16. Microwire EDM at ZEMI. (IPK and IWF handout 2004, reprinted with permission).

Micro-EDM research at IPT is focused on exploring the possibilities of an increase in the material removal rate. The gap regulation is improved to minimize the open circuit during machining (Fraunhofer IPT, 2005).

Typical micro-ECM research activities in the U.S. are being conducted at Ex One Corporation and the University of Nebraska. Contoured features, including grooves and cavities, have been machined in bores, shafts, flat surfaces, and in OD and ID conical surfaces (Figure 4.17). Typical feature depths are 2–20 μm or larger in some applications. Micro EDM does not affect static properties such as yield and ultimate tensile strength, hardness and ductility. However, a reduction in fatigue strength is experienced in a wide range of alloys including those of iron, nickel and titanium.

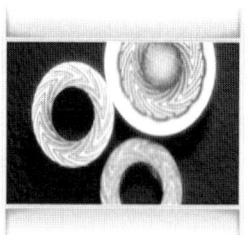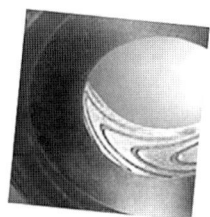

Figure 4.17. Grooves machined by micro ECM.

Electrolytic-in-process dressing (ELID) technology provides one of the best alternatives for avoiding high wheel loading when grinding with metal-bonded superabrasive wheels. The ELID system, which consists of wheel bond material, grinding fluid and a power supply, realizes a new grinding technique with high machining performance. Figure 4.18a shows the ELID grinding system mounted on a compact cylindrical grinding machine. The grinding wheel serves as the anode, and an electrode fixed at a small distance from the wheel serves as the cathode. The small interelectrode gap is supplied with chemically soluble grinding fluid and electrical current. Electrolysis is induced, and an insulating layer is generated on the dressed wheel surface. The bonding material is electrolytically removed during ELID grinding, resulting in the reduction of wheel wear (Ohmori et al., 2003). Figure 4.18b is a schematic illustration of the ELID grinding mechanism for sustaining ultra-precise finish grinding with fine abrasives.

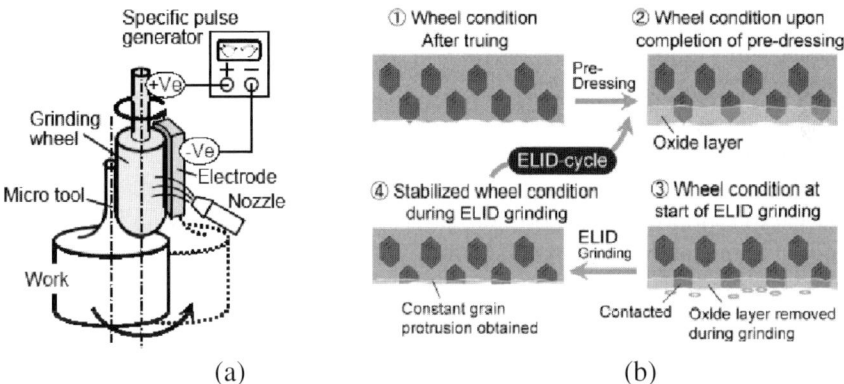

(a)   (b)

Figure 4.18. Principle and mechanism of ELID grinding (Ohmori, H. et al, 2003).

Figure 4.19. ELID machine and surface generated. (RIKEN Research Japan 2004, reprinted with permission)

A desktop cylindrical grinding machine unit (Figure 4.19) has been developed by RIKEN in Japan to improve processing efficiency. It includes an ELID grinding system and can be used for long cylindrical processing, cutting, boring, polygonal bar processing and tapering. The unit is very compact, with exterior dimensions of only 500 mm wide × 500 mm deep × 560 mm high. The most distinctive feature of this machine is the pair of opposed grinding wheel heads, which are arranged to pinch the workpiece. The two wheels can grind the workpiece from both directions and can hold the workpiece on two sides to eliminate any deformation during grinding (RIKEN Research, 2004). Dr. Ohmori of RIKEN, as well as being at the forefront of the development of novel multi-axis cutting machines such as the FANUC ROBOnano and improving the mechanical strength of micro-tools, has also pioneered ELID, which is capable of producing an unprecedented mirror finish of about 0.01 μm R_{max} (Ohmori and Nakagawa, 1995).

Energy Beam Micromachining

At the University of Tokyo, an ultraviolet (UV) laser (355 nm) is used for micromachining lithium niobate glass for optical waveguide applications. Smooth grooves of up to 10 μm depth and 300 μm wide were produced with a spot size of 9.2 μm. Depth is controlled by changing the laser energy, and width is controlled by altering the focal length/spot size. The primary advantage of this process is the ability to process fragile materials such as glass. A cabin, which measures 500 × 400 × 500 μm and has a wall thickness of 80 μm, was machined by laser cutting.

HV series laser systems at Mitsubishi Electric Corporation are able to cut very complex and meso/microparts with excellent accuracy and surface finish. They use nanocontrol technology for high-speed, high-precision and five-axis control. Accuracies of the order of 2.5 μm are achieved.

A typical application of laser micromachining is to increase the density of optical storage media such as DVDs. Currently, the track spacing on a DVD is 400 nm, yielding approximately 6 GB of data storage space. The initial target spacing for the next generation of laser micromachining systems is 100 nm, resulting in 25 GB storage capacity on a DVD. Research is being conducted at the Korean Institute of Machinery and Materials (KIMM) in the use of a variety of new lasers to generate smaller cutting widths at high speed.

In Europe, research organizations that are involved with laser micro-machining include the Laser Zentrum Hannover e.V. (LZH, Germany), the Institut für Mikrotechnik Mainz GmbH (IMM, Germany), and the Institute for Laser Technology (ILT, in RWTH-Aachen University, Germany).

LZH uses femtosecond lasers for hole drilling. Because the pulse width is extremely short, the edges produced by ablation are extremely accurate and sharp. Drilling was demonstrated by the application of injection nozzles for automotive applications as shown in Figure 4.20a. Stainless steel workpieces were laser drilled with 100–250 μm holes. Nano- and femto-second lasers have been used to drill 60 μm holes and trepan a wide variety of materials: fused silica, titanium, nitinol, alumina, and aluminum nitride.

Three-dimensional ablation was used to cut alumina carriers for airplane radar devices using excimer lasers. At IMM, laser caving of polymers as well as drilling of holes with diameters smaller than 10 μm (Figure 20b) and aspect ratio values up to 5:1 have been realized by excimer laser ablation in combination with the mask projection technique. IMM also combines LIGA and laser micromachining for building high-aspect ratio shapes, 500 μm tall with 85 μm wide features. The material was SU-8 photoresist. A large variety of materials can be laser micromachined, e.g., plastics, metals, semiconductors and ceramics (even diamond, graphite or glass).

 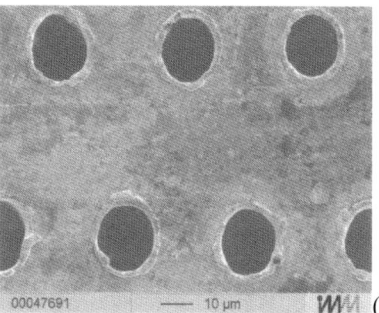

LZH laser drilling. (a)

Pore filters manufactured by laser drilling of 75 μm thick brass foil (IMM). (b)

Figure 4.20. Laser hole drilling.

ILT emphasizes technologies related to laser-supported manufacturing, including laser cutting, caving, drilling, welding and soldering as well as surface treatment, microprocessing and rapid-prototyping. Laser ablation is successfully used to fabricate meso- and microscale molding tools. Figure 4.21 shows examples of the different types of materials and tool geometries that have been processed by laser ablation. The accuracy and surface finish obtained in this process relates to the type of material. A few microns surface finish is generally descriptive of the process capability (Fraunhofer ILT, 2005).

Steel   Ceramics   Polymers

  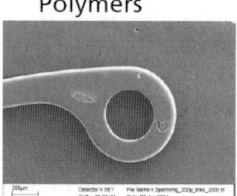

Figure 4.21. Materials and shapes formed via laser ablation process. (Sources: ILT October 2005; IPK October 2005 http://www.ipt.fraunhofer.de)

NEAR-NET-SHAPE PROCESSES

The high-temperature component processing research group in AIST has developed a new microforming technique called "reactive microforming." Its principle, as shown in Figure 4.22, is to add a small amount of binder and then form the microstructure by utilizing the generated internal heat.

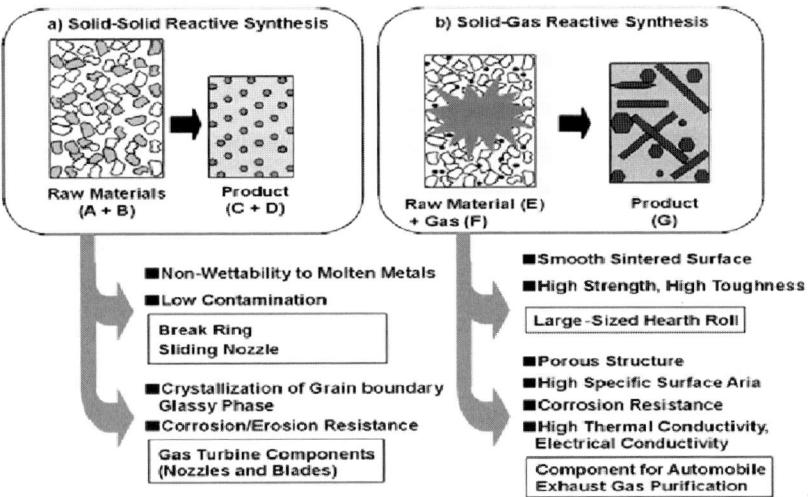

The methodology of processing research-

Figure 4.22. Reactive microforming.

A six-station micropress is used to punch a thin strip of brass (120 µm thick) to produce an air-bearing cover. Tension on the strip is supplied manually. The punched cover is deformed to produce three prongs for attachment to the housing, enclosing the spindle. Once the housing, spindle and cover are manufactured, they are assembled using a robotic microarm and microhand.

AIST has developed a system for injection molding of acrylic resin with a maximum operating temperature of 300°C. They have identified temperature control as the main challenge. Only a small amount of material can be heated above its softening point, resulting in the freezing of mold material before being injected through the nozzle. AIST is also a leader in micro- and nanoscale hot embossing of glass. Glassy carbon (GC), which is first patterned with focused ion beam (FIB) milling, serves as the stamp. Materials such as Pyrex or quartz are heated above their melting point and pressed into the full mold. Glassy carbon's lower value of thermal coefficient facilitates the removal of the part from the mold. The molds are resistance heated, and the system applies press forces and temperatures of up to 10 kN and 1400°C respectively. The process has been used to generate 3~10 µm features in Pyrex.

Metal Industries Research & Development Center (MIRDC) in Taiwan is performing extensive research on microforging, microstamping, microplastic forming, and hydrostatic extrusion processes. The produced parts are shown in Figure 4.23.

  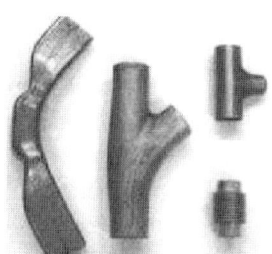

(a) Deep-drawing   (b) Bearing parts with   (c) Hydro-forming
    parts.   high-speed transfer die   parts.
                              system.

 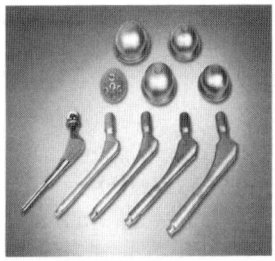 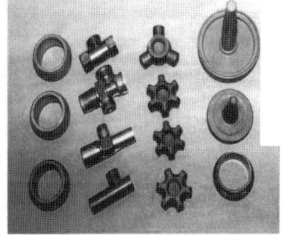

(d) Cold-forging parts   (e) Ti-alloy human   (f) Precision-forging
with progressive die.   bone.   parts.

Figure 4.23. Microforming products. MIRDC. (Source: MIRDC Research report 2004; http://www.mirdc.org.tw October 2005, reprinted with permission)

Typical research topics at MIRDC include (Reprinted with permission from MIRDC Research report, Taiwan, pages 1-21, 2004.):

1. Characteristic study of plastic forming for special alloy steel, aluminum alloy, magnesium alloy, titanium alloy, and copper alloy.

2. Precision closed-die forging, multi-action die forging, hybrid forging, cold/warm/hot forging, cross rolling, powder-metal (PIM) forging, microalloy forging, and near isothermal/isothermal forging.

3. Combined-stamping and cold-forging processes for complicated and near-net-shape parts.

4. Fine blanking, deep drawing, progressive die design, transfer dies design, combination of progressive and transfer die, and high-speed transfer die design.

5. Ultra-fine tube/wire forming, especially for wire, tube, and shape metal blanks with dimensions below 50 μm (an aluminum wire 50 μm in diameter has been successfully extruded from thick bar stock).

6. Tube and sheet metal hydro-forming for the development of automotive parts such as engine cradles, cross and side members, exhaust and intake systems, A/B/C pillars, bicycle/motorcycle components, sanitary products, bellows, and pipes and pipe fittings.

7. Simulation and analysis software for metal-forming processes.

Miniature Tool & Die Inc. conducts commercial research and application of microinjection molding technologies in the U.S. (see Figure 4.24a). A nanoinjection molding machine developed at Medical Murry Inc. can produce medical devices for disease treatment such as catheters (Figure 4.24b and c), fluid channels, balloons, and other similar devices. The parts are made of silicon rubber or thermoplastic, ranging from 0.01 to 80 mm 3 in volume and wall thickness to 120 μm.

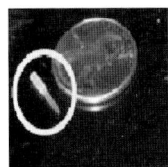

Micro-Injection-molded components
(Miniature Tool and Die Inc.)
(a) Microplastic parts.

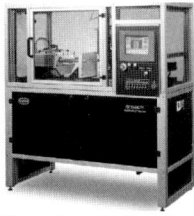

(b) Nanoinjection molding machine.

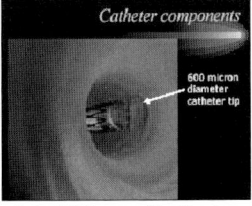

(c) 600 μm Catheter tip.

Figure 4.24. Microinjection molding in the U.S.

At the Department of Mechanical Engineering of Northwestern University, a forming assembly is used in conjunction with a loading substage to extrude micropins with a final diameter of 1 mm (Figure 4.25a, b). Theoretical or numerical solutions are proposed for optimization in the microextrusion process. A new method, Reproducing Kernel Element Method (RKEM), has been developed to simulate the microextrusion problem. RKEM addresses the limitations of FEM, while maintaining FEM's advantages. Furthermore, size effects were studied, and the individual microstructure (size, shape, and orientation of grains) and the interfacial conditions show a significant effect on the process characteristics (Cao et al., 2004).

 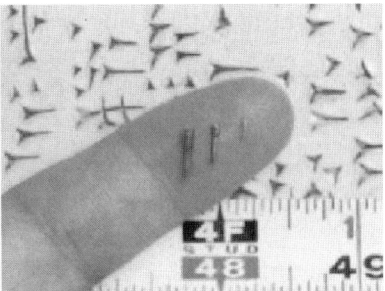

| (a) Microextrusion assembly. | (b) Extruded pin diameter: 1.2 mm, 0.8 mm, 0.48 mm. |

Figure 4.25. Microextrusion research in NWU. (Cao, J. et al. 2004, reprinted with permission)

The systematic research efforts being pursued at the Technical University of Eindhoven (TU/e, Netherlands) include theoretical and experimental work on the micromechanical modeling of polymers, size effects in microforming, and finite element models for multi-scale applications. The program has an ongoing focus on the fundamental understanding and modeling of metal forming at different length scales. The group is developing structure-property relationships by designing and implementing single-scale models, scale transitions, and embedded multiple scales in the mechanics of metals. Their approach relates the microscopic level (discrete dislocation and dislocation structures, i.e. at nanometer scale) to the mesoscale level (single and polycrystals, grain boundaries, i.e. at the micrometer scale) and the engineering macroscopic level (i.e. at the millimeter scale and above). Another program investigates the influence of grain sizes on the processing of thin metal sheets by reducing the sheet thickness (2 to 0.17 mm) at a constant grain size and then changing the grain size (0.016 to 600 μm) at a constant sheet thickness. The yield strength and the maximum load were found to decrease with a decreasing number of grains over the thickness. For grain sizes larger than the specimen thickness, the value of the yield strength increases with the grain size.

Experts at ZEMI have developed new tool and injection molding technologies to demold microstructures and optical components. The new tools for injection molding have variotherm tempering and an easy demountable mold cartridge. ZEMI is also studying the demolding behavior of components with a high aspect ratio (IPK and IWF handout, 2004).

STEAG microParts GmbH (Germany) has made some technical improvements in microinjection molding. Compression defects are reliably eliminated, and the filling of the mold has been improved. The highly precise products are in the biomedical field, including technically advanced and innovative mechanical, optical and fluidic products (Figure 4.26) manufactured from a wide range of plastics and silicon. Microinjection

molding and hot embossing enable mass reproduction plastic microproducts at low cost. The entire production process at STEAG is carried out
under strictly defined clean room conditions to minimize the contamination of particles or microbes in microsystems (IPK and IWF handout,
2004).

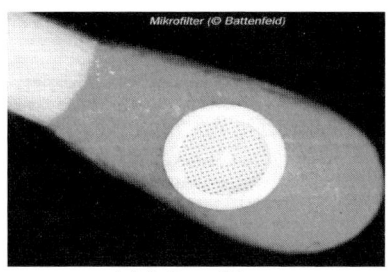 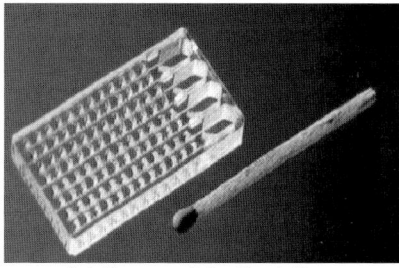

      (a) Microfilter.   (b) Porous Medical Part.

Figure 4.26. Microinjection molding parts (IPK and IWF
handout 2004, reprinted with permission).

Additive Processes

Working jointly, Ishikawajima-Harima Heavy Industries Co. and Mitsubishi Electric have developed new technology for surface coating and
cladding by small electric discharge pulses, called micro spark coating
(MSC).

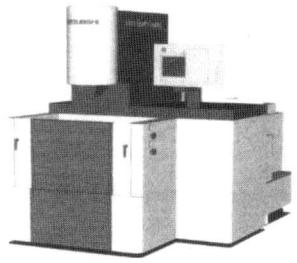 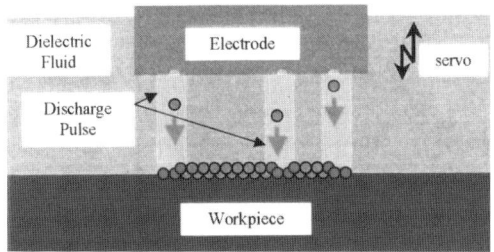

(a) Micro spark coat-   (b) Micro spark coating principle.
  ing equipment.

Figure 4.27. Micro spark coating.

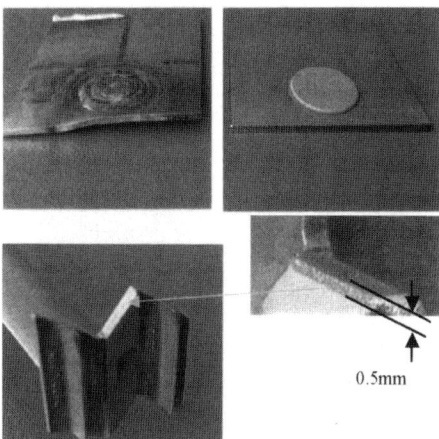

0.5mm

Figure 4.28. Micro spark coating products (Courtesy Goto,
A. et al. 2004).

Figure 4.27a shows the MSC processing equipment designed for manufacturing flow lines; the MSC principle is illustrated in Figure 4.27b. The
tool electrode is made of semi-sintered powder. Micro electric discharge
pulses, generated between electrode and part surface in dielectric fluid,
melt the electrode and part of the workpiece surface. The melted powder
moves toward the workpiece surface. Repeated discharges form strong
bonds between the coated layer and the surface. The coated layer generated
by MSC exhibits excellent characteristics when compared with plating,
welding, and plasma spray techniques. The process does not cause deformations and cracks (see Figure 4.28) because there is no heat concentration; likewise, the process does not induce peeling as does plasma spray
because the electrode material is melted onto the workpiece material. No
bench work, such as masking, is needed because the coating area is limited
only to the discharge area, which is determined by electrode shape. Lastly,
no finish machining, such as grinding, is needed because the thickness of
coatings can be controlled accurately (Goto, Magara and Moro, 1998, Goto
et al., 2001).

A metal laser sintering/milling hybrid machine, the LUMEX 25C, developed by Matsuura Machinery Corporation in Japan (see Figure 4.29),
successfully combines freeform manufacturing and high-speed milling.
The integration of laser sintering of metallic powder and high-speed cutting eliminates the finish machining operations. It layers the mixture (a
powder of 90% steel and 10% copper) over the part and then sinters it with
a 300 watt (500 watt max) CO_2 laser. The edges of the part are milled after
sintering five layers, with tolerances in the 25 μm range. For the generated
part, a surface roughness of 15 μm can be obtained. The presently achieved

hardness is about 25 Hardness Rockwell C (HRC), but their goal is to obtain 40 HRC.

Figure 4.29. LUMEX 25C. (Source: Matsuura Machinery
Corporation October 2005;
http://www.vcnet.fukui.fukui.jp/co/en/matsu
ura.html)

However, the LUMEX 25C requires cooling liquids to carry out milling processes to prevent adverse heat effects. The process suffers from porosity and relatively slow speeds. Although the process is promising, the manufacturers face significant challenges in scaling down to the micron level (Matsuura Machinery Corporation, 2005).

A real 3D microstereolithography (μSLA) process called the integrated harden polymer (IH) process was first developed by Professor Ikuta's group at Nagoya University in Japan in 1992. The polymer in the IH process is cured below the polymer/atmosphere interface by an ultraviolet (UV) beam. The beam energy is small enough to prohibit curing polymer outside of the focal point. A glass slide is placed above the cured layer. Hence the layer thickness is not limited to that usually generated by an SLA system with typical "dip and sweep" approach. An extremely thin (5 μm thick in 1992 and about 200 nm at present) layer is generated by raising the glass slide small distances above the previously solidified layer. Recently an advanced IH process (super IH) has been introduced using a HeCd laser with a wavelength of about 442 nm for higher lateral resolution. A high-viscosity polymer is used for fabricating microparts to avoid supporting structures. The team has also implemented the super IH process by using fiber optic arrays, enabling parallel generation of microcomponents. It takes about four to five minutes to fabricate a gear with an outer diameter of 15 μm.

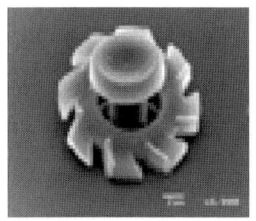
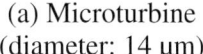

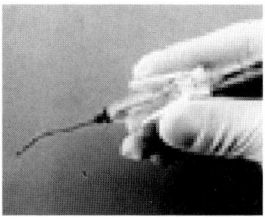

| (a) Microturbine (diameter: 14 μm). | (b) Microreactor. | (c) Micro active forceps for retinal surgery. |

Figure 4.30. Products and microsystems by IH process.
(Source: Ikuta, K. October 2005;
http://biomicro.ikuta.mec.nagoya-
u.ac.jp/~ikuta/index.html)

Applications of the IH process include movable microturbines of 15 μm diameter, microreactors, and forceps. These devices are driven by laser trapping, which generates forces by transferring momentum from light to an object and provides angular momentum to generate torque. Many other devices were developed using microfabrication technology. Though many of these devices are manually controlled, they have extremely accurate motion with excellent accuracy. Some of them can easily navigate through very small paths (e.g., blood vessels) and very long distances in the body. Typical examples of such devices are shown in Figure 4.30. The IH process requires the manual loading of the components during their various production stages, which is very labor intensive. Process automation is desired for the production versions of these components (Ikuta, 2005, Miraikan, 2005).

At Nagoya University, rapid prototyping has been used to develop an anatomical model of a human cerebral artery for use in surgery. Dr. Fukuda and his research team produced a transparent prototype with the same elastic and frictional properties and thin membranous structure as that of a human cerebral artery. The accuracy of the fabricated artery is about 1 μm. Physical models are fabricated by UV laser using a U.S.-made commercial machine, SoloidScape, under a layer thickness of 30 μm. Fabrication of a typical artery model takes about six hours. The Fukuda team has also developed a catheter-based transfer system for use in regenerative medicine. The suction cup of the catheter tip absorbs the pulsative myocardium (Research Activities Report, 2004, Fukuda Laboratory, 2005).

Dr. Jun Akedo at AIST has developed a very promising room-temperature aerosol deposition technique using a film coating technology in which the impact of solid-state particles can create a strongly adherent, high-density nanocrystalline film by gas blasting the nano-sized particles onto a surface. The deposition rate was 30 times faster than traditional deposition rates and can be carried out at room temperature. Uniform deposition rates were achieved over a 200 mm square area. The new ceramics formation mechanism is assumed to be room-temperature shock-compaction. Particles are deformed and fractured into nanocrystalline structures of 10-30 nm size upon impact. Plasma corrosion, dielectric strength, and film uniformity were all superior to sintered ceramic films. This novel Japanese process is nothing short of a technological break-through. Funding for the work was provided through a Nanotechnology Program from the Ministry of Economy, Trade and Industry (METI) and the New Energy and Industrial Technology Development Organization (NEDO). AIST and the Manufacturing Science and Technology Center (MSTC), as well as some Japanese universities, did the research and also worked together with six private industry partners: Brother, TOTO, NEC, Fujitsu, Sony, and NEC-TONKIN. Applications for this novel Japanese process include a wide range of piezoelectric, high-frequency, and optical devices.

Hitachi Chemical has developed three interesting new additive processes to solve manufacturing challenges in the production of flat panel displays, printed circuit boards, and mobile telephones. U.S. manufacturers are not facing these challenges as they are not heavily engaged in the manufacture of these products anymore.

Other products developed by Hitachi include arranged tube technology, which involves using glue to stitch very fine glass and metal wires to PC boards to act as inexpensive and elegant optical, electrical, and magnetic conduits (Hitachi Arranged Tubes Technology, 2004). Another project involves the use of lithography and lamination on a dry film photosensitive polyimide to complete fluidic structures (Photosensitive film for μTAS ME-1000 series, 2004). Anisolm, a new anisotropic conductive film developed by Hitachi, is used to connect printed circuit boards for room-temperature applications at a density of 10 lines per millimeter. The Anisolm-connected circuit boards are used in flat-panel displays (Hitachi Anisotropic Conductive Film ANISOLMTM, 2004).

There are several highlights of additive processes research in Europe. The R&D in Europe is more innovative and further advanced than it was five years ago, and the level of support is higher than in the U.S. at present.

The German Envisontec Perfactory photopolymer system has recently been commercialized. The light source for photopolymerization presents to the polymer from the bottom of the build chamber. Unique to the Perfactory is formation of the layer image using a 32 μm resolution micromirror array. The digital mask contains 1280 by 1024 pixel mirrors that are digitally controlled to allow various shades of light to expose the photopolymer. This micromirror array concept has been used to form micro-sized objects.

Integral micro-SLA process, in which a complete layer can be built using a single light exposure, has been realized at the Swiss Federal Institute of Technology in Lausanne (EPFL). The principle of integral micro-SLA apparatus is described in Figure 4.31a. Slices of a 3D computer-aided design (CAD) model are converted into bitmap files and used to drive a dynamic pattern generator, which in turn shapes a light beam. The beam is focused on the surface of a photopolymerizable liquid, which results in a selective solidification of the irradiated areas and creates a thin layer of polymer of the required shape. A complete layer is built using a single exposure. A typical exposure time of one second per layer is needed with this setup. Typical fabrication speeds of 1–1.5 mm per hour can be obtained in the vertical direction, a the superimposition of 200 to 300 layers per hour.

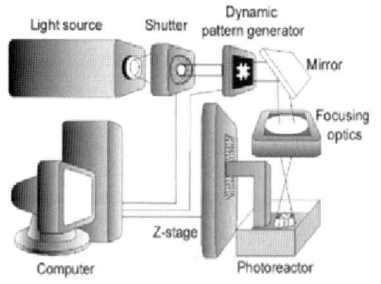

(a) Schematic diagram of integral
micro-SLA apparatus in EPFL.

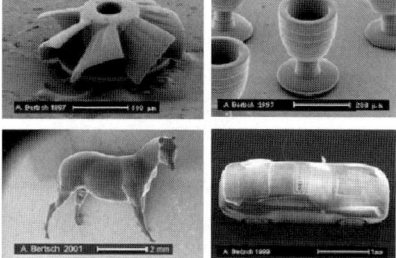

(b) Microstructures.

Figure 4.31. Integral micro-SLA principle and products.

Rapid micro product development (RMPD) is a process patented by MicroTEC Corporation (Germany). RPMD uses the mechanism of micro SLA, but with a technical innovation that increases the production speed. As in Figure 4.32a, the laser beam source is split into many beams, so this process allows the parallel production of up to 150,000 microcomponents per machine per hour. Furthermore, the component is created in thickness growth steps of less than 1 μm and resolutions of less than 10 μm. Any de sired form can be produced by RMPD, and the integration of mechanical, optical, electronic components can also be realized. Structures successfully fabricated by the RMPD method are shown in Figure 4.32b and c.

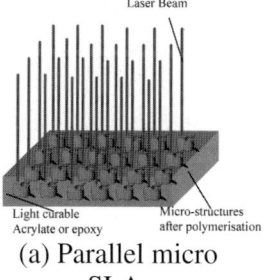

(a) Parallel micro
SLA.

(b) Microgripper with
match head.

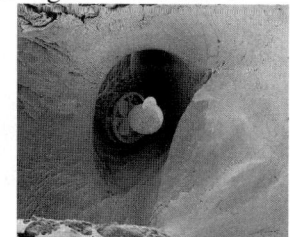

(c) Microsubmarine in
human vein.

Figure 4.32. RMPD principle and products.

LZH has developed an amazing technique called two-photon polymerization (2PP), a rapid additive technology that uses photopolymeric liquid as raw material. Exquisite parts including miniature statues (Figure 3.8) and spiders demonstrate the versatility of the technique. Part dimensions are on the order of 20–100 μm. Lateral resolution of the 2PP process is 120 nm. A design rule for 2PP is generally to build parts in a way such that the laser does not scan through cross-linked solid materials, as a slight change in refractive index occurs upon cross-linking. A new process named "Roll-based powder deposition for microlaser sintering" is also reported by LZH. This process differs from other laser sintering methods that have been developed. It is still in experimental stage; an example of a roughly fabricated structure created using this method is shown in Figure 4.33.

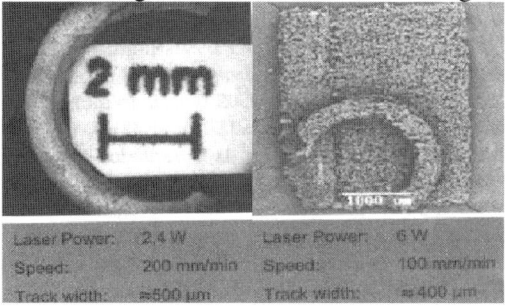

Figure 4.33. Microadditive processes in LZH. Features produced by "Roll-based" powder deposition.

In the U.S., Dr. Rosen at Georgia Tech has developed a micro-SLA device that can be used to create objects in photopolymer with features as small as 30-100 μm (Figure 4.34a). A typical micro-SLA system was made in Zhang Xiang's research laboratory in UCLA, where complex polymeric microstructures (Figure 4.34b) have been fabricated. Dr. Li at the University of Wisconsin at Madison has formed a line/grid array in photopolymer with features less than 25 μm. Gears and a square micropillar have been fabricated with overall dimensions of less than 800 μm.

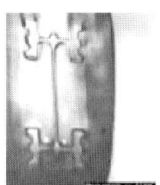 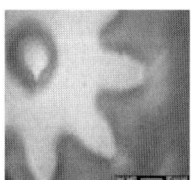  

(a) Georgia Tech (first three pictures).   (b) UCLA.

Figure 4.34. Micro-SLA products.

Electrochemical Fabrication (EFAB) is a patented technology developed by MEMGen Corporation in the U.S. It is targeted at the rapid, automated batch fabrication of high-aspect ratio metallic microstructures with arbitrary, truly 3D geometry (Figure 4.35a and b). Instant masking, a new high-speed *in situ* selective-plating technique, is used to precisely deposit an unlimited number of layers of material. The technique uses an electrochemical printing plate that contains images of the cross-sections of the microdevice to be built. EFAB can be used to form structures from any electrodepositable metal, but not from silicon. The only process constraint is that the accompanying sacrificial metal should be selectively removed (e.g., by chemical etching) after the layers are formed. Typically, nickel and copper form a desirable material system, with copper as the sacrificial material (Cohen, Zhang and Tseng, 1999).

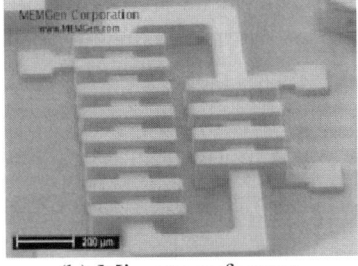

(a) Microspray nozzle array.　　　(b) Microtransformer.

Figure 4.35. Products fabricated by EFAB (Cohen, A. et al. 1999).

The Rapid Prototyping Laboratory (RPL) at Stanford University has developed a process named shape deposition manufacturing (SDM), which produces functional prototypes directly from computer models. As with other additive processes, SDM builds up a part from material layers; unlike other additive processes, however, SDM also removes support material. The additive process is accomplished with plasma or laser welding, casting, or UV curing. The subtractive step is performed by grinding, turning, or milling material using CNC machining.

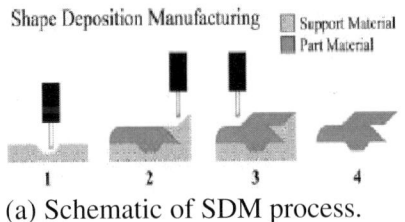 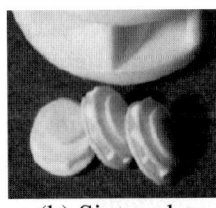 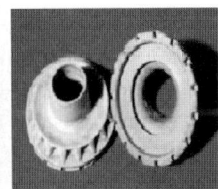

(a) Schematic of SDM process.　　(b) Sintered Al2O3 miniature turbines.　(c) Silicon nitride parts fabricated using Mold SDM.

Figure 4.36. SDM principle and products (Kang and Cooper 1999, reprinted with permission).

As shown in Figure 4.36a, repetition of the deposit and shape cycles (steps 1–3) layer by layer results in an object with a shape of the desired complexity. The sacrificial support material acts as a substrate for the part material; thus, after features such as undercuts, overhangs, and internal channels have been made on the part material, the support material can be removed by melting, dissolving, or etching, leaving a final near-net-shape object. In the building cycles, objects such as electronic circuits, sensors, or mechanical mechanisms can be embedded into each layer, to incorporate additional functions into the object. In addition, different materials can be deposited during the building cycles to produce multi-material objects with spatially varying material properties (Kang and Cooper, 1999).

MICROMANUFACTURING ISSUES

The various micromanufacturing processes discussed in this chapter vary in terms of their working principles (mechanical, thermal, dissolution, ablation, solidification, and sintering) and their material interactions (subtraction, mass containing, addition and joining). The Asian and European site visits revealed that a variety of issues—namely, those related to process modeling and simulation, process-material interactions, monitoring and control, process capabilities, tool and equipment design, metrology, economics, and application—have yet to be fully addressed (Masuzawa, 2000).

Many microscale components and products are manufactured using full-scale or downsized versions of existing macroscale precision manufacturing processes and equipment. This approach exposes the difficulties and limits inherent in using existing macroprocesses and equipment to produce microscale components and products, particularly limitations on (1) the amount of material that can be added and formed per cycle and (2) achievable precision. These limits are determined by a wide range of factors related to material properties, including: morphology; the generation and delivery of required small amounts of energy; mechanical, electrical and electronic compatibility; the effect on achievable resolution from related operational equipment; the effects of scaling on the process mechanism; process-material interactions; the removal of chips during machining processes, lubrication during forming processes, or curing during additive processes; and heat transfer.

Mechanical Micromachining

In mechanical micromachining processes, the effects of scaling on the process mechanism—including chip formation, cutting forces, vibration, process stability, and the surface integrity of the machined components—are some of the important issues that are not fully understood. Deformation caused by direct contact of the workpiece and the microtool, and the effect of deformation on accuracy, are also important topics for further investigation.

The accuracy of ultraprecision cutting is limited by the use of machine components, tools and equipment. Studies on the impact of the size effects on cutting forces and specific energy during microcutting processes have revealed that ploughing and elastic recovery of the workpiece play significant roles the more closely the chip thickness values approach the edge radii of the cutting tool. The impact of size effects on factors such as minimum chip thickness, surface generation, burr formation, built-up edge formation, and microstructure effects, has been reported (Lucca, Rhorer, Komanduri, 1991, Lucca and Seo, 1993, Lucca, Seo and Rhorer, 1994, Ikawa et al., 1991b, Ikawa et al., 1991a).

Various modeling techniques—including molecular dynamics (MD) simulation, FEM, multi-scale simulation modeling, and mechanistic modeling—have been discussed and attempted in efforts to characterize micromachining processes. Simulation technologies for machining processes such as FEM simulation of cutting, kinematical simulation of grinding, computational fluid dynamics (CFD) simulation of abrasive flow machining and CFD simulation of EDM have been used for machining at the microscale level. These methods expand the process know-how as well as help to control development costs and define process limits (Vogler, DeVor and Kapoor, 2004, Lee and Dornfeld, 2002, Waldorf, DeVor and Kapoor, 1999, Moriwaki et al., 1991, Ueda and Iwata, 1980, Liu et al., 2005).

Electro-Physical and Chemical Machining

In micro ECM—in which the material is removed by controlled anodic dissolution—the tool never touches the workpiece, nor is it consumed in the process. As a result, micro ECM offers an accurate, highly repeatable process with rapid machining times that produce final surface and edge. The continuous and uniform replacement of electrolyte in the gap and localization of anodic dissolution poses some difficulties in drilling a microhole and machining 3D complex cavities (Dutta, Shenoy and Romonkiw, 1996).

The material removal mechanisms of micro EDM make it advantageous in many microapplications. However, its use is limited with respect to tolerance and surface quality. Difficulties include poor availability of technologies and the fragmentary knowledge of material removal, tool wear mechanisms, debris removal, topography formation resulting from low discharge energy, and small gap width. To avoid arcing and short circuits at the discharge gap, gap monitoring and control systems must be developed. A comprehensive model of the phenomena at discharge gap based on the scaling effects will help optimize the process (Rajurkar and Yu, 2003, Tönshoff and Masuzawa, 1997).

Microwire EDM will be used to produce microstructured dies and molds from high-strength materials. With nearly process force-free machining, the technology's independence of hardness and strength of the workpiece material guarantee the highest accuracy and surface qualities in the manufacturing of complex geometries. Moreover, microwire EDM is exceptionally flexible and easy to automate. Therefore, it is especially suitable for small- and medium-batch production of microparts (IPK and IWF handout, 2004).

Energy Beam Micromachining

Lasers are extremely precise tools for micromanufacturing because their beams can be concentrated accurately on microscopic areas. Lasers have several advantages for micromanufacturing: high flexibility (e.g., ablation, drilling, fine cutting, welding or engraving); contactless machining without tool wear and load bearing; and the ability to machine a large variety of materials including plastics, metals, semiconductors, ceramics, and materials that are conventionally difficult to process, such as diamond, graphite and glass. Quality considerations include the avoidance of cracking, as well as clean, sharp features with no localized melting, burring or material build-up.

Femtosecond lasers have an extremely high peak power density. They are used for drilling high-quality holes. The very short influence period of intensive radiation heats up the electrons in the solid body while the lattice remains cold. Energy is not transmitted to the lattice until after a few picoseconds, resulting in sufficient energy input for direct evaporation of the material. The surrounding zones are only subject to very low heat influence; hence, precise structures in the micrometer range have been generated with minimal damage to the material. Processing results in burr-free edges, with high levels of precision and reproductability. Heat-affected zones and mechanical damage were kept to insignificant levels, resulting in almost no changes in material characteristics.

The anticipation that advanced picosecond and femtosecond laser systems will offer enhanced tailoring of optical energy density as well as improved processing strategies is expected to result in further application potential for lasers. However, issues related to the process-material interactions, heat transfer, and taper continue to present challenges that must be addressed for the successful application of laser machining technology at the microscale.

Near-Net-Shape Processes

Microforming is an appropriate technology for use in manufacturing microparts, in particular for bulk production. As with material-removal processes, microforming processes, which are based on plastic deformation, require a research and development effort to address issues related to size, shape, orientation, and grains of the material to be formed (Balendra and Qin, 2004). Flow stress, anisotropy, ductility and forming limit, forming forces, spring back, and tribology are some of the many factors that need further study for increasing the performance of microforming processes. One of the main problems encountered in microforming is the scaling effect that occurs in the tribological aspects; for example, the friction coefficient increases with decreasing specimen size. Scaling effects occur not only within the process but also throughout the forming chain. Laser techniques are being attempted to smooth the scaling effects, especially in tool manufacturing (Geiger, Mebner and Engel, 1997).

Microextrusion, a typical microforming process, is both fast and suitable for mass production. It is a bulk deformation process that produces less waste as compared to machining. However, the knowledge base of macroscale extrusion cannot be extended to microscale extrusion processes due again to size effects.

Microinjection molding is proving to be a very economical procedure for reproducing high-quality and complex plastic parts in the microscopic range. Progressive miniaturization of products such as medical instruments, nanoengines and microswitches requires flexible manufacturing procedures. However, small dimensions and high-aspect ratios of components prompt high demands on tools and machines as well as on process control. Above all, this concerns processes such as tool tempering and components demolding.

Additive Processes

Micro SLA involves laser photocuring of a photosensitive liquid photopolymer resin. A thin layer is scanned using a low-energy laser to selectively cross-link the layer. A new layer is deposited by recoating the previous layer, and the process is repeated until a part is eventually created. Similar to macroscale Rapid Prototyping, many microadditive processes also fabricate miniature features layer by layer. The fineness of the generated features, flexibilities, and related knowledge scope of the microadditive processes largely exceed that of their macro counterpart (Regenfuß et al., 2004). Issues related to material, energy, curing, staircase effects, and rigidity of manufactured parts need further studies.

Desktop Factory

Processes performed in a desktop factory (DTF) will have a dramatic impact on society, but it is not clear what direction it will take us. Sankyo Seiki believes that its DTFs might revive manufacturing in Japan and in Korea the government just started a new desktop factory project.

It is difficult to predict which strategy will enable developed countries to maintain a manufacturing base. But it is clear that, if any country wants to remain independent, especially in times of war, and maintain a high standard of living (for an appreciable percentage of its population) it must remain strong in manufacturing. This is especially true if that country is resource poor. There are several different scenarios one can think of on how to maintain a strong manufacturing edge. The bottom line, perhaps, is the need for a strategy in which manufacturing processes are only shipped abroad if new manufacturing technologies or products are in the pipeline. The latter would require a more guided economy than in the U.S., but it is paying off very well for China right now. In the popular "free market" scenario, the high level of automation of DTFs will negate their exportability, as no labor costs would be saved as a result. In the "socially responsible" scenario, promoted by MIT's Neil Gershenfeld and others, desktop manufacturing becomes as popular as desktop publishing (Gershenfeld, 2004). In the latter case, manufacturing would become much less centralized and many more people would participate. In either case, the impact of microscale manufacturing on society would be dramatic.

In the shorter term, free market forces in a global economy that is largely orchestrated by the U.S. will most likely continue to push toward low-wage countries the manufacture of highly accurate miniaturized devices that require quality control and multiple assembly. The manufacture of small-lot, high-add-on value products and items that require intense service assistance will remain in the more developed countries.

SUMMARY AND CONCLUSIONS

This chapter provided an overview of micromanufacturing processes equipment and their applications observed during the WTEC panel visits to manufacturers, universities, and research organizations in Asian and European countries. The processes and equipment include microturning, micromilling, micro electro-discharge machining, laser micromachining, near-net shape, and various microadditive processes. Applications include the components/products used in biomedical, automobile, aircraft, and microelectronic industries.

Overall Micromanufacturing activity is constrained by the limited market need. The expected breakthroughs for future emerging and perceived applications of micromanufacturing technology require extensive research and development efforts to address scaling effects, develop devices capable of generating and delivering the required energy, address precision and scale-up issues for the mass-production of products, and develop microscale machines and microfactories.

In an interesting development concerning microelectromechanical systems (MEMS) processes, Olympus has now set up a MEMS foundry. In the 80s and 90s, many engineers in the U.S. (including one of the authors of this chapter) attempted this. While experience has shown that it is very hard for a MEMS foundry to become profitable, the Japanese companies are willing not only to wait longer to be profitable, but also to learn from their clients and make new products in collaboration with them. Very few MEMS foundries currently remain in the U.S.—a rather delicate position.

REFERENCES

Balendra, R., Y. Qin. 2004. Research dedicated to the development of advanced metal forming technologies. *Journal of Materials Processing Technology* 145, 144-152.

Cao, J., N. Krishnan, Z. Wang, H. Lu. 2004. Microforming: Experimental investigation of the extrusion process for micropins and its numerical simulation using RKEM. *Journal of Manufacturing Science and Engineering* 126:4, 642-652.

Cohen, A., G. Zhang, F. -G. Tseng. 1999. EFAB: Rapid, low-cost desktop micromachining of high aspect ratio true 3-D MEMS (Micro electro mechanical systems). In *Proceedings of the 12th IEEE International Conference*, January 17-21, Orlando, USA.

Dutta, M., R. V. Shenoy, L. T. Romonkiw. 1996. Recent advances in the study of electrochemical micromachining. *J Eng Ind* 118:29, 29-36.

Ehrfeld, W., H. Lehr, F. Michel, A. Wolf. 1996. Micro electro discharge machining as a technology in micromachining. *Proceedings of SPIE* 2879, 332-337.

FANUC ROBOnano alpha-0iB machining samples, http://www.fanuc.co.jp/en/product/ROBOnano/ sample/index.html (Accessed October 20, 2005).

FANUC ROBOnano Ui, Brochure (2004).

Fraunhofer Institut Lasertechnik (ILT). http://www.ilt.fhg.de (Accessed October 20, 2005).

Fraunhofer Institut Produktionstechnologie (IPK), http://www.ipt.fraunhofer.de (Accessed October 20, 2005).

Fukuda Laboratory, Nagoya University, http://www.mein.nagoya-u.ac.jp/ (Accessed October 20, 2005).

Geiger. M, A. Mebner, U. Engel. 1997. Production of microparts: Size effects in bulk metal forming, similarity theory. *Production Engineering* 4:1, 55-58.

Gershenfeld, N. Private communication to author. December, 2004.

Goto. A., T. Magara, T. Moro. 1998. Formation of hard layer on metallic material by EDM. In *Proceedings of the 12th International Symposium for Electro-Machining*, May, Aachen, Germany.

Goto. A., T. Moro, K. Matsukawa, M. Akiyoshi. 2001. Development of electrical discharge coating method. In *Proceedings of the 13th International Symposium for Electro-Machining*, May 9-11, Bilbao, Spain.

Hitachi Anisotropic Conductive Film ANISOLM™. Brochure (2004).

Hitachi Arranged Tubes Technology. Brochure (2004).

Ikawa, N., R. R. Donaldson, R. Komanduri, W. Koenig, T. H. Aachen, P. A. McKeown, T. Moriwaki, I. F. Stowers. 1991a. Ultra-precision metal cutting: The past, present and future. *Annals of the CIRP* 40, 587-594.

Ikawa, N., S. Shimada, H. Tanaka, G. Ohmori. 1991b. Atomistic analysis of nanometric chip removal as affected by tool-work interaction in diamond turning. *Annals of the CIRP* 40, 551-554.

Ikuta, K. Biochemical Micro System Engineering Laboratory, Deptartment of Micro System Engineering, School of Engineering, Nagoya University, http://biomicro.ikuta.mech.nagoya-u.ac.jp/~ikuta/index.html (Accessed October 20, 2005).

IPK (Institut Produktionsanlagen und Konstruktionstechnik) and IWF (Institut fur Werkzeugmaschinen und Fabrikbetrieb) handout. "Future," page 3-34, Jan, 2004.

ITRI (Industrial Technology Research Institute) Annual Report "An Introduction to ITRI Nanotechnology," 2004, Taiwan, pages 1-56.

Kang, S. and A. G. Cooper. 1999. Fabrication of high quality ceramic parts using mold SDM. Paper presented at the Solid Freedom Fabrication Symposium, August, Austin, USA, 427-434.

Kawahara, N., T. Suto, T. Hirano, Y. Ishikawa, Y. Kitahara, N. Ooyama, T. Ataka. 1997. Microfactories: new applications of micromachine technology to the manufacture of small products. *Microsystem Technology* 3: 37-41.

Kitahara, T., K. Ashida, M. Tanaka, Y. Ishikawa, N. Ooyama, Y. Nakazawa. 1998. Microfactory and Microlathe. In *Proceedings of the International Workshop on Microfactories*, December 7-9, Tsukuba, Japan.

Lee, K. and D. A. Dornfeld. 2002. An experimental study on burr formation in micro milling aluminum and copper, *Transaction of the NAMRI/SME* 30, 255-261.

Liu, X., R. DeVor, S. Kapoor, K. Ehmann. 2005. The Mechanics of Machining at the Micro-Scale: Assessment of the Current State-of-the Science. Trans ASME Journal of Manufacturing Science and Engineering 126, 666-678.

Lucca, D. A., R. L. Rhorer, R. Komanduri. 1991. Energy dissipation in the ultraprecision machining of copper. *Annals of the CIRP* 40, 69-72.

Lucca, D. A. and Y. W. Seo. 1993. Effect of tool edge geometry on energy dissipation in ultraprecision machining. *Annals of the CIRP* 42, 83-86.

Lucca, D. A., Y. W. Seo, L. Rhorer. 1994. Energy dissipation and tool-workpiece contact in ultraprecision machining. *STLE Tribology Transactions* 37, 651-657.

Masuzawa, T. 2000. State of the art of micromachining. *Annals of the CIRP* 49:2, 473-488.

Masuzawa, T. 2001. Micro EDM. In *Proceedings of 13th International Symposium for Electromachining*, May 9-11, Bilbao, Spain.

Matsuura Machinery Corporation, http://www.vcnet.fukui.fukui.jp/co/en/matsuura.html (Accessed October 20, 2005).

Micro-Blast: Micropump based on liga and silicon technology, http://www.el.utwente.nl/tdm/mmd/projects/mublast/ (Accessed October 20, 2005).

Miraikan, (The National Museum of Emerging Science and Innovation), http://www.miraikan.jst.go.jp (Accessed October 20, 2005).

MIRDC (Metal Industries Research & Development) Research report, Taiwan, 2004, page 1-21, http://www.mirdc.org.tw (Accessed October 20, 2005).

Moriwaki, T., N. Sugimura, K. Manabe, K. Iwata. 1991. A study on orthogonal micromachining of single crystal copper. *Transactions of the NAMRI/SME* 19, 177-183.

Ohmori, H. and T. Nakagawa. 1995. Analysis of mirror surface generation of hard and brittle materials by ELID (Electrolytic in-Process Dressing) grinding with superfine grain metallic bond wheels. *Annals of the CIRP*, 44:1, 287-290.

Ohmori, H., K. Katahira, Y. Uehara, W. Lin. 2003. ELID-grinding of microtool and applications to fabrication of microcomponents, *International Journal of Materials and Product Technology* 18:4/5/6, 498-508.

Okazaki, K. Micromachine tool to machine micro-parts. 2000. In *Proceedings of the American Society for Precision Engineering 15th Annual Meeting*, October, Scottsdale, Arizona.

Okazaki, K. and T. Kitahara. 2000. Microlathe equipped with numerical control. *Journal of the JSPE* 67:11, 1878.

Photosensitive film for μ-TAS ME-1000 series. Hardcopy of viewgraphs from presentation on [date], [location].

Rajurkar, K.P., Z. Yu. 2003. Micro EDM and its applications. In *Proceedings of SME's Precision Micro Machining Technology and Applications Technical Conference*, June 11-12, Minneapolis, USA.

Regenfuß, P., L. Hartwig, S. Klötzer, R. Ebert, T. Brabant, T. Petsch, H. Exner. 2004. Industrial freeform generation of microtools by laser micro sintering. In *Proceedings of the Solid Freeform Fabrication Symposium*, D. Bourell, et al., eds., August 2-4, Austin, TX, 709-719.

Research Activities Report, 2003 Robotics and Mechatronics, Nagoya University, page 1-72, March, 2004.

RIKEN Research, brochure, 2004, Japan.

Tönshoff, H. K., T. Masuzawa. 1997. Three dimensional micromachining by machine tools. *Annals of the CIRP* 46:2, 621-628.

Ueda, K, K. Iwata. 1980. Chip formation mechanism in single crystal cutting of beta-brass. *Annals of the CIRP* 29, 65-68.

Vogler, M. P., S. G. Kapoor, R. E. DeVor. 2004. On the modeling and analysis of machining performance in micro-endmilling, Part I: Surface generation. *Journal of Manufacturing Science and Engineering* 126:4, 685-694.

Waldorf, D. J., R. E. De Vor, S. G. Kapoor. 1999. An evaluation of ploughing models for orthogonal machining, *Journal of Manufacturing Sciences and Engineering* 121, 550-558.

CHAPTER 5

METROLOGY, SENSORS AND CONTROL

Thomas R. Kurfess and Thom J. Hodgson

ABSTRACT

This chapter focuses on metrology, sensors and controls for micromanufacturing. In general, a variety of sensors are employed in micromanufacturing. Many of them are used for metrology, and in particular dimensional metrology. Most are used on production machines to provide feedback during production. Few metrology systems for process and product control are available. The sensors and metrology systems available for product and process analysis are, in general, slow and not readily implemented. From a metrology and sensing perspective, the currently available systems are mostly two-dimensional in nature. Most of the measurement systems are fairly slow, which makes them suitable for R&D as their measurement speed limits their utility on actual production lines. Controllers fall into two architecture categories, open and closed (standard computer numerical control (CNC) controller architecture). In most cases, commercially available systems employ more traditional closed architecture controllers, while research and development teams tend to use open architecture systems. For most manufacturing systems, process control is a desired capability. In many cases it is achieved by measuring parameters from both the process and the product and adjusting process parameters to improve the quality of the part. In general, this approach is far from being achieved in the micromanufacturing field as the actual measurement of part and process parameters in real-time is limited to a few very special instances. Furthermore once a measurement is made, the process must be tuned according to the process model. In most cases, such models do not exist. Thus, in order to effectively achieve process control and quality enhancement, improved real-time metrology systems and sensors are needed along with models that accurately describe the various manufacturing processes at the micro level such that true process control can be enabled.

INTRODUCTION

Metrology, sensors and controls are critical for addressing quality control of micromanufactured products and processes. Quality control (QC) is one of the barriers to the growth of micromanufacturing both as an enabling technology and as a new market. Quality control not only provides quality assurance (QA)/QC but also feedback to optimize the design, fabrication, and assembly processes involved in the next generation of microsystems. QC may be divided into two equally important phases: R&D and production. From an R&D perspective, metrology is necessary for an understanding of the performance of newly developed products and processes. When a new design is generated and a prototype of that design is produced, the question that must be answered is, "Does this new design meet specifications?" The specifications will be both of a dimensional and performance nature. For example, if a gear train is designed, micrometrology systems will address not only the dimensional compliance of the individual gears, but also the assembly of the gears into the gear train. Other compliance issues that must be addressed for the gear train might include material parameters, dynamic performance and overall assembly compliance. From a production viewpoint, metrology is necessary to ensure that the products are fabricated correctly. Ultimately, the assurance of producing products that are "in specification" requires that the production process be controlled. Thus, metrology at the production level is used to understand and enhance production processes to ultimately yield microcomponents and systems that meet specifications. This understanding can be used to continuously improve production processes, resulting in continuous process and product improvement. In the end, the knowledge and understanding gained from metrology systems can be used to develop next-generation product and process designs.

This chapter discusses metrology systems as they apply to the dimensional measurement of microcomponents. Sensors are then discussed as they apply to both metrology and production systems. For the purposes of this report, the relationship between the two is assumed to be that metrology systems make use of a variety of individual sensors (for example, a scanning white light interferometer uses a microscope with a charged-couple device (CCD) camera and linear glass scales—making it a metrology system). Lastly, the chapter presents a review of controls used in micromanufacturing.

Metrology

In general, metrology is usually thought of as dimensional metrology that is used to verify component geometric dimensions. From this perspective, very little capability is available for microcomponent inspection. Most inspection that is currently done is accomplished using optical microscopes, scanning electron microscopes (SEM) and tunneling electron microscopes (TEM). Metrology with microscopes is executed in a fairly primitive manner using images from the microscope in conjunction with calibrated markings. Figure 5.1 is an example of such an image. The figure shows wear at the tip of a carbide cutting insert that is on the order of 100 μm in size (wear or tip size). The "100 μm" scale at the bottom is a typical scale marking. This particular picture was generated using an optical microscope. Inspection and metrology using these means is the equivalent of conducting dimensional measurement of mesoscale parts with a ruler or other scale indicator, and is not an accurate means of part geometry validation. For smaller components, electron microscope systems are also employed. However, these systems are even more difficult to use as they must be specially calibrated to make use of scale markings. For both optical and electron microscopes, the scale markings are generally intended to provide a "feel" for the size of the image, rather than a precise tool for metrology.

Microscopes used with calibrated optics and servo stages are being used fairly effectively for micrometrology. For this application, images are taken with well-calibrated microscopes; vision processing algorithms are then used to detect part edges and to analyze part geometry in 2D. These systems are limited to the resolution of the camera's charge-coupled device (CCD) array, which is interfaced with the microscope. In general, these systems are 2D in nature, although the third dimension (height) can be resolved over the entire part using an autofocusing technique. These systems provide a trade-off between resolution and field of view. If higher resolution is desired, then a more powerful microscope objective is used. This limits the overall field of view. The microscope can be augmented with a servo stage that allows several images to be stitched, or aligned, together; however, stitching multiple images reduces the overall accuracy of the measurement. Such systems provide reasonable feedback for research and development purposes. However, at present, they are too slow to be used in-line for production and process control purposes.

Some metrology is being done with atomic force microscope (AFM) systems, but most of this work is in the early stages of development and is targeted at nano-type applications. As implied by its name, AFMs have the capability to resolve geometries at the atomic level. However, such resolution is too fine for most microcomponent applications, making them somewhat useful as a research and development tool but not as a production tool. AFMs are usually used as very high-resolution profilometers, making their measurements 2D in nature.

White light interferometry (WLI) and confocal microscopes appear to be leading metrology systems for 3D metrology in a number of research and development groups. There are a variety of WLI manufacturers that have commercially available products, although there are fewer producers of confocal microscopes. In general, U.S. manufacturers are the leading producers of WLI systems. Zeiss in Germany is the leading producer of confocal microscopes for micrometrology purposes. Both of these systems are based on the use of CCD cameras interfaced with a microscope to conduct measurements. Thus, the lateral resolution of these systems is limited by the pixel density on the CCD array. Furthermore, increased lateral resolution is realized by increasing the microscope objective power. Thus, just as with standard vision-based systems, there is a trade-off between field of view and lateral resolution. Servo stages can be used in conjunction with these systems to increase the lateral field of view; however, stitching issues reduce the effective accuracy of the overall measurement just as with microscope-based systems. Unlike microscope systems, WLI and confocal are most accurate in the vertical direction (out of plane). For example, a WLI is nominally presented as having sub-nanometer resolution. This is out-of-plane resolution and is realized by the interferometric measurement that is made on the system. The in-plane (lateral) resolution is still limited by the WLI's camera CCD pixel density and stage-positioning resolution (stitching). The other limitation of these two types of systems is that they must be used to measure relatively flat surfaces. Surfaces having angles of greater than approximately 10° do not reflect enough light to be measured.

Figure 5.1. Image from an optical microscope.

Presently, WLI and confocal systems are too slow for use in large-scale production. This is true if they are used as separate, stand-alone systems. Various research groups are looking to incorporate them directly into production machines for use in production machines. The difficulty with this approach is that such metrology systems can be very expensive. However, if micromanufacturing techniques can be used to fabricate a variety of the components necessary for these systems (e.g., lenses), then economies of scale may be affected, resulting in lower unit costs that may allow them to be more easily integrated into production systems. In particular, several institutions in Korea are building their own confocal microscopes and WLI systems in conjunction with Korean manufacturers, providing the latter with two types of experience that are critical for successful micromanufacturing. First, they are learning how to produce metrology systems. Second, they are learning how to integrate such systems into their production units. If these systems prove to be useful in production operations, Korean manufacturers will have an excellent foundation from which to launch a variety of commercial products.

There are several truly 3D systems that have been developed. One, presented by Professor Masuzawa at the University of Tokyo, is used primarily for R&D. It consists of a tactile probe that serves primarily as a sensor. This system is discussed in detail in a subsequent section of this chapter. Other, more developed systems include 3D coordinate measurement machines at Zeiss and at Japan's National Institutes of Advanced Industrial Science and Technology (AIST). The Zeiss and AIST systems are shown in Figure 5.2 and Figure 5.3, respectively. Of these two systems, the Zeiss machine, known as the F25, has recently become commercially available. The F25 has several measuring sensors integrated with it as shown in Figure 5.4, including an optical sensor that performs the same function as the microscope-based inspection systems, a tactile measurement head that is used as a coordinate measuring machine (CMM) probe in both a point-to-point mode as well as a scanning mode, and a visualizing camera that is used to aid in the positioning of the tactile probe. The F25 is approximately $1.5 \times 1.5 \times 2 \ m^3$, has a measurement volume of $100 \times 100 \times 100 \ mm^3$, a three-dimensional scanning speed of 0.5 mm/s, a maximum single-axis speed of 20 mm/s, a maximum multi-axis speed of 35 mm/s and a maximum acceleration of $500 \ mm/s^2$. The volumetric accuracy of the F25 in point-to-point mode is 250 nm. The volumetric accuracy of the F25 in scanning mode has not been determined yet. The repeatability is 50 nm, and it has a resolution of 7.8 nm. The machine is completely enclosed in its own environmental container. The projected cost for the F25 is approximately $500,000. The development of this system was partially supported by the government of Germany and is based on collaborative research with the university partners of Zeiss.

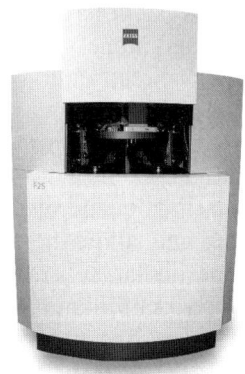

Figure 5.2. Zeiss F25.

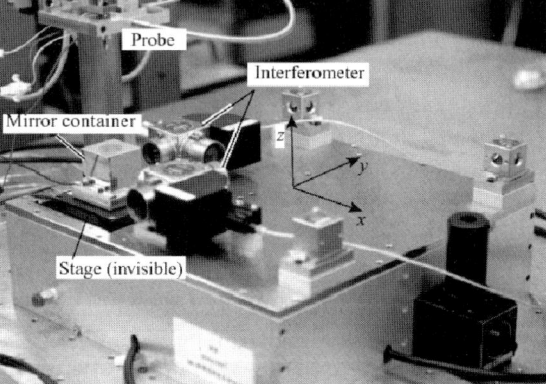

Figure 5.3. AIST CMM.

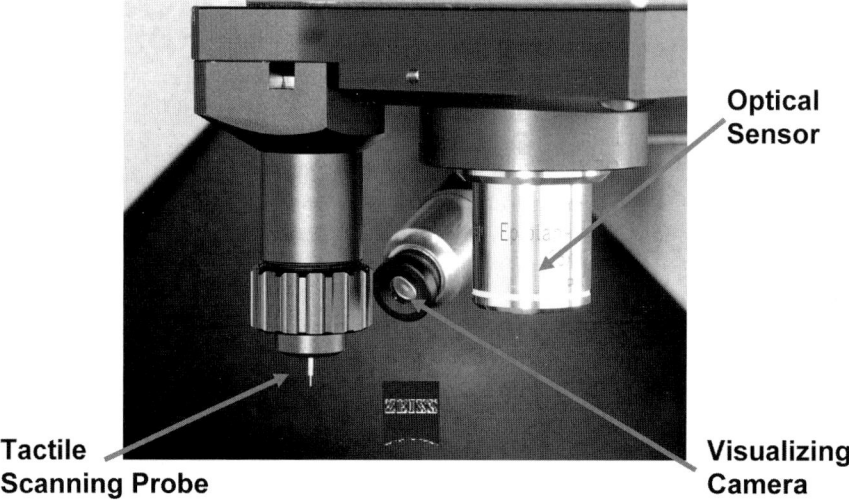

Figure 5.4. Zeiss F25 sensor heads (Courtesy Carl Zeiss
IMT Corporation).

One of the most difficult aspects of implementing the micro CMM is the probe that interfaces between the microcomponent and the CMM. Figures 5.5 and Figure 5.6 show the probes of the F25 and the AIST machines, respectively. These probes are highly sensitive and consequently very fragile. The F25's probe is placed on a silicone wafer substrate as shown in Figure 5.7. The wafer is etched to generate a flexure. This flexure is coated with a piezoresistive material that allows the probe deflection to be measured. Using this probe, the F25 is then servoed in a scanning mode in much the same fashion as other larger-scale scanning machines that Zeiss makes. Metrology software developed for other Zeiss CMMs is employed to analyze data from the F25. The smallest probe available for the F25 has a 120 µm diameter tip, with a maximum measurement depth (due to probe length) of 4 mm. Thus, the F25 is capable of measuring holes as small as 150 µm with depths of up to 4 mm. It is the only such system that can make a true 3D measurement of such holes.

 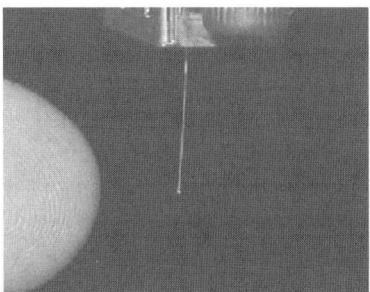

Figure 5.5. Zeiss F25 µCMM probe (Courtesy Carl Zeiss IMT Corp.).

Figure 5.6. µCMM probe (AIST).

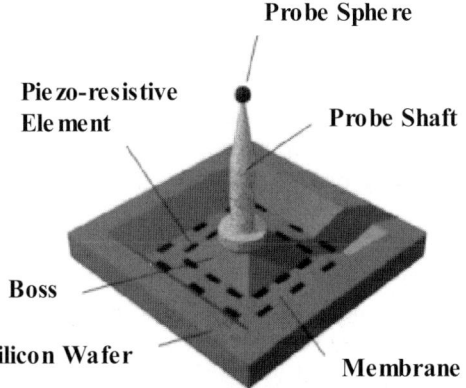

Figure 5.7. F25 µCMM probe design.

In conclusion, a variety of metrology systems are available for micro-component inspection. However, few of them are 3D in nature. Furthermore, all of them are relatively slow and expensive, making them reasonable choices for R&D but less-than-desirable choices for use in production lines. Even as faster and less-expensive systems become available, micro-metrology systems will continue to face two major hurdles: edges and standards. All components have edges, and they are used to define the geometry of a part. Edge radii of 10-100 µm are typical for manufactured components. However, such sizes may be a significant fraction of a micro-component's dimensions or those of its features. Thus, the ability to characterize and edge accurately is critical and has not been addressed to date. Finally, many of the standards that are applicable to macroscale metrology are not available for microscale metrology. Tools such as interferometers, ball bars and even gage blocks and gage balls, are not available at the micro-level for testing and calibrating micrometrology systems. As a result, the ability to calibrate and determine the capability of microscale metrology systems is limited.

Sensors

In general, sensors fall into one of two categories:

- *Machine sensors*—sensors that are used as feedback mechanisms for manufacturing and metrology systems

- *Component sensors*—sensors that are used to directly measure microscale components

The distinction between these two categories of sensors is important and must be made clear. Sensors that are used on manufacturing systems are readily available and are not necessarily "micro" in nature. For example, machine sensors measuring linear position may be 50–100 mm in length or even larger. However, they typically have a resolution of 1 nm or less. Furthermore, most of these sensors are used to measure in a single direction (example: along the X-axis), or at best need to have a single direction with an ultra-high resolution. This single high-resolution direction works well for most systems. Sensors used to measure microcomponents are those that actually take data from a component. These sensors typically must have minimal physical impact on the part. In fact, non-contacting sensors are desired. Furthermore, to acquire 3D metrology data, these sensors need to have ultra-high 3D resolution. Data from these sensors are used to verify part geometry and target dimensions. Machine sensors and component sensors are discussed separately in this section, due to the significant differences in their objectives and tasks.

Machine Sensors

The sensor of choice for providing feedback on production and manufacturing systems is the ultraprecision encoder. Nominally, the encoder comes in two variations, both of which are employed on micromanufacturing systems. The first is the linear scale used for measuring linear displacement. The second is a rotary encoder, used for measuring rotational motion. There are a variety of companies that make such encoders. In the case of the rotary encoder, which is an older and more developed technology, no particular supplier appears to dominate the market. However, Heidenhain (Germany) appears to be the dominant supplier of linear encoders, also known as linear glass scales. Most of the groups that were involved with micromanufacturing or micrometrology systems employed scales produced by Heidenhain when such feedback devices were used. Sony was also used or mentioned by a fair number of groups in Asia. FANUC indicated that they would use whatever type of linear scales a customer specifies on their ROBOnano, although they indicated the majority of systems were produced with Heidenhain scales, with Sony as the next most popular choice.

Both rotary and linear optical encoders have the advantage that they can be non-contact in nature if care is taken in their incorporation into the machine metrology system. As shown in both Figure 5.8 and Figure 5.9, the read head does not contact the optical encoder. Thus, if the read head is mounted on an independent frame, also known as a metrology frame, it can be mechanically, thermally and electrically isolated from the scale. In many instances the use of a separate metrology frame enables designers to produce machines with superior accuracy and resolution.

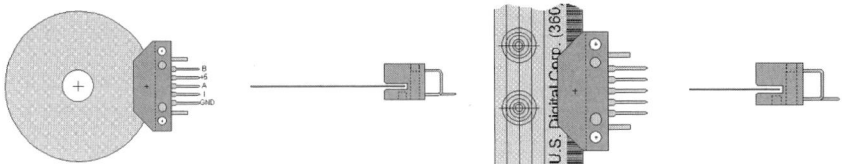

Figure 5.8. Rotary encoder.   Figure 5.9. Linear encoder.

The other machine sensor that is heavily employed in the microfactory is the interferometer. Two types of interferometer are typically used in micromanufacturing applications. The first is the white light interferometer. This device is typically used to characterize the surface of a component, and is discussed in the ensuing section on component sensors. The second type of interferometer is the heterodyne laser interferometer. Similar to a linear encoder, this device measures relative motion of a device with ultra-high precision. The laser interferometer uses interference of laser light (typically HeNe at a wavelength, $\lambda=632.8$ nm, although less-stable solid-state lasers are sometimes employed). The laser interferometer can be used as a positional feedback mechanism in the same fashion as the linear encoder. It is also a non-contacting sensor that can be used quite effectively with a separate metrology frame to greatly enhance the overall machine accuracy. However, the laser interferometer is an expensive piece of equipment ($100,000 for a basic set-up), and is much more difficult to employ than a linear scale. Thus, the linear glass scale is preferred over the laser interferometer in all but the most demanding applications. Both the interferometer and the scales (linear and rotary) output a quadrature signal. This allows similar interface hardware to be employed when using both types of devices.

While laser interferometers are significantly more difficult to employ than glass scales, they can be used in conjunction with a machine's scales to remove the machine's repeatable scale errors via calibration. This procedure is known as error mapping. As the name suggests, the interferometer is used to map scale errors for each linear axis on the machine. The scale error is the difference between the position of the readout scale and that of a known reference (the laser interferometer) along the axis of motion. The procedure is relatively straightforward, the machine is moved a specific distance (e.g., 1 mm increments), and the error for each step is recorded. These errors are entered into the machine controller, which has the capability of compensating for the error. Typically, only scale errors are mapped on machining systems and metrology systems, although errors such as angular, straightness and squareness (29 total for a three axis (X, Y, Z) machine) can also be mapped. For example, the accuracy of the FANUC ROBOnano is 2.5 µm before error mapping and is reduced to 10 nm with positional error mapping. Calibration is done at the factory, and the machine can be recalibrated on-site by an experienced professional. As most controllers are capable of error mapping scale errors, the only additional cost for this significant improvement in accuracy is the calibration. The machine requires no extra hardware. Error mapping is a major tool in achieving the target volumetric accuracy of a micromanufacturing system in a cost-effective manner.

Two primary difficulties must be addressed when using machine sensors with high resolution and relatively high-feed velocities: (1) the speed at which position increments must be counted, and (2) the number of position increments that must be counted. These issues are tightly related to the controls area and are discussed in more detail in that section of the chapter.

Sensors for Measuring Components

A vast number of new sensors have been used to measure components. A variety of other sensors have been specifically developed to measure components. In general, most sensor metrology applications target dimensional/geometric metrology (e.g., size, geometry, surface finish). However, there are a few sensors that are also targeting the ability to measure other quantities such as temperature, reflectivity, and a variety of material properties (e.g., Young's modulus and the coefficient of thermal expansion). With one exception, all of the sensors that were seen during this visit were used for dimensional metrology. This section may be considered a bit redundant with respect to the metrology systems presented in the previous section. However, a bit more detail is presented in this section from a sensor perspective.

Many of the sensors for measuring the components have been discussed in the background report. In general, the sensing systems for geometric metrology can be divided into two sets of categories. The first category set is units that employ vision systems, cameras or devices that use pixel-type arrays, and those that do not. Vision-based systems include, but are not limited to white light interferometers, scanning electron microscopes and optical microscopes. X-ray tomography may also be grouped into this category, as X-ray sensors are comprised of pixel-based elements. Such devices rely on taking a "picture" of the device and, subsequently, performing some type of analysis (usually edge detection) on the data to determine quantities like size. Nominally, the resolution of such systems is limited by a combination of the magnification and the pixel density in the CCD array on the camera used to capture the data. In many instances, high resolution is achieved by using high magnification (e.g., using a higher powered optic in a microscope). In such cases, the range of the sensor is typically reduced. Thus, several data sets must be taken and stitched together. In this instance, it is the registering of each data set to the others that is typically the limiting factor of the system's resolution. Thus, the increase in resolution generated by magnifying a part is typically offset by the resulting reduction of range (field of view).

To better illustrate this issue, Figure 5.10 shows a microgear that was inspected using a Zygo WLI. The major diameter of the gear is approximately 0.7 mm, with feature (tooth) sizes on the order of 100 μm. The direction of maximum sensitivity for the WLI is the Z direction, which is in and out of the plane of the image shown on the left part of Figure 5.10. The dimensions of interest in this particular inspection are the gear dimensions that are in the plane of this image. The high resolution direction of the WLI does show that the gear's top surface is curved. This can easily be seen in the 3D image shown in Figure 5.10. This information is important, as it demonstrates that the lapping process employed to finish this particular part rounded the edges of the part. However, the part dimensions of interest such as major and minor diameter do not make use of the WLI's interferometric resolution of less than 1 nm. Rather, the quantification of these dimensions is based on the lower lateral resolution of the WLI. Furthermore, the sidewalls of the part cannot be measured using the WLI, so no information for them is available at all. Thus, the resulting inspection can be considered 2.5D at best.

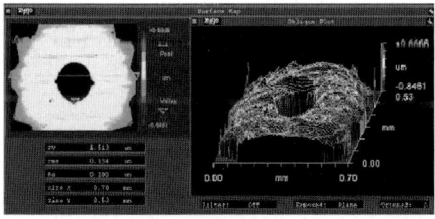

	Contact	Non-Contact
2D		
3D		

Figure 5.10. Microgear.

Figure 5.11. Contact and non-contact.

Sensors systems not employing vision type systems include contact probe systems such as AIST's micro CMM (Figure 5.10).

The second category of sensor systems can be further divided into sensors that are nominally designed for 2D inspection, and those that are truly 3D in nature. In some instances, 2D systems can be used to measure 2.5D but are not fully capable of measuring in 3D. An example of this is the WLI, which is primarily intended to measure surface finish characteristics but has been used to generate 2D measurements. In the case of the WLI, the part's edge is detected, and it is this edge that is used for metrology. As previously discussed, edge detection is another issue limiting metrology systems' capabilities that is discussed in the metrology section. Inspection of a part's top surface is not a problem, but the fundamental principals that govern the operation of the WLI do not permit it to inspect the side walls of a part. In general, all 2D systems used for inspection are fundamentally limited to 2D inspection and cannot be employed for 3D inspection. The matrix shown in Figure 5.11 presents the breakdown of the two category sets for sensors and sensor systems used for dimensional metrology.

A variety of contacting cantilever-sensing devices has been developed for use in metrology. Four such microcantilever devices were demonstrated at Seiko Instruments, Inc. (SII) by Professor Matsuzawa: (1) the self-sensitive cantilever for displacement, (2) the nanoprobe for electrical resistivity, (3) the self-sensitive dual cantilever for multi-resolution displacement, and (4) the thermal cantilever for temperature measurements. The self-sensitive cantilever, which has been commercialized and used in dynamic force microscope (DFM) and AFM applications, is an alternative to the standard probes used on these systems. Standard probes employ an optical displacement measurement arrangement and require a laser and associated optics. The self-sensitive cantilever replaces the optical deflection measurement with strain measurements at the base of the cantilever by a piezoresistive element connected in a Wheatstone bridge configuration along with a reference resistance (Figure 5.12). The principal advantages are reduced overall size, the ability to be operated in the dark field, and ease of use and setup. Calibration is performed by using standard samples. The size of the cantilever is approximately W50 x L120 x T5 μm, with frequency response 300 kHz. The Nano4probe, fabricated by focused ion beam (FIB) technology, uses the four-point resistivity measurement method implemented by four narrowly pitched (200 nm) electrodes. The dual cantilever displacement probe uses a bimorph to switch between the large and small area scanning probes. The large area probe is used when ultra high-resolution measurements are not necessary. Thus, the dual cantilever displacement probe addresses the issues of the AFM having too great of a resolution for the inspection of most microcomponents. However, the high-resolution (small area) probe can be employed when the full resolution of the AFM is required.

The thermal cantilever for temperature measurements is one of the few sensors that were observed that were not developed for dimensional metrology; rather, this sensor is used to measure the temperature of microcomponents. Due to calibration issues, it would be difficult to use a thermocouple to measure the temperature of the part directly. This sensor employs a thermopile (a large number of thermocouples) to detect temperature differences between the probe (which is heated) and the surface of the inspected part. The thermopile provides feedback to the sensor as to the temperature difference between a calibrated and controlled temperature on the sensor and the microcomponent. The thermopile indicates the temperature difference and the direction of that difference (e.g., higher or lower); it does not provide a value for the actual temperature difference. The probe reference temperature is modulated until there is no difference between the reference temperature and the component temperature. Then a thermocouple that has been calibrated for use on the probe's reference temperature generator is used to measure the probe's reference temperature.

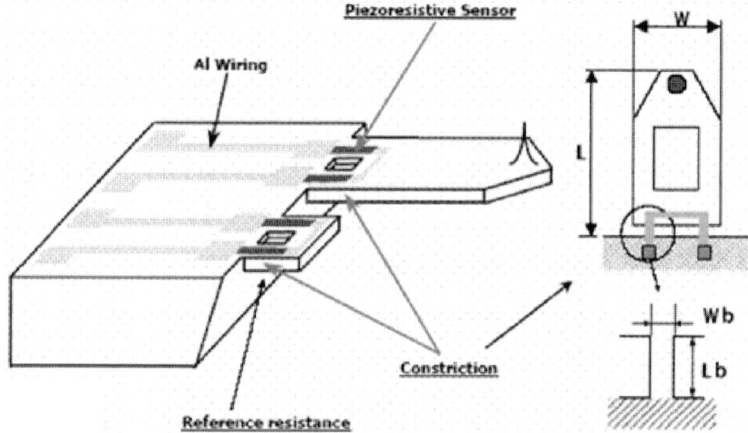

Figure 5.12. Structure of self-sensitive cantilever.

The metrology work done by Professor Masuzawa's group at the University of Tokyo specifically targets the measurement of hole sidewalls. While this device is not a complete 3D measurement system, it does provide metrology information that is not available from most other measuring systems. Two major hurdles for most micrometrology systems are the measurement of side walls and the measurement of holes. As already discussed, measuring side walls is a difficult task as most optical-based systems do not receive enough light from a reflection of the side walls to make a meaningful measurement. Secondly, measuring a small hole requires that a probe be sent down the inside diameter of the hole, again a very challenging task (one that only the Zeiss F25 is currently capable of performing). Professor Masuzawa's system addresses both of these two major issues. The system consists of a vertical probe as shown in Figure 5.13. The probe has a piezoresistive material on it that can be used to measure deflection. This is much the same as the probe shown to us at Zeiss and SII.

To make a measurement, the probe is servoed into the part as shown in Figure 5.14. The output of the probe is also shown in Figure 5.14. While the probe is not contacting the part, the horizontal line in the plot is generated (about 0–4 μm on the plot). Once the probe contacts that part, the voltage output of the probe increases linearly for small deflections as is shown in Figure 5.14 (4–8 μm). Figure 5.15 is a picture of the machine. Figure 5.16 is a close-up of the head. Basically, this approach generates two lines. These lines are fit and the intersection of these two lines is used as the contact or measurement point. This approach is much more robust but slower than the SII technique. Robustness comes from the fact that the intersection of the two lines is used. If the sensitivity of the piezoresistive material changes, the slope of the second line will change; however, the intersection of the two lines will not. If there is noise, its effect is reduced by the fitting of the lines. This eliminates the need for the environmental add-on in the SII unit. The one disadvantage over the SII probe (Figure 5.12) is that the SII probe will be much faster, generating significantly more data per unit time than Masuzawa's probe. From a CMM analogy, the Masuzawa approach is like a trigger CMM, going point-to-point. The SII probe is more of a scanning probe.

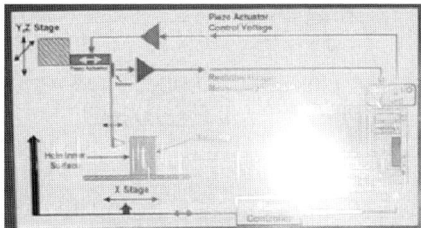

Figure 5.13. Masuzawa
vertical probe concept.

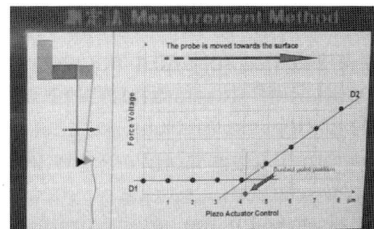

Figure 5.14. Masuzawa
measurement method.

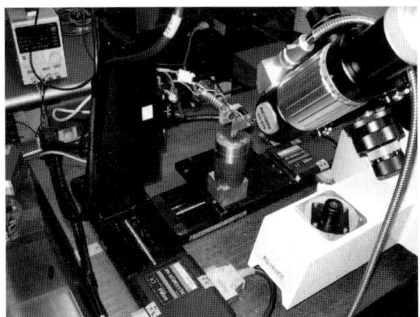

Figure 5.15. Hole meas-
urement machine.

Figure 5.16. Probe for hole
measurement machine.

In conclusion, a variety of sensors are available for microcomponent metrology. Many of them circumvent calibration issues, which are difficult to address, by either incorporating their own standards, or by employing procedures that generate data that can be interpreted without the need for precise calibration. One of the major issues still facing these sensors for dimensional metrology is edge detection and analysis. Most of the sensors are extremely high precision and cannot be used to measure part edges. Rather, they are used to measure geometric aspects of components that do not have high-frequency spatial variations. Also, all of the sensors that were demonstrated were used for measuring static parts and did not address the need to measure operational dynamic systems. Finally, many of the sensors presented can be implements in an array fashion. Such an implementation will likely generate a significant amount of data that will overwhelm most sampling and analysis systems. Thus more advanced techniques for parallel sampling and processing of large quantities of data will be required to fully utilize the capabilities of these new sensors as they become more readily available.

Controls

This section presents issues related to controls, including critical considerations for micromanufacturing as its requirements relate to controls, current controls implementation of micromanufacturing systems, and specific needs in the systems, product and process controls areas.

System Controllers

In general, there are a variety of controls used for micromanufacturing. These controls can be broken into two fairly distinct groups: open and closed architecture. In general, machine tool manufacturers appear to be utilizing two closed architecture controllers, the Fanuc 15i Controller and the Mitsubishi CNC 700 series controller. Both of these controllers have high-speed reduced instruction set computing (RISC)-based processors and are capable of controlling a wide variety of systems. The CNC 700 series features nanocontrol technology (RISC-CPU and high-speed optical servo network) high-speed, high-precision 5-axis control. These systems are closed architecture in nature, and the producers of these controls do not anticipate moving to a more open architecture.

The Bosch Rexroth IndraMotion MTX (Figure 5.19) is a new industrial controller that is entering the market and appears to be more open architecture in nature. Bosch intends to make use of these controllers for controlling their high-precision grinding and machining applications that produce a wide variety of microcomponents for their fuel injector and fuel pump divisions. The Siemens 840 series CNC controllers also have a somewhat open architecture capability. Both the Bosch and Siemens controllers will permit more advanced process control beyond the standard servo position control discussed above. For example, process control based on external inputs such as power, force or temperature may be executed with these controllers.

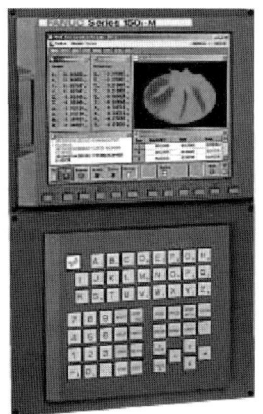  

Figure 5.17. FANUC   Figure 5.18. Mitsubishi CNC 700
15i controller.   controller.

Most research and development groups visited employed open architecture systems produced by Delta Tau, D-Space, and National Instruments. The processing of data for these controllers is generally done with a combination of LabVIEW and MATLAB; although other programming languages are also employed. Few research groups employ any of the standard CNC controllers.

The FANUC series 15i control is used to control the ROBOnano ultra high-precision multi-function machine tool. It can function as a five-axis mill, a lathe, a five-axis grinder, a five-axis shaping machine and a high speed shaper. It uses friction free servo systems (all linear slides and screws are static air bearings). For milling, an air turbine spindle is employed for rotational speed up to 70,000 rpm. Shaping is done using a high-speed shuttle unit capable of producing three grooves per second, and a 3 kHz fast tool servo using a PZT actuator. Cutting is nominally accomplished with a single-crystal diamond tool. Error mapping on the 15i control is limited to positional (axial) errors only. The resolution of the linear axes is 1 nm and 0.00001° for the rotational axes. The design of the machine is such that it experiences no backlash and stick-slip motion. The size of the ROBOnano is 1270 x 1420 x 1500 mm. This does not include the CNC locker or the thermal management unit. However, in total, the machine is very compact. The range on the X-, Y- and Z-axes of the machine is 200 mm (horizontal, left and right) by 20 mm (horizontal, in and out) by 120 mm vertical, respectively. The maximum feed speed for the X- and Z-axes is 200 mm/min, and 20 mm/min for the Y-axis. The development of ROBOnano was partially supported by the government of Japan.

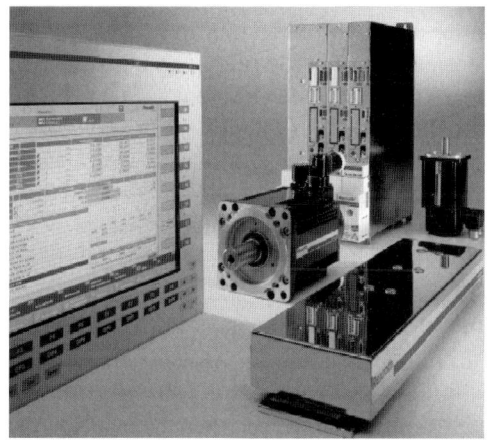 

Figure 5.19. Bosch Rexroth IndraMotion MTX. (Source: Rexroth Indramotion MTX October 2005)

Figure 5.20. FANUC ROBOnano machine tool.

Issues Related to Control

Several major issues are specifically related to the control of micromachine tools. They include higher speed controllers, position counters having higher resolution capabilities, spline interpolation, higher order motion control and environmental control. Increased controller speed is critical for micromachine implementation for several reasons. First the higher speed allows for smoother interpolation. Second, servoing at ultra-high resolution (on the order of 1 nm) demands that the controller is capable of tracking encoders at very high rates. For example, if a machine has 1 nm resolution and is traveling at federate of just 1 mm/s, the controller must track an encoder pulse at a rate of 1 pulse/μs, resulting in a 1 MHz clock requirement. If the system is traveling at 1 m/s, an encoder pulse is generated every ns, requiring a 1 GHz clock. If this same machine has a one-meter range, the controller must be able to track a billion encoder counts, requiring a 30-bit (minimum) counter. The speed at which the encoders must be read in conjunction with the range of the encoder counts is significant and outside of most standard controller specifications. Thus, only very high-end controllers are used in such applications.

In order to generate smoother trajectories, spline interpolation is used in conjunction with high-resolution, high-speed controllers. The spline interpolation results in improved surface finish, especially over shorter-range motions. Furthermore, trajectory planning and control are accomplished on the machine's position, velocity, acceleration, and jerk (the time derivative of the acceleration). This differs from typical machine control where only the position, velocity and acceleration are considered. The incorporation of jerk into both the trajectory planning and control results in smoother and more consistent machine motion trajectories. Error mapping is employed on almost all of the micromachine tools (and measurement systems). For all systems reviewed, only positional (axial) error maps were developed and used. Thus, angular and straightness errors were not considered. Positional error mapping appears to be sufficient to achieve the desired machine accuracies at the present time. As an example of the utility of positional error mapping, the accuracy of FANUC's ROBOnano was increased by a factor of 100 via error mapping.

Finally, given the size of the micromachine tools and of CMMs, many of them are enclosed in their own environmental chambers. This enables tight control of the machine tool's temperature, reducing variations due to thermal expansion and enhancing the machine's accuracy. Typically the environmental control uses an air shower temperature control unit, whose temperature is controlled to 0.1°C. To demonstrate the importance of thermal control, the following example is presented. The thermal coefficient of expansion of steel is approximately 12 parts per million per degree C. Thus, a bar of steel 1 m in length grows 12 μm for every degree Celsius that it is heated. This is a significant amount of change when producing microcomponents. Coolant temperatures are also controlled to an even tighter level.

Open Architecture Controls / Control Flexibility

There is a clear dichotomy regarding the utilization of open architecture controls in the micromanufacturing area. Most of the research and development is done on an open architecture and flexible platform. This is true at academic institutions, government laboratories and in industry. A variety of controllers were seen including PC-based controllers operating with real-time systems, single-board motion controllers (usually employing a PC as the interface for programming), and field programmable gate arrays (FPGA). However, from an industry/implementation perspective, closed architecture controllers using hardware based on application specific integrated circuit (ASIC) systems are employed.

Even with the "closed architecture" controls such as the FANUC or Mitsubishi controllers, a single controller can run multiple types of machines. For example, ROBOnano can be configured for milling, turning, shaping and several other operations using the same FANUC 15i controller. While these controllers are flexible, they are not open architecture. It appears that the mainstream control companies in Japan such as FANUC and Mistubishi will not move into the open architecture market. The German control companies such as Siemens and Bosch appear to provide a more open architecture platform; however, these controllers are still not as open as the DSP and FPGA controllers.

Process Control

The ultimate goal of controlling a micromachine tool is to control the machining process. While significant work is being conducted in the realm of servo control of machine tools, little is being done regarding process control. For example, positioning accuracy and repeatability were amply demonstrated. However, the effect of a cutting process, such as milling or turning, on the overall machine performance was not demonstrated. For a variety of ultra high-production processes, the actual cutting dynamics must be considered so that the process can be controlled well enough to achieve the desired results. Thus, the ability to control parameters other than tool position trajectory will be important in micromachining. Furthermore, process control implies that there is a process model available to use in actuator trajectory generation. For most micromachining processes, models do not exist. Furthermore, it has been shown for a variety of processes that micro-level process models are not accurately represented via macro-level models. Thus, new process models for the micro-level must be developed before process control can be properly executed.

In conclusion, a variety of controllers that can adequately control micromachine tools are available. These controllers are flexible in nature and can be either open or closed architecture depending on the application. The systems are capable of controlling a variety of process parameters, resulting in some of the most accurate machine tools available on the market. However, to fully control various processes, improved and more accurate process models are required. These will have to be developed before true process control can be achieved.

SUMMARY AND CONCLUSIONS

In general, a variety of 2D and 2.5D metrology systems are currently available. They are relatively expensive and slow, but are useful for initial microcomponent and system R&D. From the standpoints of both robustness and inspection speed, metrology systems lack the ability to be used in a production setting from both a robustness perspective as well as an inspection speed perspective. There are a few truly 3D metrology systems. These systems are quite capable, but are expensive and, again, are not built for a mass-production environment. The sensors that are available for micromachines are adequate for single-axis position control. Other sensors are currently being developed but are not yet well enough developed for industrial-scale use. In particular, a sensor is needed to interface metrology systems to the small-scale microcomponents. Ideally, this sensor would be non-contact, have high resolution and high bandwidth. This would enable high-speed, high-precision dynamic measurements of microcomponents and systems.

Controllers are evolving to address a number of issues raised in controlling micromanufacturing systems. In particular, they are operating at higher speeds, communicating with drives and sensors at higher speeds, generating smoother trajectories (position, velocity, acceleration and jerk) that result in smoother and more accurate surfaces. They have faster quadrature decoders with higher counting capabilities to allow them to properly track high resolution motions over the machine tool's range. Controllers are also becoming more flexible to address the variety of processes that are being used in the microfactory. While they are becoming more reconfigurable—for example, controlling a lathe and a mill—major control system manufacturers in Asia are not looking towards open architecture controllers. These companies control the majority of the CNC market, and their customers are not requesting controllers that make use of open architectures. Thus, there is no great incentive to move in that direction. Those using open architecture controllers in Asia are typically using single-board open architecture controllers fabricated in the U.S. and Europe. While machine control has progressed quite well, process control has not. This is primarily due to a lack of models, process understanding, and experience. Thus, significant efforts are needed in developing micromanufacturing process models and the controllers and control algorithms to utilize these models to improve the overall process and, ultimately, the product.

REFERENCES

CNC Family SINUMERIK, http://www2.automation.siemens.com/mc/mc-sol/en/701ecff2-0611-47eb-8be8-73d4be9f33cd/index.aspx (Accessed October 20, 2005)

Rexroth IndraMotion MTX, http://www.boschrexroth.com/country_units/america/
    united_states/en/products/brc/a_downloads/
    71237AE0104_final.pdf (Accessed October 20, 2005)
Super 5-axis Nano Machine FANUC ROBOnano alpha-0iB,
    http://www.fanuc.co.jp/en/product/robonano/index.htm (Accessed October 5, 2005)

CHAPTER 6

Non-lithography Applications

Marc Madou

ABSTRACT

For the WTEC study, the mutually agreed-upon working definition of non-lithography machining included, (1) mechanical (traditional) machining and, (2) non-mechanical (non-traditional) machining. In addition to non-lithography-based micromachines, the study panelists were also interested in establishing the impact of microelectromechanical systems (MEMS) and nanoelectromechanical systems (NEMS) on non-lithography-based machining. Examples include the use of MEMS to make a micromold for plastic micromolding, nanoimprint lithography (NIL) and the fabrication of fibers using MEMS spinnerettes.

The panel agreed that lithography-based MEMS and NEMS advances are highly oversold in the most public relations-hungry universities and government institutes in the U.S. Although less advertised, non-lithography micromachining, practiced mostly in highly competitive, private companies such as Sankyo Seiki, Samsung, and Olympus is most likely to continue to lead to more practical products faster. These products include lenses for telephone cameras, flat panel displays, automotive parts, microfuel cells, microbatteries, micromotors, and desktop factories (DTFs). Based on the state-of-the-art and current investment levels, both private and government, Germany, Switzerland, Japan, and Korea will gain the most from developments in non-lithography-based machining, given their long tradition with and heavy investment in this field. The U.S. over the last twenty years has emphasized lithography-based MEMS with outstanding research results and a dominant market position, but as many MEMS products have become commodity products, Asian countries stand to reap more benefits in the near future from it. Actually, even MEMS

111

foundries, which are very hard to make profitable in the U.S., are moving more and more to Asia; Olympus in Japan has already the largest MEMS foundry in the world. During our Asia trip we gained, in general, more from our industrial visits than from the visits to academic institutions—this is understandable as micromanufacturing is very applied and product-driven, and academia is not. We believe that to succeed in non-lithography-based machining a stronger-than-usual link with industrial partners and academia is required. In this regard we are now behind in the U.S., although it was in the U.S. that the trend of academia/industry collaborations started. The links between industry and academia are now better in both Europe and in Asia. It was speculated that technology transfer offices in U.S. academia have become so unwieldy that they prevent smoother and better collaboration with industry.

In some showrooms of the Asian hosts, the panel came to realize that none of the products on display were manufactured in the U.S. anymore. As noted in Chapter 4, new product demands are stimulating the invention of new materials and processes. The loss of manufacturing goes well beyond the loss of one class of products. If a technical community is dissociated from the product needs of the day, say those involved in making larger flat-panel displays or the latest mobile phones, communities cannot invent and eventually cannot teach effectively anymore. Chapter 4 lists several such new manufacturing processes. A yet more sobering realization is that we might invent new technologies, say in nanofabrication, but not be able to manufacture the products that incorporate them. It is naïve to say that those new products will still be designed in the U.S. because the latest manufacturing processes and newest materials need to be understood and used in order for a good design to be developed.

To stem the hollowing out of the manufacturing bases within their countries, the governments of many developed countries have made huge investments in the miniaturization of new products, including MEMS and NEMS, and in the miniaturization of manufacturing tools such as DTFs. These efforts are intended to regain a manufacturing edge. To illustrate this point, Olympus' Haruo Ogawa (the leader of their MEMS team) says that MEMS may help rebuild Japan's power as a manufacturing nation. Sankyo Seiki believes that its DTFs might revive manufacturing in Japan. In Korea the government just started a new DTF project. Finally, in some quarters in the U.S., nanotechnology is seen as a means for the U.S. to remain a high-technology innovator.

It is difficult to predict which of these strategies will enable developed countries to maintain a manufacturing base. But it is clear that if any country wants to remain independent—especially in times of war—and maintain a high standard of living, it must remain strong in manufacturing. This is especially true if that country is resource-poor. There are several different scenarios one can think of to maintain a strong manufacturing edge. One approach is a strategy in which manufacturing of products is only transferred abroad if new manufacturing technologies or products of equal or higher value are in the pipeline. The latter would require a more guided economy than we are used to in the U.S., but it is a strategy that seems to be paying off very well for China now. In another scenario, DTFs will become very automated and require very few human operators, so that exporting these machines does not make sense anymore, since no labor costs are saved; this is a free-market approach. Yet another popular scenario sees DTFs becoming as popular as desktop publishing; this could be called a socially responsible approach. This has been promoted by, for example, MIT's Neil Gershenfeld, but may be somewhat naïve given the current political climate. In this scenario manufacturing becomes much less centralized, and many more people get involved. In any of these cases, the impact on society would be dramatic.

APPLICATIONS OF NON-LITHOGRAPHY MACHINING

European and Asian countries are investing significant amounts of money and effort into improvements in mechanical machining. Often this involves small incremental improvements in projects that one could not get funded in the U.S., as they would not be deemed innovative enough. One could argue that these small improvements are the most secure way to improve manufacturing skills and product quality and yield. Higher accuracy machining is indeed needed to provide better computer memory disks, and optical mirrors and lenses with accuracies to a fraction of the wavelength of light. Non-lithography, ultraprecision manufacturing is still the commercially preferred technique for the production of computer hard discs, mirrors for X-ray applications, photocopier drums, commercial optics such as polygon mirrors for laser-beam printers, consumer electronics such as mold inserts for the production of compact disc reader heads and camcorder viewfinders, in addition to high-definition television (HDTV) projection lenses and VCR scanning heads.

Figure 6.1 shows a scanning electron microscope (SEM) photograph of a single-crystal diamond cutting tool. The radius of the edge on this tool is estimated at less than 0.05 μm; with such a tool submicron grooves equivalent to those produced by silicon micromachining can be generated (Figure 6.2). Using such a diamond tip and numerical control (NC) rice grain-sized cars were machined at Nippondenso (one car was 4.5 mm long, and another was 7.5 mm long) (Figure 6.3). The shell of the car was made by the sacrificial mold technique. A piece of aluminum (Al) was NC machined and plated with 30 μm thick electroless nickel (Ni); after cutting off the lower part of the body with electro-discharge machining (EDM), the Al

was removed by heated potassium hydroxide (KOH) etching. Finally, the
Ni car shell was gold- (Au) coated to protect it from oxidation. The
stainless chassis and wheels were also made by NC machining, and the
core shaft and coil of the motor were made by NC machining and EDM.
The zirconia wheel shaft was machined down to a 250 μm diameter. The
electromagnetic step motor is driven by an external magnetic field. The
car's permanent barium ferrite rotor runs at a maximum speed of 100
mm/sec, developing a torque of about 10^{-6} Nm (at 3V and 20 mA). To assemble the various microparts into a complete car, a mechanical micromanipulator, ordinarily used for handling biological cells, was employed.

Figure 6.1. Diamond ultra-high precision tools.
Single-crystal diamond tool. Angle of
tool is 20°. Maximum depth of cut is
500 μm (Bier, W. et al. 1991).

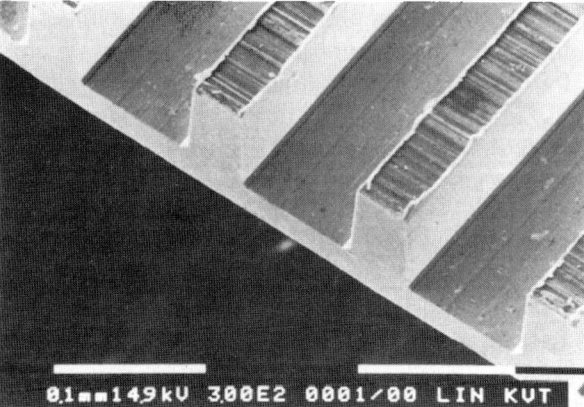

Figure 6.2. Microgroove produced by diamond machining a 100 mm thick Al foil. Width is 85
μm, depth is 70 μm (Bier, W. et al. 1991).

Figure 6.3. Microcar and rice grains (*left*) and microcar on
sandpaper (grain size = 200 μm) (*right*). The car
is 7 mm long, 2.3 mm wide, and 3 mm high
(Teshigahara et al., © [1995] IEEE, reprinted
with permission).

Although obviously not a commercial product, this little car demonstrates the manufacturing prowess of Nippondenso, a company that was
unfortunately not on our list of visited companies. Most of the compa-
nies/universities we visited in Europe and Asia were consistently better
equipped than the U.S. labs in this specific area. They also had similar
skills and all the required tools to do as well as Nippondenso in pushing
the limits of traditional mechanical machining.

The single-crystal diamond tool refinements are not the only reason for
the high precision achieved with mechanical machining today. Submicron
precision is also being achieved through high-stiffness machine beds, air
bearings (with a rotational precision of 0.01 μm and better available now),
and measurement systems such as laser interferometry. Furthermore,
highly precise instruments such as servomotors, feedback devices, and
computers have been implemented. Many types of machine tools are now
equipped for computer numerical control, further improving precision and
reproducibility of the manufactured parts. Progress in machine control and
feedback in the U.S. and Europe is often based on modeling and theory,
whereas in Asia it is more based on trial and error and practical experience.

FANUC's ROBOnano U_i is an ultraprecision micromachining station
capable of making such high-precision parts as the mold for diffraction
gratings or aspheric lenses (RoboNANO, 2005). This super-precision micromachining equipment consists of a diamond milling tool rotating at
high speed on a super-precision positioning table with nanometer resolution. Using this setup, FANUC succeeded in relief engraving a minute,
mirror-surfaced Noh mask in copper with a "diamond end mill," which results in a surface smoother than a mirror (Figure 6.4). The roughness of the
forehead (R_{max}) of the mask is 60 nm. Thus, FANUC has realized a super-

precision micromachining technique that allows mirror finishing in any direction except in a narrow groove.

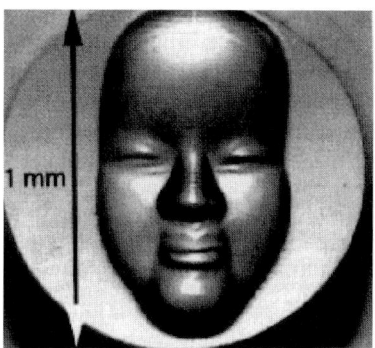

Figure 6.4. Japanese NOH mask. (Source: Fanuc, RO-
BOnano alpha-0iB,
http://wwwl.fanuc.co.jp/en/product/robonano/sa
mple/index.html)

This machine is not currently available in the U.S., as FANUC says it could not provide the support for the machine abroad. During the time this machine is only available to Japanese industry, substantial new manufacturing strength can be garnered in that country.

Interesting and important applications of sophisticated mechanical machining concepts were demonstrated, for example at Phillips. This was in the area of large optical surfaces, where a collection of optical features working in unison is used to achieve an optical function. For example, a very large Fresnel lens was machined using a custom-made circular lathe.

FLEXIBLE MANUFACTURING AND DESKTOP MANUFACTURING

Despite all of the progress in mechanical machining reported above, the fact that it often takes a two-ton machine tool to fabricate microparts where cutting forces are in the milli- to micro-Newton range is a clear indication that a complete machine tool redesign is required for the fabrication of micromachines. Along this line (in Japan the concept is called desktop flexible manufacturing systems (DFMS) or DTF) an effort to build micromachines was launched in the early 1990s. Similar efforts were also initiated last year in Korea and Europe, but there is currently no such coordinated effort in the U.S. The manufacturing units in this approach would be tabletop size and include universal chuck modules to which workpieces are continuously clamped through most of the manufacturing process. A significant fraction of this current WTEC study was geared towards measuring progress in desktop manufacturing.

The most advanced application of this concept was found at Sankyo Seiki. The DTF machines developed at Sankyo Seiki have modular units that carry out different functions. Modules we saw in operation include: oven baking of parts, glue application to parts, cleaning of parts, assembly of two parts, a direct input (DI) unit, a small high-efficiency particulate air (HEPA) filter for atmosphere conditioning, drilling and ID turning. Some of these modules were shown to work together. The DI unit was the largest unit of a series of connected DTFs as shown in Figure 6.5.

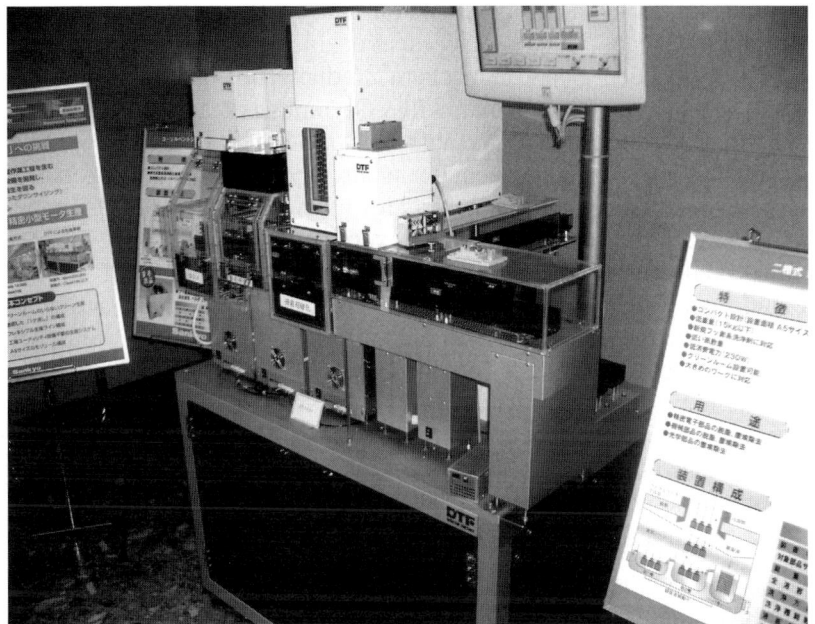

Figure 6.5. Sankyo Seiki DTF units.

The connection between the modules is with an automatic carrier vehicle, which was the size of a hand. Thirty-eight patents have been applied for (domestic and international) around this concept. The cleaner module has been available to outsiders since the end of September 2003. Eighteen enterprises have composed the DTF project. And some research organizations and universities, such as AIST, are participating in this activity.

Important components in the different modules are particle monitors, robot hands (instead of arms), vacuum pumps, compressors, linear motors, direct-drive motors, auto-guided pallet, and DI water supply. By the end of next year integrated systems for assembly will be available. Machining systems will be a third-generation effort and will come later. The DTF is connected to a personal computer (PC) environment and an Ethernet connection. This was a very impressive, well-planned and executed effort by a company that has been manufacturing precision components since 1946.

The extension to micromanufacturing is not easy at this point, since parts are typically larger than 1 mm. Currently available positioning technology limits effort to extend into microscales.

At Olympus the microfactory concept is seen as a means to save energy, hand labor, materials and cost. Some lenses handled today with a DTF at Olympus are 300 μm in diameter, and at this size, even skilled workers have trouble with assembly at high efficiency. Production is envisioned in an island type of modality where each worker controls a substantial number of manufacturing/assembly steps on one small workstation rather than in a production line. Desktop equipment enables a small local environment that is easy to keep clean, and where products can be assembled with a joystick without needing special skills. Olympus is also building a microfactory that will build and assemble MEMS. For now their key focus is microassembly. One microassembler is already in use to put together lens modules for cellular phones and parts for medical endoscopes. Olympus is the market leader in endoscopy—the examination of the stomach or colon for suspected cancerous tissue and the treatment or removal of the tissue. The endoscopy technique is being adapted to machine repair and a smart capsule endoscope is under development. The latter will be swallowed by the patient and will make its way to the designated organ, identify the troublesome tissue, and either sever it or deliver a drug to eliminate it. Optical microscanners will be attached to the tips of endoscopes to identify diseased tissue at the molecular and cellular level. Olympus projects this to be on the market by the end of 2008.

The benefits of a microfactory that Olympus sees are:

1. Improvement in assembly accuracy
2. Improvement in productivity per square meter (small foot print)
3. Energy and materials savings
4. Easier to keep clean

MEMS FOUNDRY

It will be interesting to see if Asian countries can make the MEMS foundry concept a commercial success. Most such U.S. foundries are now out of business. The Olympus MEMS foundry has a fabrication facility in Nagano that has a four-inch line, and expansion plans for six and eight inches. It has over a 100 experienced engineers along with design and marketing located in the Tokyo headquarters at Shinjuku-ku. Services started in February 2002. Today many projects involve either single-process or optical MEMS devices. The foundry already has built inkjet printer heads for a customer with hundreds of tiny heads crammed onto the surface of each head. For other companies they have been building MEMS light switches (Email: MEMS_Lab@ot.olympus.co.jp).

SUMMARY

Non-lithography methods are preferred for obtaining truly 3D and complex shapes, better relative tolerances in all dimensions, and smoother surfaces in all directions (X, Y and Z). For very small absolute tolerances and 2D shapes, lithography is the best approach. Small, incremental improvements in non-lithography techniques have given Europe and Asia an important edge over the U.S., where MEMS is more heavily emphasized and leap-frogging technology funded more extensively. MEMS has mostly become embedded in commodity products that are mostly manufactured abroad.

The most obvious realization from this study is that U.S. machining today is mostly forced into niche markets with high add-on value products. However, the time for even those more sophisticated engineered products—including bearings, silicon (Si) pressure sensors, dynamic random access memory semiconductors (DRAMs), and lenses—to become commodities is steadily shrinking. U.S. activities appear increasingly relegated to early intellectual property (IP) exploitation, product design, and services. This is a very alarming situation, as the U.S. is quickly losing ground in developing new manufacturing processes and materials. Without the knowledge of product challenges, it is hard to be innovative in new products; even if the U.S. is innovative, it might not be able to implement its innovations anymore.

An approach for the U.S. would be to launch a concentrated effort in very advanced, new manufacturing techniques and reintroduce the societal merits and value of actually making things. With the information technology sector depressed, and high-paying jobs still scarce, this is a good time to launch such an effort. The current WTEC study could be a first attempt towards this goal. Hybrid manufacturing approaches, incorporating top-down and bottom-up machining approaches, could be a key to attract a new generation of motivated engineers and scientists into the science and engineering of manufacturing.

A good start would be to start courses in comparative machining methods (i.e., where do you use what machining option?) scaling laws, continuous manufacturing methods, bottom-up manufacturing methods, self-assembly; small portable machining tables, and wet/dry machining method combinations.

A change in funding policy might be necessary. If new technology in the U.S. continues to be more and more military driven, as it is now, we run the risk of repeating a Soviet-type economic disaster: a country that can develop very sophisticated, very expensive military technology but does not know how to mass produce anything anymore. Earlier this year a meeting in Tsukuba, Japan was held on Biosensors for Homeland Security. The American scientists talked about bioterrorism and finding weapons of mass destruction (WMD); in contrast, the Japanese hosts talked about smart toilets! This anecdote illustrates an urgent issue that the U.S. needs to address: how to extract the commercial value out of U.S. government investments in military hardware lest the fruits of all that work only benefit countries that still know how to manufacture. In addition, the country should find ways to balance funding for "blue-sky" work such as nanotechnology with support of incremental development of more traditional manufacturing options. Not only do they both have merit; neither can survive without the other.

REFERENCES

Bier W., G. Linder, D. Seidel, K. Schubert. 1991. Kernforschungszentrum Karlsruhe. *KfK Nachrichten* 23:2-3, 165-173.

ROBOnano   alpha-0ìB,   http://www.fanuc.co.jp/en/product/robonano/sample/index.html (Accessed October 20, 2005)

Teshigahara, A., M. Watanabe, N. Kawahara, Y. Ohtsuka, T. Hattori. 1995. Performance of a 7-mm microfabricated car, *J Microelectromech Syst* 4: 76-80.

CHAPTER 7

BUSINESS, EDUCATION, THE ENVIRONMENT, AND OTHER ISSUES

Thom J. Hodgson

ABSTRACT

This chapter reviews elements of the study that are not central to the particular technologies studied, but that nonetheless affect both the development and the efficacy of micromanufacturing. Facets of the educational systems, business potentials and practices, governmental policies, and cultural characteristics in all of the countries visited drive and/or enable the technological development necessary for moving micromanufacturing forward. While this chapter makes some effort to compare the U.S. with other countries on these issues, readers can easily conduct similar comparisons from their own perspectives.

BUSINESS

The potential of micromanufacturing from a business perspective—and, collaterally, from the societal perspective as well—is significant. Medical applications, to consider one area, currently abound. With the movement toward minimally invasive surgery, very high-precision miniaturized devices are needed to perform procedures within the body without the collateral damage inherent in conventional surgery. There are also numerous military applications. The evolution of combat into a highly focused effort that inflicts damage on combatants and not civilians necessitates the increased use of miniaturized sensors and delivery devices. Allowing an increase in the extent that the U.S. is already dependent on foreign sources for micromanufacturing may not serve the country's security needs well. One of the important areas of application for both military and commercial use is miniaturized optics. A very visible recent application is electronic cameras in cell phones.

Many Asian companies are concerned about losing manufacturing "off-shore" to China. The response of many companies is to increase their investment in research and development to maintain their competitive edge; for example, Samsung (South Korea) plans to more than double its rate of investment in research and development to about 8½% of the corporate budget, while FANUC personnel estimate that their firm currently invests up to 10%. There is a belief that this policy will work as long as there is a concentration on bringing high-tech, short-lifetime products to market, thereby keeping all others out of the loop. As soon as a product is on the market and others understand both the product technology and the manufacturing technology, then they will have the potential to jump into the market. In the short term, at least, this may be the most viable strategy.

Concern over the movement of manufacturing offshore did not seem to be as strong in Europe. That may be because, as in the U.S., this is a phenomenon that has been going on for a much longer time. Europe seems to have developed its own response. For example Kugler, GmbH, which is a small German manufacturer of very high quality ultraprecision machine tools, has a normal research and development budget of 15–20%.

Outsourcing is an issue that affects the Far East as well as the U.S. When questioned, the response at Samsung was well-structured in that outsourcing was focused on mature products that had reached or were reaching maturity—personal computers (PCs), computer mice and keyboards, small cooling fans, TV inverters, and conventional TVs—and had attained commodity status. Again, these firms appear to see their future as dependant on maintaining their lead in the technology race. However, by controlling the outsourcing, they continue to control their market segment, and maintain a share of the profit.

In Japan and South Korea there seemed to be a large number of young PhDs at several of the industrial laboratories that the panel visited, and there were clearly plans in place to hire more PhDs at several of the laboratories. However, at one university an increasing numbers of PhDs with industrial experience were looking for jobs at universities. Apparently, this is because some companies are downsizing their research organizations. One expects then that the response in Japan and Korea to manufacturing moving offshore is as varied as the response is in the U.S.

In Japan and Korea it appeared that in some cases, industry was willing to invest in enabling technology for micromanufacturing, even though the total market for a particular machine technology might be on the order of 10-12 units. RIKEN is commercializing machines that it develops and selling them to industry. Those machines are used to produce ultra high-precision tools that are used, in turn, for micromanufacturing operations, thus supporting a relatively wide range of micromanufacturing efforts with only a few specialized machines.

The president of The Metal Industries Research & Development Centre in Kaosiung, Taiwan, made some very interesting comments during discussions. First, the probability of success for micromanufacturing projects is quite high when compared with MEMS projects because they are industry-motivated and application-driven. Second, virtually all materials used in micromanufacturing are conventional materials. There has been little need to develop new materials for micromanufacturing. Innovations in materials must be performance-driven since material volumes are too low for either cost or economics to drive the development. However, since the actual volume for micromanufactured parts is so small, very expensive materials can be economically justified. A lesson to be learned here for U.S. funding agencies is that independent materials research that is not strongly connected to industrial needs may not be appropriate for funding.

EDUCATION

In Japanese universities, and to some extent in Korean, Taiwanese, and European universities as well, many of the individual academic departments had names that defined them uniquely (e.g., precision engineering, precision machinery engineering, microsystem engineering, mechanical engineering and intelligent systems, engineering synthesis, information storage engineering, mechanical systems engineering) in addition to a conventionally named department of mechanical engineering. In the U.S,, the functions represented by these dfferent departments would typically reside in a single mechanical, or perhaps a mechanical and aerospace engineering, department. Taken together, these departmental groups can be larger than any single mechanical engineering department in the U.S. These uniquely named departments typically represent multi-disciplinary educational efforts on a scale not seen in the U.S., where such efforts are usually found in much smaller cross-disciplinary centers. There is clearly a commitment to multi-disciplinary work, which is necessary for both manufacturing and manufacturing research, that well exceeds the commitment in the U.S. This, in part, may explain the excellence in micromanufacturing in these countries.

When questioned, both Asian and European industry personnel noted that, while initial training ranging from six months to two years was necessary, new engineering graduates typically had a good background for their technical careers. This indicates that they are making use of the traditional technologies taught in the universities. Like many U.S. companies, they find that their new engineering hires may not have all of the communication skills and team-building skills that they need to excel in industry.

The Japanese Government funded an interesting effort to bring technology to the younger generation. Professor Ikuta at Nagoya University designed, implemented, and now supports a microsystems interactive experience for children at the National Museum of Emerging Science & Innovation. It targets microsystems that are big enough for children to see under a standard microscope. The set-up at the museum is capable of downloading new designs from Professor Ikuta's laboratory so that new designs can be generated for the children to experience. The web site for the museum is http://www.miraikan.jst.go.jp. The museum hosts a microart contest for children to use the "art-to-part" process. This is one way that the Japanese government is insuring that the next generation of Japanese citizens is motivated to pursue careers in technology.

As noted in the previous section, a large number of young PhDs were found working at some of the Asian industrial research laboratories visited. Up to 30% or more of those PhDs were acknowledged to come from outside Japan or South Korea, including the U.S. The output of PhDs from the universities in Europe and the Far East individually now exceeds the U.S. In addition, in most U.S. engineering schools, the majority of PhD students are foreign born. While many of those students stay in the U.S. for their professional careers, many go back to their own countries since an increasing number of opportunities may be found there. The U.S. is now seventeenth in the world in the per capita output of engineering and natural science degrees. In 1975, the U.S. was third, behind Japan and Finland. China's output of engineers alone far exceeds the output of U.S. engineers. This does not bode well for the future competitiveness of the U.S. in terms of its ability to support the U.S. high-tech manufacturing effort. The reasons for this are many, and go well beyond the scope of this report.

A number of effective educational efforts are being employed throughout Asia and Europe to develop the skills necessary for micromanufacturing. An extremely creative educational effort was observed at Professor Hisayuki Aoyama's microrobotic laboratory at the University of Electro Communications in Tokyo. He deals primarily with graduate students. His students use off-the-shelf components (available in the "electronic city" Akihabara area in Tokyo). Professor Aoyama motivates his students by making their projects fun. Projects include building a desktop factory, "robot mating" using artificial insemination (a fish egg was actually fertilized by his students' robots), "microdrop mixturing" (mixing very small quantities of liquids, potentially for drug applications), and a micromanipulator driven by a Sony Play Station controller. The cost of each of the projects was less than $1 million, exclusive of the laboratory PCs that were used. It was clear that Professor Aoyama's students were ready and able to move into industry and to be productive very quickly.

ENVIRONMENT

During site visits, the WTEC panel saw no processes that would present a problem in terms of pollution. Some of the very fine-machining processes can involve the potential danger of particulate ingestion to manufacturing personnel. In each case a simple venting system and/or filters were sufficient to limit exposure. While the potential for hazardous materials is certainly present, most of the processes observed were scaled down in size from the conventional manufacturing counterpart. This results in, at least, lowering the volume of pollutants. In the case of the Korean Institute of Machinery and Metals (KIMM), one of the justifications for micromanufacturing was that it results in reduced industrial pollution. However, a number of the sites visited did actively consider environmental issues throughout product life.

Some companies, such as Mitsubishi, have ongoing environmental management systems that monitor the effects of manufacturing activities and products on the environment, and attempt to prevent pollution before it occurs. Japan has national environmental laws, local ordinances, and agreements. Mitsubishi has set voluntary standards that promote energy conservation, clean energy, CO_2 emission reduction, waste reduction, restricted use of dangerous chemical substances, and environmentally friendly designs that support product life cycles. Mitsubishi's environmental policies are well-documented and open to the public. All employees are familiarized with it. The approach seen here is not necessarily directed at micromanufacturing. However, the environmental characteristics of micromanufacturing noted above certainly make it a viable alternative to more conventional technologies.

The Institute of Reliability and Micro-Integration, at the Fraunhofer Institute Berlin (IZM), takes an integrated environmental approach in the early stages of product design and process engineering. They work with industry partners to analyze electronic devices from an environmental point of view to foster eco-efficiency. IZM, in collaboration with the University of Tokyo, has developed a joint eco-design program.

GOVERNMENT POLICIES

At KIMM, the panel saw an excellent example of government policy directed at the mainenance of industrial competitiveness. The emphasis there was not on fundamental issues but rather on developing the ability to produce next-generation systems such as white light interferometers (WLIs), and precision motion and machining capabilities. They also have an excellent history of working with industry to transfer technology. They appear

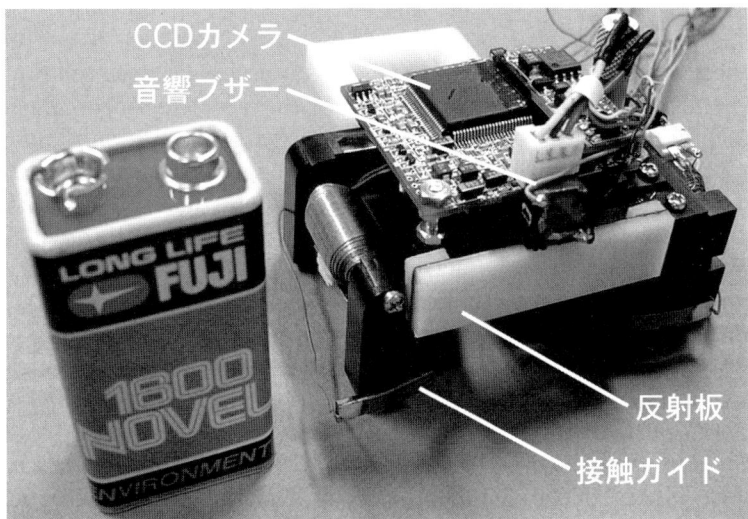

Figure 7.1. Female robot used for robot artificial insemination project.

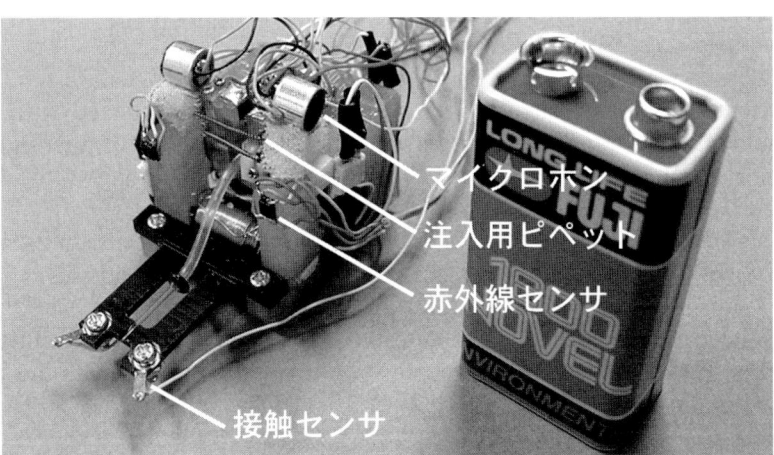

Figure 7.2. Male robot used for robot artificial insemination project.

to be poised to provide enabling technologies to Korean industry as their various micromanufacturing projects develop and mature.

Some Japanese government research laboratories may be feeling the effects of the global economy in that there appeared to be a change of mission in progress that did not seem to be well understood, even by the personnel in the laboratories.

The DTF project of the Fine Manufacturing Systems Group at AIST is shown below. There is a government-sponsored effort to make small machines to make small things, and this is one example of that effort. Figure 7.3 (left) is an overall view of the factory. Figure 7.3 (right) is a picture of the small press.

The Taiwanese National Science Council (NSC), part of the executive branch of the government, was established in 1959. The Department of Engineering & Science (DES) within the NSC funds academic research and places equal emphasis on fundamental and applied research. They support multi-year integrated and individual PI projects. In 2003, they received about 9,000 proposals, and were able to fund approximately 5,000 projects (about ~55%—compared to the ~15% funding rate in the Design, Manufacture, and Industrial Innovation Division at the National Science Foundation in the U.S.). The average level of funding for projects is $175,000 - $350,000 per year. Overhead on projects is limited to 8% (maximum), as compared to the roughly 50% that most U.S. universities charge. There is no specific program in place at the NSC to support micromanufacturing.

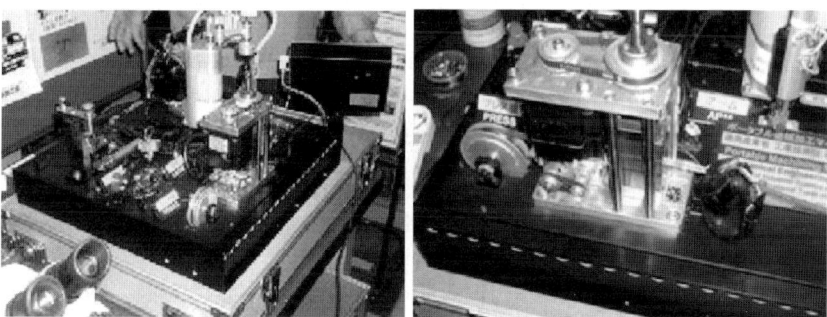

Figure 7.3. AIST desktop factory project.

The German Fraunhofer Institutes enjoy an extremely good relationship with German industry. Many times during conversations with German industry representatives, the subject of the Institutes came up. In every case, it was clear that the industries depended on the Fraunhofers for technologies, and that the relationship between the two was very solid. From the other side, it was clear that the Fraunhofers take their relationship with industry seriously and are dedicated to being a critical resource for industry. The result is that it is relatively easy for even a small German company to stay abreast of technological advances in their field. This is a tremendous advantage in a rapidly changing technological environment.

Besides providing research and laboratory training faculties to university students, the Fraunhofer Institutes offer industry-oriented training programs for the technologies in which they have developed expertise. They organize workshops and conferences as well as one-day technology forums for technology transfer to industry. The institutes also organize company-specific workshops.

Other Issues

One of the more impressive characteristics that ran across almost all of the industrial R&D laboratories that were visited was that there was a focus on making sure that a product and its associated enabling manufacturing technologies were brought to the factory floor very quickly. Laboratory personnel appeared to be well-connected to their counterparts in manufacturing engineering.

Perspectives on FANUC

FANUC indicates that it will not move its production overseas for several reasons. They do not feel that they are producing systems in quantities that are significant enough to warrant a transition overseas. They feel that it is important for R&D to be close to manufacturing. It may be that their high level of automation necessitates this, although this was not stated.

They are working hard to have a minimal impact on the environment, but major concerns appear to be not cutting trees down when they build new facilities and minimizing packing materials. They do not appear to consider environmental issues from a total product life perspective. For example, they appear to have not considered how to recycle their used robots.

The staff of FANUC indicated that they hired all engineers, with about 70% being mechanical engineers and 30% being electrical engineers. It was indicated that some training was necessary for new engineers, but they nominally had a good background for their technical careers at FANUC. As in the U.S., there was some concern regarding communication and team skills for new engineers. Among their engineering staff, three were women. This is an improvement from the past when no female engineers were included in their ranks.

FANUC has participated in a few government-funded projects, for example, the Intelligent Manufacturing System project that was recently finished.

Intellectual Property and Commercialization

The issues of commercialization and intellectual property were discussed with personnel at many of the university and government laboratories that were visited. As a general observation, the South Korean university educational community appears to be more closely tied to the U.S. university educational community than is the case for the Japanese. The South Koreans more typically have been educated in the U.S. An entrepreneurial spirit connects them with their industry. It appeared that Japanese educators were not particularly pressing the patent process, whereas Korea and Taiwan were. In fact, it seemed that in many instances in Japan, getting a patent might have been a fairly painful experience for academic personnel as they could be forced to commercialize material that was developed using government funding, and that they would not necessarily reap any economic benefit from the commercialization.

The Europeans seemed to be comfortable with the intellectual property and commercialization process. The relationships between universities, government laboratories, and industry appeared to be strong, with all players in this three-sided relationship appreciative of the contributions of the others. One could observe that the intellectual property and commercialization process appears to be considerably more mature in Europe than in the U.S.

The Importance of Human Resources in Micromanufacturing

With all of the focus on the technology of micromanufacturing, one tends to forget that there are real people that run the high-precision and highly automated manufacturing processes. This was brought home in the following way. Til Kugler, the son of the founder of Kugler, GmbH (Germany) commented that it took a special kind of worker to succeed in a high-precision environment, and hiring the right workers was critical. They had to be very neat and well organized. He said that he could tell in 15–20 minutes of conversation whether or not a potential worker would be able to succeed at Kugler.

An anecdote will serve to demonstrate the extent that individual human requirements in a high-performance micromanufacturing environment are challenging and that certain personality traits are highly correlated with the ability to perform well in the micromanufacturing environment. The panel next visited Zumtobel Staff (Austria). Here similar comments were heard. In this case, it was pointed out that there was one worker whose performance exceeded any of the other workers. Although the thesis of the comments was that it took a special person to succeed in micromanufacturing, there was no characterization given of the traits necessary for success.

Panel members were later given an opportunity visit a work area where three workers were actively involved in their trade, including the high-performance individual in question. As the panelists approached the individual's work area, it was clear that they were invading his space in that he immediately stopped tending his turning center and came over to us. The panelists stopped by a worktable that held the recent output from his work center. There was absolutely nothing out of place on the table. Recently manufactured high-precision parts were lined up perfectly on the table, as were tools. A quick inspection of the work spaces of the other two individuals working in the area revealed that their individual work areas were likewise neat and well organized. However, they paled in comparison to the work area of the high-performance individual. His was at a completely different level in this respect. Knowledgeable management, in some cases, is able to identify these personnel, and do so in comparison to other highly trained technicians. Clearly, the selection of workers is a very important element of maintaining a successful micromanufacturing enterprise.

APPENDIX A. PANELIST BIOGRAPHIES

Kornel F. Ehmann
(Panel Chair)

Dr. Ehmann currently serves as James N. and Nancy J. Farley Professor in manufacturing and entrepreneurship in the department of mechanical engineering at Northwestern University. Dr. Ehmann's main research interests are in the interrelated areas of machine tool structural dynamics, metal cutting dynamics, computer control of machine tools and robots, accuracy control in machining, and micro/mesoscale manufacturing. Dr. Ehmann has published over 150 articles and supervised over 35 Master of Science (MS) and 35 PhD students. Dr. Ehmann is currently the technical editor of the American Society of Mechanical Engineers (ASME) Journal of Manufacturing Science and Engineering (formerly Journal of Engineering for Industry), an associate editor of the Journal of Manufacturing Processes and is serving as president of the North American Manufacturing Research Institution of the Society of Mechanical Engineers (NAMRI/SME). He is a former chair of the Manufacturing Engineering Division of ASME.

David Bourell

Dr. Bourell is associate chairman of administration and research and Temple Foundation endowed faculty in the mechanical engineering department. Dr. Bourell's research interests include deformation processing, powder processing, failure analysis, nanocrystalline materials, the kinetics of sintering, and selective laser sintering. He is the 1999 recipient of the Lockheed Martin Aeronautics Company Award for Excellence in Engineering Teaching. He has served on the College of Engineering faculty since 1979. Dr. Bourell has had two years of industrial experience and has published more than 100 technical articles and reports. Dr. Bourell is also a member of the University of Texas at Austin Texas Materials Institute.

Martin L. Culpepper

Dr. Culpepper is the Rockwell International Assistant Professor of mechanical engineering at the Massachusetts Institute of Technology (MIT). His research interests include precision machine design, six-axis compliant mechanism design, reconfigurable compliant mechanism design, and the design of instruments and equipment for micro- and nanomanufacturing. Dr. Culpepper is the recipient of a 2004 PECASE Presidential Early Career Award for Scientists and Engineers (sponsored by NSF's Nanomanufacturing Program), two R&D 100 awards (1999, 2003), a 2004 TR100 award, and a Joel and Ruth Spira Teaching Award. Dr. Culpepper received his Bachelor of Science (BS) in mechanical engineering from Iowa State University in 1995. He completed his MS and PhD at MIT in 1997 and 2000, respectively.

Thom J. Hodgson

Dr. Hodgson is the James T. Ryan Professor, a Distinguished University Professor, and the Director of the Integrated Manufacturing Systems Engineering Institute at North Carolina State University. His areas of interest include scheduling, logistics, production and inventory control; manufacturing systems; and applied and military operations research. He served as head of the industrial engineering department at NCSU ('83–'90); director of the division of design and manufacturing systems at the National Science Foundation ('91–'93); professor of industrial & systems engineering at the University of Florida ('70–'83); operations research analyst at Ford Motor Company ('66–'70); and an officer in the U.S. Army ('61–'63). He is a Fellow of IIE and INFORMS, and a member of the National Academy of Engineering. He is the author or co-author of over 70 journal articles and book chapters. He served as associate editor, departmental editor ('81–'84; '88–'91), and editor-in-chief ('84–'88) of IIE Transactions. He received a

BS in science engineering in 1961, an MBA in quantitative methods in 1965, a PhD in industrial engineering in 1970, all from the U. of Michigan.

Thomas R. Kurfess

Dr. Kurfess is a professor and BMW Chair of manufacturing in the mechanical engineering department, as well as Director of the Carroll A. Campbell Jr. Graduate Engineering Center at Clemson University. Previously he served as professor of mechanical engineering at the Georgia Institute of Technology. His research there focused on the design, fabrication and control of precision manufacturing and metrology systems. Dr. Kurfess began at Georgia Tech as an associate professor in the fall of 1994. Prior to that he was an assistant and associate professor at Carnegie Mellon University. He received his PhD from MIT in 1989. Dr. Kurfess has served as an associate technical editor for the ASME Journal of Dynamics Systems, Measurement and Control, and is currently an associate technical editor of the ASME Journal of Manufacturing Science and Engineering (formerly Journal of Engineering for Industry), associate editor of the Journal of Manufacturing Processes and an associate editor for the International Journal of Engineering Education. He is the secretary of the North American Manufacturing Research Institution of the Society of Mechanical Engineers (NAMRI/SME), and is on the Advisory Board for the journal Mechatronics.

Marc Madou

Dr. Madou is a Chancellor's Professor of mechanical and aerospace engineering at UC Irvine, CA, in addition to professor of biomedical engineering and materials concentration. He also serves as professor of integrated nanosystems research facility (INRF) and as a NASA Ames Research Center associate. Dr. Madou's research focuses on DNA arrays, CD fluidics-sample preparation, polymer actuators, carbon microelectromechanical systems (C-MEMS), and novel lithography techniques. He previously served as vice president of advanced technology at Nanogen in San Diego, CA, and as center for materials research (CMR) scholar (endowed chair) at Ohio State University (OSU). He was professor of materials and chemistry at OSU. From 1989 to 1996 he served as visiting scholar at Louisiana State University, visiting Miller professor at UC Berkeley and visiting scholar at UC Berkeley.

Kamlakar Rajurkar

Dr. Rajurkar is a Distinguished Professor of engineering, director of center for nontraditional manufacturing research and interim associate dean at the college of engineering and technology at the University of Nebraska-Lincoln. Dr. Rajurkar's research focuses on modeling and analysis and control of traditional and nontraditional manufacturing processes. He earned his bachelor's degree in engineering from Jabalpur in India, his MS in 1978 and PhD in 1981 from Michigan Tech.

Richard E. DeVor (Advisor)

Dr. DeVor is a college of engineering Distinguished Professor of manufacturing and research professor at the University of Illinois at Urbana-Champaign (UIUC). His research interests include: engineering statistics and quality control; environmental engineering; manufacturing systems; nano-, micro-, and mesotechnology. He earned his BS, MS and PhD from the University of Wisconsin, Madison in 1967, 1968 and 1971, respectively. DeVor started his career as assistant professor in the department of mechanical and industrial engineering at UIUC in 1971. Since then he rose to associate professor in 1974, professor in 1984, associate head in 1987, executive director of the institute for competitive manufacturing in 1989, director of the NSF/DARPA Machine Tool Agile Manufacturing Research Institute (MT-AMRI) of the department of mechanical and industrial engineering in 1994 and Grayce Wicall Gauthier Professor in the department of mechanical and industrial engineering in 1995.

Traveling Team Members

Other team members traveling with the group included sponsors Sreeramamurthy Ankem (NSF), Jian Cao (NSF), George Hazelrigg (NSF), and Kershed Cooper (ONR), as well as Hassan Ali (WTEC), Gerald Hane (Q-Paradigm), Roan Horning (WTEC) and Eriko Uemura.

APPENDIX B: QUESTIONNAIRE FOR SITE VISITS

INTRODUCTION TO THE QUESTIONNAIRE

In this study we are focusing on microscale manufacturing processes and systems, and we distinguish this area from what is often referred to as microelectrocmechanical systems (MEMS). Our interest in this study is in processes and applications that may contain 3D (freeform) surfaces, employ a wide range of engineering materials, and have relative accuracies in the 10^{-3} to 10^{-5} range. Our interest in this study also includes the "microfactory" manufacturing paradigm and the development of miniaturized processes for the manufacture of microscale components and assemblies. In the U.S., MEMS is meant to include processes and applications that are based primarily on exploiting silicon planar lithography as the core technology. Hence, these processes and applications are 2D to 2.5D and are limited in the engineering materials employed and involve relative accuracies (tolerance-to-feature size) in the 10^{-1} to 10^{-2} range. While our interests in this study range from the nano-level to micro-level and even beyond, we are excluding those processes and applications that are more commonly associated with the semiconductor industry.

The following questions are provided in the spirit of helping you, the hosts, to prepare for the visit of the U.S. delegation. We apologize for the large number of questions. This is a measure of our interest in learning about your work and outstanding accomplishments. We do not expect detailed answers to each of these questions. Please feel free to examine the list and determine which questions would be most appropriate for your organization. We hope that discussion on issues related to these questions will lead to a productive exchange of views during and after our visit.

CHAPTER 1

Executive Summary, Introduction

1. Are you currently developing and/or using technologies that are specifically designed to address the manufacture of miniature components and products that require high relative accuracies, have 3D features, and involve a wide range of engineering materials? What are these products and technologies?

2. What do you believe are the most important factors influencing the trend toward the miniaturization of manufacturing equipment for microsystems applications?

3. What do you think are the key application areas that will benefit the most from the use of the microfactory idea for manufacturing systems? For example, transportation, telecommunication, biomedical, and consumer goods.

4. What do you believe are the most important problems/barriers (scientific, technological, and economic) to overcome in the realization of a fully functional microfactory?

5. What is your perception of the importance of micromanufacturing in comparison to standard manufacturing and nanoelectromechanical systems (NEMS)/MEMS manufacturing? In particular, do you feel that micromanufacturing will dominate, from an economic perspective, one or the other?

6. Do you consider this class of technologies to be paradigm changing? If so, would you consider it to be a technology push or is the industry demanding products and processes that require the development of these new microscale manufacturing technologies?

CHAPTER 2

Design

1. What do you perceive as the major challenges/difficulties in designing and analyzing miniaturized components, assemblies and products? How applicable are existing national/international standards in relation to miniaturized product development?

2. How applicable are conventional design methods, principles and software tools for miniaturized product development? What do you see as the main distinguishing features between design methodologies for macro, micro/meso and the NEMS/MEMS realms?

3. Are you currently engaged in any R&D effort to develop methodologies and tools for the design of micro/mesoscale components and assemblies? If so, what are the goals of your program?

CHAPTER 3

Materials

1. Have you generated any materials innovations in micro- and nano-manufactured parts and components? And if so, can you describe them?

2. Can you report on any research relating to fundamental understanding of material structure and properties at the micro- and/or nanoscale?

3. What challenges remain relative to materials in micromanufacturing? Are you or others that you know of exploiting novel new materials in micromanufacturing, including but not limited to carbon nanotubes or Bucky Balls, nanocrystalline particulates, photonic materials, nanocomposites, single-cell or biological materials?

CHAPTER 4

Processes

1. What is the status of modeling and simulation efforts for the processes you work with? To what extent and how have you accounted for the scaling effects?
2. What new processes, including hybrid processes, do you believe will be required to better deal with microsystems manufacturing in the years to come?
3. What kind of surface integrity studies have you conducted for these processes? What instruments do you use? What are the resolutions of those instruments?
4. What is the state-of-the-art in terms of feature size, absolute and relative tolerances, accuracy and repeatability, and material usage? What are the projected limits in the next five years? Next 10 years?

CHAPTER 5

Systems Integration, Assembly, Metrology and Controls

1. What target accuracies, ranges and repeatabilities are you striving towards? What types of measurement systems do you most commonly use for these assessments?
2. What control models and algorithms do you use? What types of hardware and software systems do you use to implement those control models and algorithms?
3. How important do you feel parallel processing, data transfer, networking and reconfigurability will be to micromanufacturing? What are the trends in this area that are useful to your work? Where is research needed in these areas?

CHAPTER 6

Applications

1. Are you looking at using MEMS/NEMS to create tools for more traditional machining? For example, molds for plastic molding, spinneret nozzles, and MEMS for electro-discharge machining (EDM) tools. Are there commercial examples of products made this way?

2. How much overlap are you seeing between mechanical manufacturing and lithography-based manufacturing, if any?

3. In the fabrication of protein and DNA arrays, advanced drop delivery systems are used and in the manufacture of glucose sensor strips, continuous web-based manufacturing is used. What are the latest developments in your country on advanced drop delivery systems for array manufacture and on continuous web-based manufacturing, in general? In this context are you aware of efforts on more advanced web-based processes: e.g., continuous lithography and continuous sputtering.

CHAPTER 7

Broader Issues

1. How large is the capital investment to bring microscale manufacturing online? How long would it take to bring this technology to commercial viability? How long do you believe it will take to "break even" on your overall investment?

2. Does microscale manufacturing require organizational changes within the company? What are the most important factors and safeguards that need to be put in place? What are your concerns about protecting the processes/machines and/or technology you develop? Is this an issue?

3. Have you conducted any studies on process economics and/or the environmental impact of the processes you employ? Have you conducted any studies on process economics, energy consumption, waste disposal and/or environmental impact of the processes you use?

4. Is the present educational system able to train the kind of individuals that you need in order to drive the kinds of technologies that you are developing? For example, to staff the manufacturing facilities necessary for microscale manufacturing technology?

5. In what ways can industry and universities work together on joint development? How could the educational system be changed to best serve your needs in technology development and training?

6. What government policies have helped you in your development of technology? What government policies have hindered you in your development of technology?

7. Does the manufacturing process create waste products that require special handling and/or disposal? What are the end-of-life disposal issues for the final product?

8. Has the government been helpful in dealing with environmental issues? How?

9. How would success in commercialization of the new technologies benefit in any of the following categories; a) lower unit cost of manufacturing; b) developing new supply chains; c) increased revenue opportunities; d) potential spillover applications; and e) societal and quality-of-life benefits? Can you quantify your projected benefits from these areas?

10. How would the development and commercialization of such technologies provide you with or maintain your leadership in a global marketplace?

APPENDIX C: SITE REPORTS—ASIA

Site: National Institute of Advanced Industrial Science
                      and Technology (AIST)
                      Namiki, 1-2-1 Tsukuba, Ibaraki 305-8564, Japan
                      http://unit.aist.go.jp/imse/finemfg/

Date Visited: December 6, 2004

WTEC Attendees: D. Bourell (Report author), M. Madou, T. Hodgson,
                      K. Cooper, R. Horning, E. Uemura

Hosts: Dr. Yuichi Okazaki, Leader, MJSPE, MASPE, Fine
                      Manufacturing Systems Group, Tel: +81-29-861-
                      7224, Fax: +81-29-861-7201,
                      Email: okazaki-u1@aist.go.jp
                      K. Ashida, Tel: +81-29-861-7155,
                      Fax: +81-29-861-7201, Email: ashida.k@aist.go.jp
                      M. Takahashi, Tel: +81-29-861-7186,
                      Fax: +81-29-861-7167,
                      Email: m.takahashi@aist.go.jp
                      N. Mishima, Tel: +81-29-861-7227,
                      Fax: +81-29-861-7201,
                      Email: n-mishima@aist.go.jp
                      T. Kurita, Tel: +81-29-861-7204,
                      Fax: +81-29-861-7201, Email: t.kurita@aist.go.jp
                      J. Akedo, K. Kitano

BACKGROUND

National Institute of Advanced Industrial Science and Technology
(AIST) is a national laboratory founded in 1999. The lab has almost 2,400
researchers with over 4,000 visiting researchers per year. Over half of the
funding comes from Federal allocation. There are six research fields, one
of which is nanotechnology, materials and manufacturing, the branch we
visited. Under the leadership of Dr. Okazaki, a number of excellent, world-
class micromanufacturing processes and machines have been developed
and demonstrated over the last ten years. A notable development is the microfactory, a series of micromanufacturing processes fitting into a suitcase
for creation and assembly of a three-piece ball bearing just over 1 mm
long. This represents one of six research areas in nano- and micromanufacturing at AIST.

RESEARCH AND DEVELOPMENT ACTIVITIES

Ceramic Coating of Microdevices (J. Akedo)

High-density ceramic coating on surfaces has been demonstrated using a variant of powder cold spraying. Ceramic coating materials include lead zirconate titanate (PZT) and alumina. Substrate materials are silica, soda glass and copper. The approach is to force a carrier gas, either helium or dried air, through a submicron-sized powder mass to create an aerosol. The aerosol is transported to the work area where it is accelerated to approximately 200 m/s. It is flooded through a previously prepared mask created using laser cutting or a photoresist technique. This takes place in a low-vacuum (0.1 Torr) work chamber. The particulate impinges and embeds onto the surface of the substrate mounted on a four-axis fixture. This process has been used to deposit multi-layer materials onto a polymer circuit board, creating a capacitor array. Deposits vary in width according to the mask geometry, but a width of several hundred microns is achievable. Deposit thickness varies with time, but values on the order of 10–20 µm are possible.

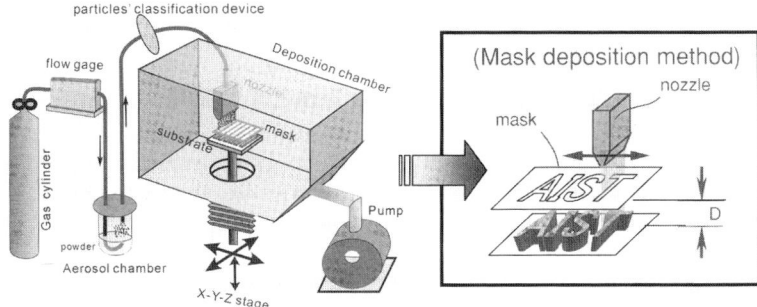

Figure C.1. Ceramic coating of microdevices.

The Microfactory (Y. Okazaki)

The microfactory is a complete micromanufacturing facility that fits in the size of a regular suitcase. It consists of three machines and two handling robots. The demonstration part to be manufactured in the microfactory is a 3D air bearing composed of three components, all brass. The first component is a 1 mm long and 0.5 mm diameter cylindrical bearing housing. The second component is a spindle, a cylinder with >1 mm long wires feeding out thereof. The third component is a cover that encapsulates and retains the spindle inside the housing.

The Microlathe (Professor Kitahara)

This miniature machine tool includes a diamond wedge cutting tool. The starting stock is 2 mm in diameter. Ends of the material are machined with a depth of cut of 100 µm. No metrology is applied to the machining process.

The Micromill (N. Mishima)

A 700 μm diameter carbide end mill is used to machine the air bearing housing. Starting with 900 μm diameter brass, a cylinder is faced and bored to accept the spindle.

The Micropress (K. Ashida)

A thin strip of brass 120 μm thick is punched in a six-station micropress to produce the air bearing cover. Tension on the strip is supplied by hand. The cover is a flat punched piece of brass deformed to produce three prongs for attachment to the housing, enclosing the spindle.

Once the housing, spindle and cover are manufactured, they are assembled using a robot microarm and robot microhand (tweezers/"chopstick").

The "Shalom" Ultra High-Speed Milling Machine (Y. Okazaki)

A 100–500 μm diameter carbide cutting tool is rotated at 300,000 RPM and is used to micromachine a number of materials including aluminum and hardened steel. Part features as small as 30 μm tall walls have been demonstrated. The advantage of this process is high jerk (up to 2.1 g acceleration in 30s), excellent surface finish due to fine chip formation resulting in minimal surface damage as a result of machining. Depths of cut on the order of 20 μm ensure low cutting force, a prerequisite to successful micromachining without fracture of the cutting tool.

The Hybrid Micromachine (T. Kurita)

A major problem in multi-stage microfabrication is "setting" or registration errors resulting from moving the workpiece from station to station. The hybrid micromachine alleviates this problem by maintaining the workpiece in fixed mode while changing out the tools which have fewer degrees of freedom. A large number of interchangeable process fixtures have been designed and constructed, including milling, drilling, cutting, grinding, polishing, electro-discharge machining (EDM), electro-chemical machining (ECM), laser machining, laser treating, laser-assisted milling, and EDM- and ECM-grinding/polishing. Parts are in the 3–10 mm size range.

Microinjection Molding (K. Ashida)

A project is underway to perform injection molding of acrylic resin. The molder has a maximum operating temperature of 300°C. The initial part is a 14 mm diameter optic lens. The main challenge is temperature control. A small charge of material can be heated above its softening point. Since the volume is so small, the thermal mass is miniscule. This causes the charge to freeze before being injected through the nozzle.

Micro/Nano Hot Embossing of Glass Fabricated by FIB Etching (M. Takakhashi)

A glassy carbon mold cavity is machined using focused ion beam (FIB) milling. Either pyrex or quartz is heated above its melting point and pressed into the full mold. The glassy carbon is resistant to wetting by the glass. Since the thermal coefficient of glassy carbon is small (2.2×10^{-6} °C^{-1}), removal of the part from the mold is facilitated. The molds are resistance heated. Maximum press forces and temperatures are 10 kN and 1400°C, respectively. Traditionally, fluidic or optical microelectromechanical systems (MEMS) processes are used, but they suffer from low etching rate and high cost. The ion source is Ga, and the FIB etching depth is linearly proportional to the accumulated beam current dose. Features in pyrex are on the order of 3-10 µm and include rectangular structures and pyramidal shapes. Additionally, parallel grooves/lines have been created, such as 1 µm wide lines 0.8 µm high on 1.5 µm centers, and 100–300 nm lines 400 nm high on ~400 nm centers. The machine operating conditions were 600°C temperature, 0.22 MPa pressure and molding time of 60 seconds.

SUMMARY AND CONCLUSIONS

AIST represented one of the most extensive micromanufacturing efforts. The realization of the concept of the microfactory, ultra high-speed machining, ceramic coating of microdevices, hybrid micromachining, micro-injection molding and hot embossing of glass using focused ion beam (FIB)-etched glassy carbon tools embrace the width and breadth of the technology. A desire was expressed to clarify the mission of the AIST. What are the goals and mission, and how do the micromanufacturing efforts relate?

REFERENCES

Akedo, J. 2004. Aerosol deposition and its application. *J Surface Sci Soc Japan* 25:10, 635-641 [in Japanese with English abstract].

Akedo, J. An Innovative Ceramics Integration Technology Using Compaction of Nano-Crystals at Room Temperature. Brochure.

Akedo, J. and M. Lebedev. 2001. Influence of carrier gas conditions on electrical and optical properties of Pb(Zr,Ti)O3 thin films prepared by aerosol deposition method. *J Jap Appl Phys* 40:9B, 5528-5532.

Akedo, J. and M. Lebedev. 1999. Microstructure and electrical properties of lead zirconate titanate (PBbZr52/Ti48)O3 thick films deposited by aerosol deposition method. *J Jap Appl Phys* 38:9B, 5397-5401.

Akedo, J. and M. Lebedev. 2000. Piezoelectric properties and poling effect of Pb(Zr,Ti)O3 thick films prepared for microactuators by aerosol Deposition. *Appl Phys Ltrs* 77:11, 1710-1712.

Furuta, K. 1999. Experimental processing and assembling system (microfactory). In, *Proceedings of the 5th International Micromachine Symposium*, October, Tokyo, Japan.

Gaugel, T., H. Dobler, B. Rohrmoser, J. Klenk, J. G. Neugebauer, W. Schäfer. 2000. Advanced modular production concept for miniaturized products. In *Proceedings of 2nd International Workshop on Microfactories*, October, Fribourg, Switzerland.

Hollis, R. and A. Quaid. 1995. An architecture for agile assembly. In *Proceedings of ASPE 10th Annual Meeting*, October, Austin, USA.

Kawahara, N., T. Suto, T. Hirano, Y. Ishikawa, N. Ooyama and T. Ataka. 1997. Microfactories; New applications of micro machine technology to the manufacture of small products. *Microsystem Technologies* 3:2, 37-41.

Kitahara, T., Y. Ishikawa, K. Terada, N. Nakajima, K. Furuta. 1996. Development of microlathe. *Journal of Mechanical Engineering Laboratory* 50:5, 117-123.

Lebedev, M., J. Akedo, Y. Akiyama. 2000. Actuation properties of lead zirconate titanate thick films structured on Si membrane by the aerosol deposition method. *J Jap Appl Phys* 399B, 5600-5603.

Mishima, N. and K. Ishii. 1999. Robustness evaluation of a miniaturized machine tool. In *Proceedings of ASME/DETC99*, DETC/DAC-8579, September, Las Vegas, NV.

Nakada, M., K. Ohaashi and J. Akedo. 2004. Electro-optical properties and structures of (Pb,La)(Zr,Ti)O3 and PbTiO3 films prepared using aerosol deposition method. *J Jap Appl Phys* 43:9B, 6543-6548.

National Institute of Advanced Industrial Science and Technology Company Profile and Research, Brochure.

Okazaki, Y. et al. 2001. Desk-top NC milling machine with 200 kprm spindle. In *Proceedings of the ASPE 2001 Annual Meeting*, November, Crystal City, USA.

Okazaki, Y., N. Mishima, and K, Ashida. 2004. Microfactories: Concept, history and developments. *J Manufacturing Science and Engineering* 126, 837-844.

Portable Machining Microfactory. Brochure. 2004.

Reshtov, D. N. and V. T. Portman. 1998. *Accuracy of Machine Tools*, New York: ASME Press.

Taguchi, G. and S. Konishi. 1988. *Quality Engineering Series* Vols. 1-4, New York: ASI Press.

Takahashi, M., Y. Murakoshi, K. Sugimoto and R. Maeda. 2004. Micro/nano hot embossing Pyrex glass with glassy carbon mold fabricated by focused-ion-beam etching. In *DTIP of MEMS & MOEMS*, 12-14 May, Montreaux, Switzerland.

Tanikawa, T., H. Maekawa, K. Kaneko and M. Tanaka. 2000. Micro arm for transfer and micro hand for assembly on machining microfactory. In *Proceedings of the 2nd International Workshop on Microfactories*, October, Fribourg, Switzerland.

Tanikawa, T. and T. Arai. 1999. Development of a micro-manipulation system having a two-fingered micro-hand. *IEEE Transaction on Robotics and Automation* 15:1, 152-162.

Site: **Asia Pacific Microsystems, Inc. (APM)**
                             No. 2 R&D Road 6,
                             Science-Based Industrial Park
                             Hsinchu, Taiwan
                             http://www.apmsinc.com

Date Visited: December 14, 2004

WTEC Attendees: K. Ehmann (Report author), D. Bourell, G. Hane, K.
                    Rajurkar

Hosts: Dr. Star R. Huang, Chief Technical Officer, Tel:
                             +866-3-666-1188 Ext. 1401,
                             Fax: +866-3-666-1199, Email: star@apmsinc.com

BACKGROUND

Asia Pacific Microsystems, Inc. (APM) was established in 2001 with a capital investment of over $50 million. It was created by a team of engineers with pioneering research and development expertise in microelectromechanical systems (MEMS) technology. APM provides fabrication technology and services in the integrated device manufacturing (IDM) arena, specifically in the wireless, sensors, broadband optical telecommunication, inkjet head microstructure and BIO-MEMS sectors. APM is also becoming a leading MEMS foundry supplier by providing a variety of groundbreaking technologies and manufacturing know-how and services. Today it employs about 300 people, including over 100 in engineering. All employees own a share in the company. APM owns over 30 patents. Its revenues are 80% derived from services and 20% from products.

RESEARCH AND DEVELOPMENT ACTIVITIES

The visit to APM lasted less than an hour and consisted of a presentation by its chief technology officer, Dr. Huang, who is one of the founders and still holds a faculty position at National Tsing Hua University in Hsinchu.

Microsystems/MEMS Platform Technologies

APM's core competencies include design, manufacturing and testing vertical integration for IDM, original equipment manufacturer (OEM) and original design manufacturer (ODM) products. It also offers technology development and manufacturing services for foundry customer's products. APM has expertise in silicon (Si)-bulk micromachining, membrane stress control, thick structural layer and deep trench, sacrificial layer etching, through-wafer interconnect, electrochemical etching and wafer-level chip size package (CSP), and logic IC integration, among other things. APM is also aggressively pursuing advanced MEMS packaging technologies.

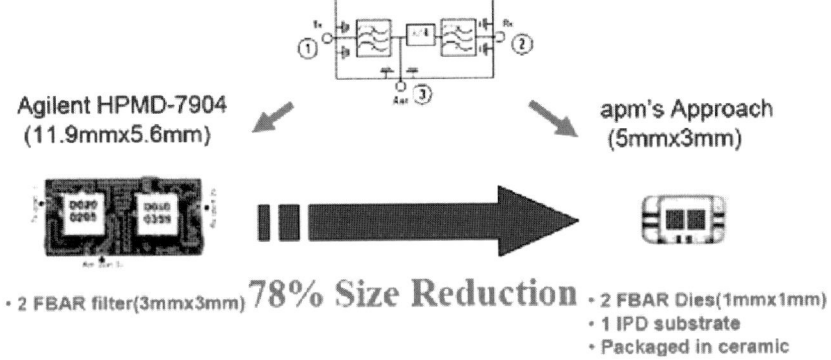

Agilent HPMD-7904
(11.9mmx5.6mm)

apm's Approach
(5mmx3mm)

• 2 FBAR filter(3mmx3mm) 78% Size Reduction • 2 FBAR Dies(1mmx1mm)
• 1 IPD substrate
• Packaged in ceramic

Figure C.2. FBAR duplexer.

Products

APM's product line includes pressure sensors, Bluetooth modules, different wireless products, film bulk acoustic resonators (FBAR), FBAR duplexers for code-division multiple-access (CDMA) (see Figure C.2), variable optical attenuators, optical switches, digital blood pressure monitors, and wireless tire pressure monitoring systems.

SUMMARY AND CONCLUSIONS

APM is a small, fast-growing company with a growth rate of about 30% annually since its inception. It caters to the microsystems market (in a broader sense than MEMS only) and unlike typical IC manufacturers is a small-volume, wide-variety and high-margin manufacturer and service provider.

REFERENCES

Asia Pacific Microsystems, Inc. http://www.apmsinc.com/en/htm/mainindex.asp (Accessed October 20, 2005).

Site: **FANUC, FA & Robot**
                             Oshino-mura, Yamanashi Pref, 401-0597, Japan
                             http://www.fanuc.co.jp/eindex.htm

Data Visited: December 8, 2004

WTEC Attendees: T. Kurfess (Report author), B. Allen, J. Cao, K. Eh-
                    mann, G. Hane, K. Rajurkar

Hosts: Dr. Yoshiharu Inaba, President and CEO, Tel: +81-55-
                             584-5555,
                             Fax: +81-55-584-5512, Email:
                             inabu.yoshiharu@fanuc.co.jp
                             Mr. Atsushi Watanabe, Executive Vice President, Sen-
                             ior Development Engineer
                             Dr. Shinsuke Sakakibara, Senior Development Engi-
                             neer, Tel: +81-55-584-5363,
                             Fax: +81-55-584-5879, Email:
                             sakakibara.shinsuke@fanuc.co.jp
                             Mr. Tomohiko Kawai, Manager, Tel: +81-55-584-
                             5847, Fax: +81-55-584-5526

BACKGROUND

FANUC's businesses were founded in 1956 when it started to develop numerical controls (NCs) and servo systems. FANUC Ltd. was established in 1972 when the Computing Control Division became independent from Fujitsu. They are a publicly traded company. FANUC has 10 major divisions with operation revenue of $2.2 billion. The divisions visited by the team focused primarily on the nanomachining center and the robotics lab. The entire set of facilities at the Oshino-Mura location occupies 1.3 million square meters, employs 400 workers and uses 1,000 robots for manufacturing tasks. FANUC holds approximately 55% of the market share in robotics (Siemens is second with less than 30%), mostly for automotive manufacturers. They have two joint ventures with the U.S.; a controls joint venture (JV) with General Electric (GE) and a robotics JV with General Motors (GM). It was estimated that they spend approximately 10% of their revenue on R&D. Their mission is to supply the best solutions to their customers.

RESEARCH AND DEVELOPMENT ACTIVITIES

There are a number of research and development activities being conducted at FANUC; however, this report will focus primarily on the ROBOnano nanomachine tool and control systems while only somewhat on robotics.

ROBOnano

The ROBOnano (Figure C.3) is an ultra high-precision multi-function machine tool. It can function as a five-axis mill, a lathe, a five-axis grinder, a five-axis shaping machine and a high-speed shaper. It uses friction-free servo systems (all linear slides and screws include static air bearings). For milling, an air turbine spindle is employed with rotational speed up to 70,000 rpm. Shaping is done using a high-speed shuttle unit capable of producing three grooves per second, and a 3 kHz fast tool servo using a lead zirconate titanate (PZT) actuator. Cutting is nominally accomplished with a single-crystal diamond tool. Control is achieved using a FANUC series 15i control, with error mapping for positional (axial) errors only. The resolution of the linear axes is 1 nm and 0.00001° for the rotational axes. The design of the machine is such that it experiences no backlash and stick-slip motion. The system is enclosed in an air shower temperature-controlled unit, whose temperature is controlled to 0.01°C. This eliminates errors due to thermal growth. The size of the ROBOnano is 1,270 mm by 1,420 mm by 1,500 mm. This does not include the computer numerical control (CNC) locker or the thermal management unit. However, in total, the machine is very compact. The range on the X-, Y- and Z-axes of the machine is 200 mm (horizontal, left and right) by 20 mm (horizontal, in and out) by 120 mm vertical, respectively. The maximum feed speed for the X- and Z-axes is 200 mm/min, and 20 mm/min for the Y-axis.

This is an extremely impressive machine and is truly a multifunctional next-generation micromachine tool. It is targeted at a relatively small market including optics manufacture (gratings, liquid crystal display (LCD) light-guiding panels, and small lenses), micromold manufacture and other small ultraprecision parts. Most of the parts that were shown to us required approximately five minutes to machine. Even the most intricate parts did not take longer than 30 minutes to machine. The size of the system was established based on the ability to make components up to the size of a personal digital assistant (PDA) or cell phone. The cost of the ROBOnano is $1 million. Much of this cost is due to the manual assembly of the system as well as the significant engineering and design time that has been invested in developing this machine. Approximately 30 individuals spent 17 years developing this machine.

Controls

FANUC is a major manufacturer of controllers. They are not moving toward open architecture controllers, but their controllers are flexible enough to support reconfigurable systems such as ROBOnano (e.g., milling, grinding, turning, and shaping). They have just released their next-generation controller (one generation is approximately 10 years; during each generation they have three to four upgrades). The major advance on the new generation controller (of which the 15i is a member) is smoothness of the tool trajectory. To accomplish this, various techniques are employed including non-uniform rational b-spline (NURB) interpolation. Another issue considered in the controller is the first three derivatives of the tool position (e.g., velocity, acceleration and jerk). These considerations (as well as others) allow for smooth tool trajectories while machining at the micro and nano levels. FANUC personnel indicated that generating a smooth surface was a major obstacle that had to be overcome in the development of the ROBOnano. The 15i also employs error mapping to improve the ROBOnano's performance. However, error mapping is only done on the positional errors of each axis (not angular errors). Error mapping improves the positional accuracy of each axis from approximately 2.5 µm to 10 nm. It should be noted that the 15i controller is also being employed by other micromachining system manufacturers such as Toshiba.

Robotics

FANUC employs robots for assembly, so everything is designed with automated assembly in mind. Figure C.4 shows a typical FANUC robot in their robot laboratory. They are developing new sensors and sensor-processing capabilities such as vision and force control to allow robots to perform tasks that require more flexibility. They have used these sensors to reduce the reliance of the robots on well-defined part locations that often require expensive fixtures. An excellent example of this is their third-generation production facility where the robot automatically identifies the orientation of a part and grasps the part based on its newly determined orientation. The robot then loads the part onto a fixture that is subsequently loaded into a machine tool for an operation. In the past, a person loaded the part onto the fixture for the robot to load into the machine tool. Such an approach may be important for material handling in future microfactories.

Currently, they have defined teaching the robot as a major bottleneck in productivity. While they still teach the robot with teach pendants and program in their own robotic language, they are developing new interactive video systems that will enable programming of robots in a video game-like environment. Programming will be accomplished in a highly user-interactive computer graphic environment. This approach will be used to develop course robot programs. Sensors such as vision units and tactile units in conjunction with the interactive video programming tools will then be used to refine the robot programs. This could be quite useful in developing programming schemes for next-generation and multi-functional micro-factory systems.

Figure C.3. ROBOnano.

Figure C.4. Robotic laboratory.

Other Pertinent Issues

The staff of FANUC indicated that their hires are all engineers with 70% being mechanical engineers and 30% being electrical engineers. Even their technicians are specifically from technically oriented preparatory schools that are similar to high schools. They indicated that some training was necessary for their new engineers but that the new engineers nominally had a good background for their technical careers at FANUC. This indicates that they are making use of traditional technologies that are being taught in the university systems. There was some concern regarding communication and team skills for their newly hired engineers.

Future Issues

FANUC indicated that it will not move its production overseas for two reasons. First, they do not feel that they are producing systems in quantities that are significant enough to warrant a transition overseas. And second, they feel that it is important for research and development to be close to manufacturing. Also, they are not moving towards open architecture controls, as the bulk of their market (50% automotive) does not require this. Future technologies at FANUC at their basic research lab include robots with the dexterity of humans and nanometric precision.

SUMMARY AND CONCLUSIONS

This was one of the most impressive tours. Clearly, FANUC is an industry leader in the micromachine tool area. Their advances in control and micromachining indicate that they are a well established leader in the micromachining area. It appears that they are targeting lower volume ultraprecision industries for their ROBOnano as opposed to producing the machine in large quantities for more generalized production usage. While we did not visit their machine tool and motor production facilities, these additional areas of expertise give FANUC a significant edge in terms of developing technology for the microfactory.

REFERENCES

FANUC Global Network. http://www.fanuc.com (Accessed October 20, 2005).

Site: **Hitachi Chemical R&D Center in Tsukuba**
                    48 Wadai Tsukuba-shi
                    Ibaraki 300-4247, Japan

Date visited: December 8, 2004

WTEC attendees: T. Hodgson (Report author) D. Bourell, K. Cooper, R. Horning, M. Madou, E. Uemura

Hosts: Dr. Shigeru Hayashida, Hitachi Chemical Co., Ltd., R&D Center, Tel: +81-29-864-4000, Fax: +81-29-864-4008
                Higeharu Arike, Chief Researcher
                Masaaki Yasuda, Mgr., Lab. for Elect. Packaging & Technology
                Masahiko HIRO, Staff Researcher
                Dr. Takumi Ueno, R&D Director

BACKGROUND

Hitachi Chemical is primarily a materials company, producing, for example, display-related materials, circuit board materials, advanced functional films, and automobile parts. Materials improvement enables improved manufacturing processes. We were presented with some excellent examples, such as higher resolution dry photo-resist (e.g., negative and positive polyimides, anisotropic conductive film, micro-arranged tube technology, optical wave guides).

RESEARCH AND DEVELOPMENT ACTIVITIES

Polyimides

Polyimides were developed in partnership with Dupont. By tailoring photosensitive and non-photo-sensitive polyimides, more and more applications have been opened up. These include stress isolation, heat insulation, electrical insulation, and optical wave guides.

μ-ARTS

In μ-ARTS, microtubes and metal wires are arranged on a flat polyimide sheet by using ultra-sonic to plastic welding (based on a U.S.-invented technology).

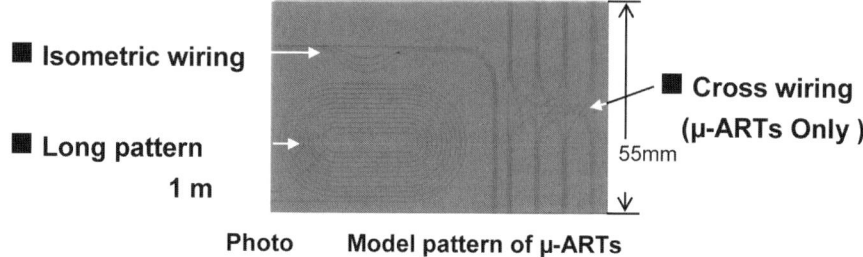

■ **Isometric wiring**

■ **Long pattern**

1 m

■ **Cross wiring**

(μ-ARTs Only)

55mm

Photo   Model pattern of μ-ARTs

Figure C.5. Model pattern of μ-ARTs.

The μ-ARTS device motion to deposit the material resembles that of a sewing machine. The tubes and wires are glued to the substrate using an adhesive.

Applications that are envisioned for this new technology are a heat exchanger for microelectronics, a motherboard for microreactor arrays, and capillary electrophoresis. The examples mentioned are all based on hollow glass tubes, but also fiber optic glass can be deposited without fracture. Typically, a 190 μm outer diameter and 50 μm inner diameter fiber optic cable size is deposited. This product can also be used for integrated circuits and hybrid applications.

Anisotropic Conductive Film (ACF)

This film's market name is ANISOM™. It is an insulating polymer matrix doped with metal-coated polymer balls (usually 5 μm in diameter) which allows conductivity in the Z-axis when the film is compressed between components. Printed circuits with density of up to 10 lines per millimeter can be connected (Hitachi Chemical Company, 2004). 3M has a competing product. One technological challenge is to improve the line density of the film.

Polymer Optical Waveguide

This is a modification of the polyimide composition by fluorination, which creates a heat-resistant optical waveguide with optical loss of 0.4 db/cm.

μ-TAS

This is a dry photo resist film to laminate and create microfluidic structures on glass.

SUMMARY AND CONCLUSIONS

This is a company that is improving material properties that, in turn, enable new micromanufacturing processes.

REFERENCES

Hitachi Chemical Anisotropic Conductive Film ANISOLM, ND. Brochure (n.d.).
Hitachi Chemical Company, Shinjuku-ku, Tokyo, Japan, 2004.
Hitachi Chemical Polymer Optical Waveguide, Tokyo, Japan, 2004.
Photosensitive film for μ-TAS ME-1000 series, R&D Group, Photosensitive Material Division. Presentation at WTEC panel visit, December, Tokyo, Japan, 2004.

Site: **ITRI – Mechanical Industry Research**
                             Laboratories (MIRL)
                             Industrial Technology Research Institute
                             A000, MIRL/ITRI
                             Bldg. 22, 195, Sec. 4, Chung Hsing Rd.
                             Chutung, Hsinchu, Taiwan 310
                             http://www.mirl.itri.org.tw

Date Visited: December 14, 2004

WTEC Attendees: K. Ehmann (Report author), D. Bourell, G. Hane,
                             K. Rajurkar

Hosts: Dr. Bill S.Y. Tsai, Vice President and MIRL General
                             Director, Tel: +886-3-591-6503,
                             Fax: +886-3-582-0235, Email: sytsai@itri.org.tw
                             Dr. Ivan T.C. Wu, Deputy General Director
                             MIRL/ITRI, Tel: +886-3-591-8099,
                             Fax: +886-3-582-0235,
                             Email: tungchuanwu@itri.org.tw
                             Dr. Ben Sheng Lin, Director MEMS Division,
                             Tel: +886-3-591-6630, Fax: +886-3-582-0043,
                             Email: bslin@itri.org.tw

BACKGROUND

Industrial Technology Research Institute (ITRI) was founded in 1973 as a nonprofit R&D institution with the mission to lead the development for Taiwan's technology-based industries (semiconductor, opto-electronics, displays, advanced materials, and precision machinery) through timely technology transfers and spinoffs. ITRI today employs over 6,000 employees with a budget exceeding $500 million (50/50 split between government- and industry-sponsored commercial contract services).

The Mechanical Industry Research Laboratory (MIRL) was founded in 1969 under the name of Metal Industrial Research Institute. It was subordinated to ITRI in 1973, and changed to its current name in 1983. MIRL employs about 750 people, over 50% of whom hold MS and PhD degrees. The annual budget is $70 million, over 40% of which is derived from industry contracts. The remaining 60% comes from the Ministry of Industry Affairs. About $5 million is generated from license fees—the largest amount in ITRI. MIRL applies for about 200 patents a year.

MIRL's R&D activities cover: micro/nanomachine technology; precision manufacturing; precision machine and control technology; vehicle and power technology; and optoelectronic and semiconductor process equipment technology.

RESEARCH AND DEVELOPMENT ACTIVITIES

The visit lasted only slightly over one hour and consisted of a general overview of activities, given by Dr. Wu, followed by a tour of the facilities. The emphasis during the tour was placed on MIRL's activities in the mechanical micro/nanotechnology area.

Mechanical Micro/Nanofabrication

MIRL focuses on biology, information technology (IT) and energy-related developments. The general approach is to cover the whole spectrum of technology development from design, tooling, and processing to equipment and system integration. Current product developments in the bio area, for example, include bio-chips, drug delivery systems and motion sensors, and in the IT area, next generation backlight displays, microoptical structure technology and image and display elements (e.g., see the hybrid optic film for LCD back light in Figure C.6) and 3D wireless mice. In the energy area, product development activities include mechanical micro-generators and miniature heat spreaders. The latter two are of particular interest here since they include miniature mechanical components. The microgenerator, for example, is of the self-generating motion type used in the past in watches. ITRI's concept involves stacking a number of thin miniature heat spreaders on a vapor chamber produced by X-ray lithography, electroplating and molding (LIGA). This alleviates "hot-spot" problems within the central processing unit (CPU) chipset. The key component of the 3D mouse is a small mechanical gyroscope that is being jointly developed with Carnegie Mellon University.

Figure C.6. Hybrid optic film for LCD back light module.

Laboratory Tour

The quick tour of the laboratories highlighted MIRL's LIGA capabilities, single-point diamond machining performed on equipment that was modified in-house to meet the nano- and microscale feature machining capabilities required for their optical applications (e.g., microlens arrays, gratings) and direct E-beam writing facilities. MIRL also utilizes the Nanotechnology Research Center (NTRC) common facility. Ductile mode machining is routinely used for brittle materials.

A particularly interesting development, from the standpoint of this study was the miniaturized injection molding machine shown in Figure C.7. The most advanced miniaturized injection molding machine, developed at MIRL, is a five-ton, all-electric miniaturized machine. It is designed as a complete intelligent molding cell that integrates injection, part handling, and vision inspection into one machine in its own clean-room environment. The specifications are: 1–5 ton clamping force, 2,500 bar injection pressure, 800 mm/sec injection speed, and molded product weight less than 0.01g. The machine was developed in conjunction with industrial partners and may soon appear on the market.

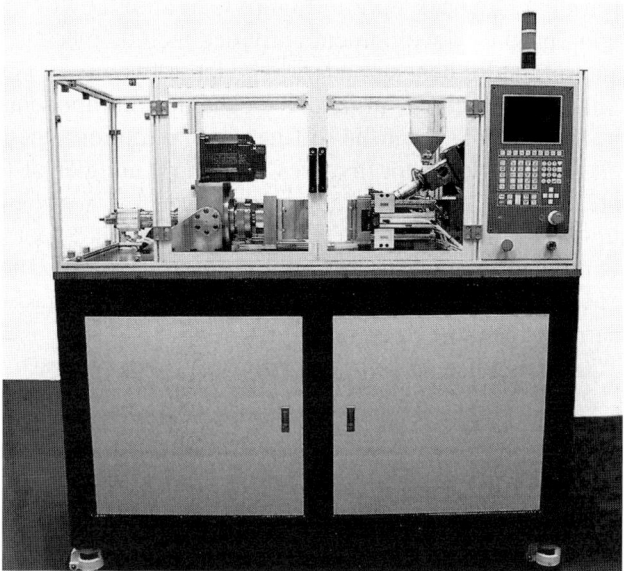

Figure C.7. Miniaturized injection molding machine.

SUMMARY AND CONCLUSIONS

The visit to MIRL offered primarily an overview of micro- and nano-scale-related efforts that focus on the current needs of Taiwan's industry. The short amount of time prevented the group from looking deeper in

MIRL's precision engineering developments for which they are also very well known. A particularly noteworthy item was the miniaturized injection molding machine that offered the configuration and capabilities the chairman of Sansyu had expressed a significant need for, in particular, for future microinjection molded parts. Dr. Tsai has indicated that plans are being considered for the initiation of a more focused micro/mesoscale manufacturing R&D program.

REFERENCES

Mechanical Industry Research Laboratories. http://int.mirl.itri.org.tw/eng/index.jsp (Accessed October 20, 2005).

Site: **ITRI – Nano Technology Research Center (NTRC)**
                             Industrial Technology Research Institute
                             0A00, NTRC/ITRI
                             Bldg. 67, 195, Sec. 4, Chung Hsing Rd.
                             Chutung, Hsinchu, Taiwan 310

Date Visited: December 14, 2004

WTEC Attendees: K. Ehmann (Report author), D. Bourell, G. Hane,
                    K. Rajurkar

Hosts: Dr. Tsung-Tsan Su, General Director,
                                 Tel: +886-3-591-7787, Fax: +886-3-591-0086,
                                 Email: Tsung-Tsan_Su@itri.org.tw
                             Dr. Ping Ping Tsai, Senior Researcher,
                                 Program Manager, Tel: +886-3-591-7651,
                                 Fax: +886-3-591-0086,
                                 Email: pptsai@itri.org.tw

BACKGROUND

Industrial Technology Research Institute (ITRI) was founded in 1973 as a nonprofit R&D institution with the mission to lead the development for Taiwan's technology-based industries (semiconductor, opto-electronics, displays, advanced materials, and precision machinery) through timely technology transfers and spinoffs. ITRI today has over 6,000 employees with a budget exceeding $500 million (50/50 split between long-term research from government contract services and mid- or short-term research contract services from either the government or the private sector). The private sector by itself accounts for about 30% of ITRI's total budget.

The Nano Technology Research Center (NTRC) was founded in 2002. It constitutes the focal point of ITRI's nanotechnology activities by performing three important functions: strategic planning, management and liaison with the outside world, and operation of a cutting-edge common research facility. In 2003 NTRC had a budget of $53 million for about 130 projects under 11 major thrusts that involved multi-disciplinary teams of over 550 researchers. The National Nanotechnology Program has been budgeted $616 million for the period 2003–2008. The program is led by ITRI, which is responsible for about 65% of the resources for the industrialization portion of the budget.

RESEARCH AND DEVELOPMENT ACTIVITIES

The visit lasted about 90 minutes starting with a presentation by Dr. Su on the general philosophy, capabilities and accomplishments of NTRC and the ITRI nanotechnology program, and was followed by a tour of the common facilities lead by Dr. Tsai.

R&D Programs

The R&D activities focus on four areas: (1) nanoelectronics applications, (2) nanomaterials applications, (3) process and equipment development, and (4) biomedical applications. The guiding philosophy of the Center follows the 20/60/20 formula in terms of resources devoted to immediate applications and technological frontiers for maintaining Taiwan's competitiveness and cutting-edge research for future applications, respectively.

Some of the examples of technologies developed include: magnetoresistive random access memory (MRAM); near-field read/write technology; master disk fabrication technology; low-temperature bonding for advanced packaging; mesoporous materials; nano glass-like hybrid substrates for active matrix displays; nanostructure fabrication equipment; microscope objective type atomic force microscope; nanoscale 2D laser encoder (Figure C.8); nanoparticles for *in vivo* diagnostic imaging; nanostructural simulation, and many others.

Specifications:

Item measured	ITRI	Heidenhain PP 281 R
Area coverage	100 × 100 mm² or custom-designed	68 × 68 mm² or custom-designed
Resolution	1 nm w/ 100-fold interpolation	10 nm w/ 100-fold interpolation
Signal output	A, B phase	A, B phase
Measuring speed	20 mm/sec	NA

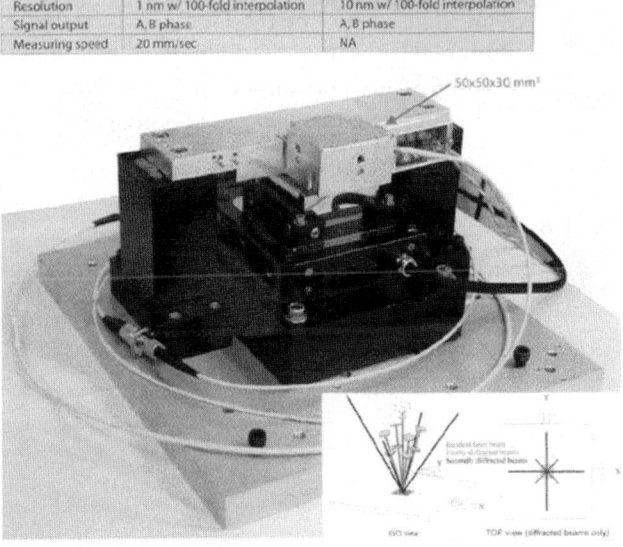

Figure C.8. Nanoscale 2D laser encoder.

Nanotechnology Common Facility Center

This facility is intended to provide a world-class R&D environment. It occupies 11,220 m^2 that includes two clean rooms (1,100 m^2) with several zones, ranging from class 1 to 10,000. The facility offers analysis, measurement and fabrication with extensive capabilities. Most of the equipment is less than two years old and 85% of the laboratories are fully equipped. A complete list of available equipment and the operating procedures of the Center can be found at http://www.ntrc.itri.org.tw/eng/index.jsp (or http://www.ntrc.itri.org.tw/eng/business/index.jsp#sharing-of-facilities). Achievable precision for nanostructure fabrication reaches feature sizes of 100 nm and critical dimensions of 20 nm. With high-resolution field emission transmission electron microscopy (FE-TEM), thin film interface microstructures can be created with dimensions under 3 nm, spatial resolution less than 1 nm and image resolution of less than 0.2 nm. For surface analysis, chemical state analysis is applied with depth profile resolution of less than 0.5 nm. Composition analysis capability is 1 ppt. The total investment into the equipment in this facility exceeds $30 million. NTRC also connects ten existing satellite ITRI facility centers. A certification system for outside users is also in place.

SUMMARY AND CONCLUSIONS

ITRI's Nano Technology Research Center is the focal point that integrates nanotechnology-related research and development by overseeing and maintaining a common Nanotechnology Facility Center and conducting research in critical areas. Most of the Center's resources are directed toward major technological frontiers crucial to the competitiveness of Taiwan's technology industries and aimed at "orders of magnitude" advances. The Center also coordinates and seeks collaboration with outside institutions and industries.

REFERENCES

Nano Technology Research Center. http://www.ntrc.itri.org.tw/eng/index.jsp (Accessed October 20, 2005).

Site:	**Instrument Technology Research Center (ITRC)**
	National Applied Research Laboratories (former
	   Precision Instrument
	Development Center (PIDC), National Science
	   Council, Executive Yuan)
	20 R&D Road VI
	Hsinchu Science Park
	Hsinchu 300, Taiwan

Date Visited: December 14, 2004

WTEC Attendees: K. Ehmann (Report author), D. Bourell, G. Hane,
                          K. Rajurkar

Hosts: Dr. Chien-Jen Chen, Director General,
                          Tel: +866-3-577-9911 Ext. 100; Tel: +866-3-579-
                          5873, Fax: +866-3-578-1226,
                          Email: cjchen@itrc.org.tw
                       Dr. Jyh-Shin Chen, Research and Development
                          Division, Tel: +866-3-577-9911 Ext. 314,
                          Fax: +866-3-577-3947, Email: jschen@itrc.org.tw
                       Dr. Fong-Zhi Chen, Tel: +866-3-577-9911 Ext. 200,
                          Fax: +866-3-577-3947, Email: chen@itrc.org.tw
                       Dr. Chi Hung Huang, Researcher,
                          Tel: +866-3-577-9911 Ext. 557,
                          Fax: +866-3-577-3947,
                          Email: chhwang@itrc.org.tw

BACKGROUND

Instrument Technology Research Center (ITRC), formerly PIDC, was established in 1974 by the National Science Council (NSC), and reorganized into a nonprofit research center on January 16th, 2005. The mission of the Center was to establish and develop the technology and manufacturing capabilities for precision instruments. Today, the Center employs about 150 people (28 PhD and about 60 MS). Its activities can be categorized into: (1) R&D consisting of remote sensing technology, nanotechnology and vacuum technology, (2) manufacturing and maintenance, and (3) technical services. The Center's core competencies are rooted in precision optics, nano/microstructure fabrication and vacuum technologies. The annual budget of the Center is about $8 million from which about $1 million is spent on major equipment development.

RESEARCH AND DEVELOPMENT ACTIVITIES

The visit to ITRC lasted about 3.5 hours and consisted of a presentation by Dr. Jyh-Shih Chen, followed by a brief discussion and a tour of the facilities. A topical summary of ITRC's capabilities and accomplishments follows.

Micro- and Nanofabrication

ITRC defines their work in this area as belonging to three distinct categories, viz., microstructures, nanostructures, and integrated microsystems. In the area of microstructures, emphasis is placed on lithography, electroplating and molding (LIGA)-like fabrication techniques that extend to focused ion beam (FIB) processes (see Figure C.9 for an example). They have been working on microoptical devices that, among other elements, contain microcylindrical lenses (see Figure C.10). As a particular accomplishment, Dr. Chen has pointed to the example of dry etching of high-aspect ratio structures by the inductively coupled plasma (ICP) process. The achieved aspect ratio is 1:40 with a remarkably high surface finish of the walls of Ra = 8 nm in the absence of ripples (see Figure C.11), which are almost always otherwise present. Other examples presented included various optical switches.

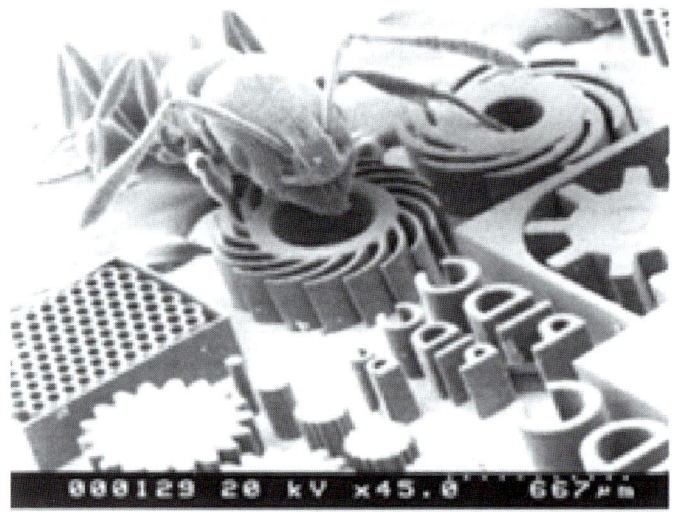

Figure C.9. Example of UV-LIGA.

ITRC's nanostructure fabrication program is inspired by nature. Attempts are being made to emulate butterfly wing structures for color control, moth eyes for high optical transmission capability, and lotus leaf surfaces for creating hydrophobic properties. Fabrication methods used include gray-scale lithography, E-beam, ICP and FIB processes.

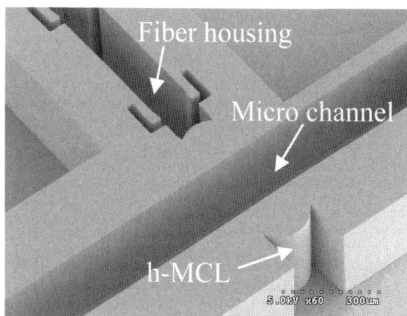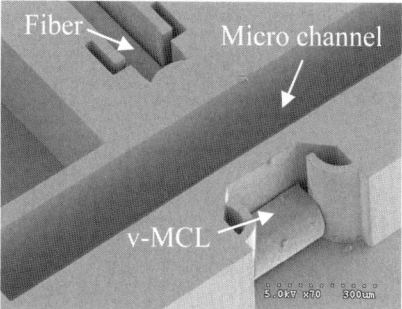

Figure C.10. Example of optical device with microcylindrical lenses (MCL).

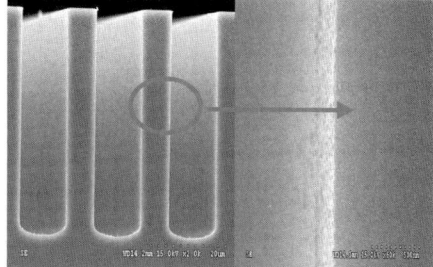

Figure C.11. ICP process with sidewall mirror finish.

Further activities include system integration and nanodroplet manipulation and mixing by passive methods as well as by the use of magnetic fields acting on nanobeads (a few nm in size) in the droplets (about 3 μL or ϕ1.3 mm). One of the aims of this research is to ultimately use this manipulation capability to develop a droplet-based bio-detection chip whose conceptual block diagram is shown in Figure C.12.

Optics Manufacture

The Center is uniquely equipped and qualified to design, manufacture and characterize ultra high-precision optics. They are capable of producing optical components of 30 cm aperture with less than 1 μm form error. ITRC's optics-related expertise is an important component in their major remote sensing development activity. The remote sensing activities are aimed at developing airborne (tested and implemented) and space-based (not launched yet) terrain imaging capabilities.

Micromanufacturing

Capabilities are centered on micro-electro-discharge machining (μEDM) applications, and optical component manufacture. Diamond turning capabilities are expected to be added this year, (2005). The focus is on processing technologies for optical devices.

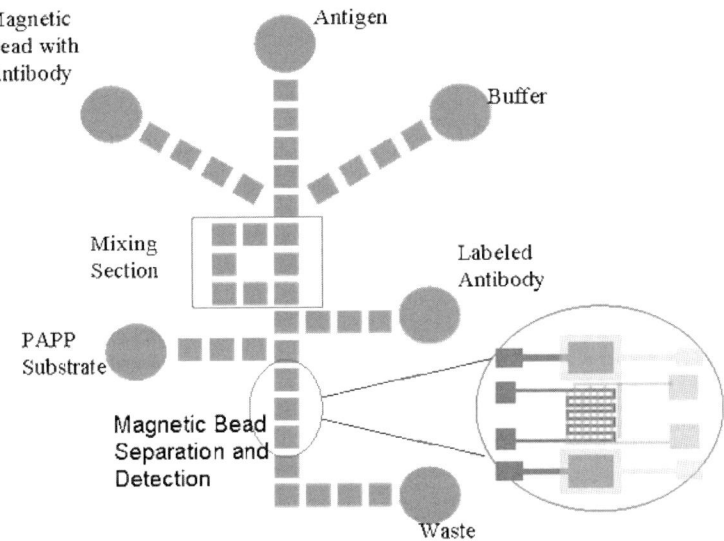

Figure C.12. Droplet-based bio-detection chip.

Laboratory Tour

The tour has included a visit to the remote sensing laboratory, optics fabrication and characterization and of the microfabrication facilities. Microfabrication capabilities included a clean room, hot embossing, excimer laser, photolithography, ICP deep reactive ion etching (RIE), and electroplating.

Other Items

ITRC supports the work of about 40-50 PhD and MS students from different universities who make use of their laboratories and expertise in conducting their research. Students are selected in response to a country-wide announcement and receive a small stipend of about $350 for PhD students and $250 for MS students.

ITRC's R&D agenda is formulated by both a top-down and bottom-up approach. The proposed program is reviewed by experts in the field before adoption. Ongoing projects are evaluated on a monthly basis. The overall performance criteria include the number of patents and publications, technology transferred to industry, and funding generated from industry. Starting this year (2005), the Center will enter a transition period in which it has to switch from full NSC support to a nonprofit research center.

SUMMARY AND CONCLUSIONS

The Instrument Technology Research Center, formerly Precision Instrument Development Center (PIDC), is a government-supported research center with the mission to establish and develop specific technology and manufacturing capabilities for precision instruments in Taiwan. ITRC's core capabilities are in electro-optical systems, vacuum technologies, precision measurement and control, opto-electro-mechanical system integration. Many of these are dependent on their nano/microstructure fabrication developments.

REFERENCES

Cheng, S.C. et al. 2003. Nanofabrication of surface plasmon polariton element by focused ion beam. In *Proceedings of The 7th Nano Engineering and Micro System Technology Workshop*, November 20-21, Taipei, Taiwan.

Hsieh, J. et al. 2005. Realization and characterization of SU-8 micro cylindrical lenses for in-plane micro optical systems, *Microsystem Technologies Journal* 11, 429-437.

Hsieh, J., et al. 2004. Toward an improved micro spectrometer by integrating micro cylindrical lens and discrete vertical grating. In *Asia-Pacific Conference of Transducers and Micro-Nano Technology (APCOT MNT2004)*, July, Sapporo, Japan.

Instrument Technology Research Center. http://www.itrc.org.tw/index-e.php (Accessed October 20, 2005)

Lin, Y.H., et al. 2003. Nano scale high aspect ratio silicon molds for PDMS low cost and batch production. In *Proceedings of The 7th Nano Engineering and Micro System Technology Workshop*, November 20-21, Taipei, Taiwan.

Lin, M.-Y., et al. 2004. A spontaneous droplet manipulation system for glucose and triglycerides colorimetric measurement. In *Biomedical Engineering Annual Symposium*, A-I-2, December 17-18, Tainan, Taiwan.

Yin, H.L., et al. 2003. Fabrication of Fresnel lens using electron beam gray-scale lithography for miniaturized fluorescence detection system. In *Proceedings of The 7th Nano Engineering and Micro System Technology Workshop*, November 20-21, Taipei, Taiwan.

Yin, H.L., et al. 2003. Rapid fabrication of nano-pillar arrays using electron beam lithography in SU-8. In *Proceedings of The 7th Nano Engineering and Micro System Technology Workshop*, November 20-21, Taipei, Taiwan.

Yu, C.-S., et al. 2004. Self-alignment optical detection system for droplet-based biochemical reactions. In *Proceedings of IEEE Sensors 2004 Conference*, October 24-27, Vienna, Austria.

Yu, C.-S., et al. 2003. Spontaneous moving phenomenon of droplets. In *Proceedings of The 7th Nano Engineering and Micro System Technology Workshop*, November 20-21, Taipei, Taiwan.

Site:	Korean Advanced Institute of Science and Technology (KAIST) 373-1, Guseong-dong, Yuseong-gu Daejeon 305-701, Korea

Data Visited: December 14, 2004

WTEC Attendees: T. Kurfess (Report author), B. Allen, J. Cao,
K. Cooper, T. Hodgson, M. Madou

Hosts:

Professor Seung-Woo Kim, PhD,
Tel: +82-42-869-3217, Fax: +82-42-869-3210,
Email: swk@kaist.ac.kr
Professor Dae-Gab Gweon, Tel: +82-42-869-3225,
Fax: +82-42-869-3210,
Email: dggweon@kaist.ac.kr
Professor Dong-Yol Yang, PhD,
Tel: +82-42-869-3214, Fax: +82-42-869-3210,
Email: dyyang@kaist.ac.kr
Professor Min-Yang Yang, Tel: +82-42-869-3324,
Fax: +82-42-869-3210,
Email: myyang@kaist.ac.kr
Dr. Hyungsuck Cho, Tel: +82-42-869-3213,
Fax: +82-42-869-3210,
Email: hscho@lca.kaist.ac.kr
Mr. Kang-Jae Lee (Peter), PhD,
Tel: +82-42-869-3264, Fax: +82-42-869-3210,
Email: za_e@kaist.ac.kr

BACKGROUND

The Korean Advanced Institute of Science and Technology (KAIST) is Korea's leading domestic educational institution. It was established in 1971 and has a charge to be an international leader in science and technology education by focusing on research. Two specific goals listed by KAIST are: to produce a Nobel Prize winner, and to be a driving power for national economical growth and industrial development. Furthermore, the education is oriented primarily for graduate students. KAIST consists of 287 Professors, 67 associate professors, 40 assistant professors, 3,000 enrolled undergraduates, 2,000 enrolled graduate students, and 2,300 enrolled PhD students.

The team specifically visited the Division of Mechanical Engineering Department in KAIST. Since 1971 Mechanical Engineering has graduated a total of 3,531 students (undergraduate: 970, graduate: 1,885, doctor: 676). The Division of Mechanical Engineering offers majors in the following areas: nano/microsystems technology, IT-based intelligent mechanical systems, thermofluid and energy systems, biomedical system engineering, mechanics and design innovation, and pro-human engineering. These major areas provide an excellent foundation to the students and researchers in micromanufacturing.

RESEARCH AND DEVELOPMENT ACTIVITIES

The two-hour tour of KAIST focused on the facilities in the Mechanical Engineering Division. There are a number of facilities and lab groups in the division; however, there was only time to visit four of these labs. The details of the four lab tours are given in this section.

Nano-Opto-Mechatronics

The Nano-Opto-Mechatronics lab is working on two major projects: precision positioning and optical metrology. The precision positioning project is targeting the development of an advanced stage for use in an atomic force microscope (AFM). Of course, there are a wide variety of other applications for this system. The stage uses a laser interferometer for position feedback. It has 1 nm resolution and 100 μm range. The accuracy of the stage has yet to be determined. A lead zirconate titanate (PZT) actuator with a flexure hinge is used to drive the stage. In the near future, the team will employ an X-ray interferometer for improved resolution. This X-ray interferometer work will be done in conjunction with Korea Research Institute of Standards and Science (KRISS).

The optical metrology project targets the use of a confocal microscope (that they have built) to inspect microscale objects. The confocal microscope is shown in Figure C.13. They are working with Samsung on the development of this system. The target applications for this work are optical media (CD / DVD) storage inspection and wafer inspection. The out-of-plane resolution of the system is on the order of 10 nm, and the lateral resolution is approximately 250 nm. They are investigating the use of bire fringent optics to improve the lateral resolution to approximately 70 nm. They have also integrated a heterodyne laser interferometer with the confocal microscope to eliminate pi (phase wrapping) problems that are encountered by the interferometer on discontinuous surfaces.

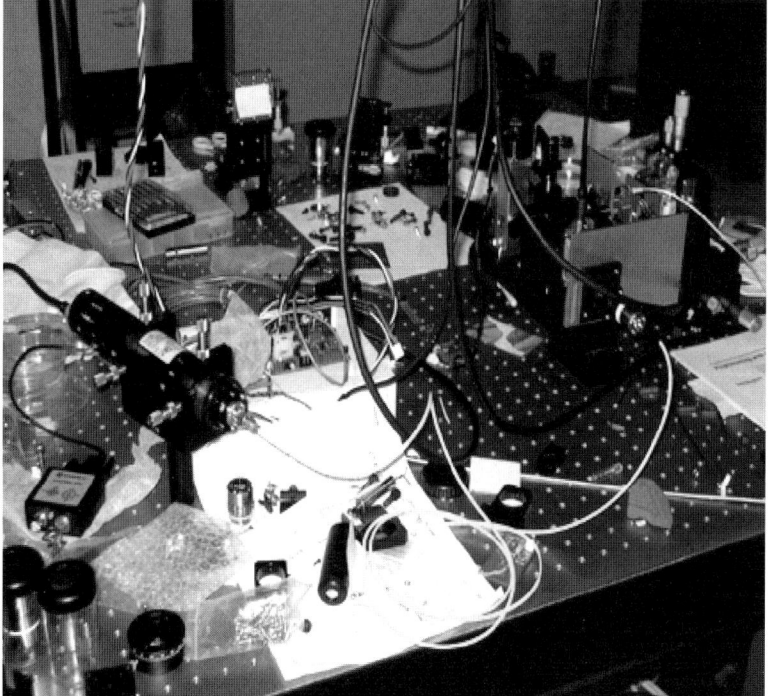

Figure C.13. Confocal microscope.

Microassembly

This group is developing systems to assemble microsystems. The current system employs two cameras to provide 3D information. The two cameras are of different resolutions; this is done in an effort to reduce the cost of the system. Currently, the unit is capable of inserting a micropeg into a microhole. The target component size for the assembly unit is on the order of 100 μm and larger. The minimum clearance that can be assembled at present is 29 μm. The current system utilizes four axes of motion on a six-axis stage. In the future, the system will employ all six degrees of freedom (DOF). The controller employed by this system is a U-Mech controller (off the shelf). The stage is a standard stage made by Physik Instrumente (PI). The system speed is limited by the speed of the vision-processing algorithms, rather than the servo velocities or sensor bandwidths. Thus, as processing power increases and new, more efficient vision algorithms are developed, the speed of the system will increase. Force sensing is currently not used, but in the future force will be monitored and controlled. This will help to address inaccuracies in the system.

Figure C.14. Six-axis microassembly system.

Micromechatronics

The Micromechatronics group demonstrated their nanogripper project. Carbon nanotubes are used as the tips of the gripper. The tip length is set by trimming the carbon nanotubes; however, this is not a simple and highly repeatable task. The grippers are controlled manually using visual feedback from a scanning electron microscope. A pico-motor with 10 nm resolution, or a PZT tube with 4 nm resolution, is used to servo the nanogripper in the workspace. The demonstration was done in 2D (e.g., just moving the grippers in a single plane, X, Y). Alignment in the third direction (in and out of plane) is accomplished manually. It appears that this is accomplished by ensuring that the target and gripper tips are both in focus. One problem that was identified by the team is that it is difficult to drop the part once it is grasped. The nano-aligning system is shown in Figure C.15.

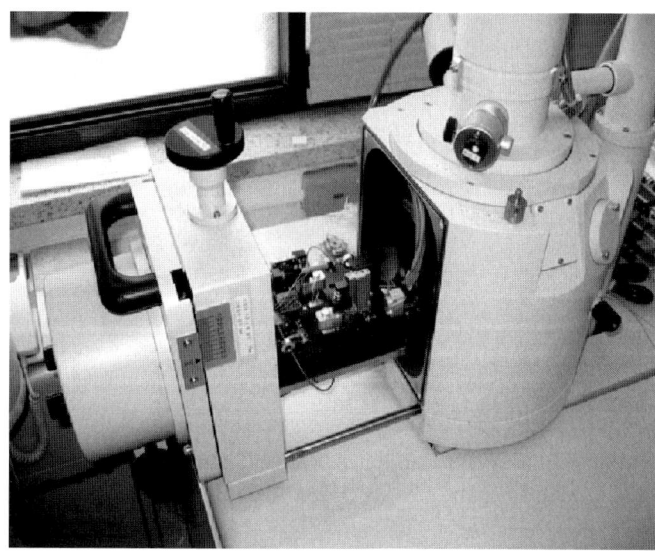

Figure C.15. Nano-aligning system.

Billionth Uncertainty Precision Engineering

This group has several projects underway, and is funded for work in the precision 3D metrology area. In particular, the group is developing the ability to measure parts within a volume of 1m × 1m × 1m with an accuracy of 1 nm. The group is headed by Professor S.-W. Kim and has 13 PhD candidates, four MS candidates and four researchers. The lab has a funding level of $500,000 a year. There are several interesting research projects being conducted in this group. They include a volumetric interferometer that provides 3D information via laser interferometry. This system differs from traditional laser interferometry in that it is capable of 3D measurement as opposed to 1D measurements using a standard laser interferometer. With this technique additional dimensions are measured using laser beams that are guided via fiber optics to multiple charged-couple device (CCD) targets. The CCD targets act as receivers for the beams. Basically, the system operates in much the same fashion as the global positioning system (GPS). Calibration is required for this system, and calibration procedures have been developed by Professor Kim's team. Figure C.16 shows a coordinate measuring machine (CMM) fabricated at KAIST that employs the volumetric interferometer for 3D measurement.

Figure C.16. Robotic laboratory.

This lab is also developing probe capabilities for measurement systems. These probes are the measurement point of a micro CMM. Two types of probes are being investigated. The first is based on an AFM tip, and detects when the tip contacts the target surface. This approach is similar in nature to typical trigger probes used on conventional CMMs. The second design is based on a confocal microscope approach, where the focal distance is the distance between the probe and the target surface. The application of this probe is similar, conceptually, to a variety of non-contact scanning probes used by standard CMMs.

Finally, the group is studying inspection of large-scale optics (10–100 mm) using interferometry. In particular, they are studying the use of three interferometric techniques for these measurements: a Fizeau interferometer using an oblique fiber source, an oblique fiber optic diffraction interferometer, and a phase-shifting, grating diffraction interferometer. All of these approaches have generated promising results.

This project concentrates heavily on hardware development. Specialists in the area of statistics are not involved in this project; however, they might be involved in the future.

SUMMARY AND CONCLUSIONS

There was no overall formal discussion of KAIST or the Mechanical Engineering Division at the meeting; rather, a series of lab tours was given.

The overall impression from the lab tours is that first-rate research is being conducted at KAIST. They are targeting some excellent fundamental issues and are developing systems that will provide the basis for more pragmatic production systems in the future.

REFERENCES

Billionth Uncertainty Precision Engineering. http://pem.kaist.ac.kr/bupe/ (Accessed October 20, 2005).

Korea Research Institute of Standards and Science, http://www.kriss.re.kr/new2004/english/index.jsp (Accessed October 20, 2005).

Nano Opto-Mechatronics Laboratory. http://nom.kaist.ac.kr/ (Accessed October 20, 2005)

Site: **Korean Institute of Machinery and Materials**
                    (KIMM)
                    171 Jang-Dong, Yseong-Gu
                    Daejeon 305-343, Korea

Data Visited: December 14, 2004

WTEC Attendees: T. Kurfess (Report author), B. Allen, J. Cao,
                    K. Cooper, T. Hodgson, M. Madou

Hosts: Dr. Jong-Kweon Park, Principal Researcher, Machine
                    Tools Group, Tel: +82-42-868-7116,
                    Fax: +82-42-868-7180, Email: jkpark@kimm.re.kr
                    Mr. Seung-Kook Ro, Senior Researcher, Machine
                    Tools Group, Tel: +82-42-868-7115,
                    Fax: +82-42-868-7180, Email: cniz@kimm.re.kr
                    Dr. Eung-Sug, Lee, Director, Intelligence & Precision
                    Machine Department, Tel: +82-42-868-7140,
                    Fax: +82-42-868-7150, Email: les648@kimm.re.kr
                    Dr. Doo-Sun Choi, Principal Research Engineer,
                    Nanoprocess Group, Tel: +82-42-868-7124,
                    Fax: +82-42-868-7149, Email: choids@kimm.re.kr

BACKGROUND

The mission and function of the Korean Institute of Machinery and Materials (KIMM) is to promote the advancement of Korea's science and technology, and to provide technical support to industry. KIMM, as the only national laboratory in Korea, conducts research and development, test and evaluation, and provides technical support in the fields of machine systems and material science. The total budget for KIMM in 2004 was $91 million ($38 million from government, $51 million from industry and remaining from "other sources"). There are 168 researchers, 17 engineers, 40 technicians and 24 administrators at KIMM. Their facilities are well-equipped and have an area of 53,400 m^2.

KIMM is the lead institution for a new five-year microfactory program for Korea sponsored by the government. KIMM will coordinate all efforts in this program targeting micromanufacturing. This program was initiated in October 2004. The project has five major groups, some located at KIMM and others housed at various other institutions. The five major groups are: micro/mesomechanical processing systems (at KIMM), assembly/integration control technology for reconfigurable microfactory (at KIMM), microelectrochemical exclusion systems (at Yonsei University), vision/inspection systems (at Pusan University), and ultra-fine microplastic processing systems (at the Korea Institute of Industrial Technology (KITECH)). The project is sponsored by the government for five years at $2 million per year. There are three planned phases for the program during its duration. From 2004–2006 they will be looking at components for micromachines. From 2007–2008 they will be putting machines together. During this period, funding will also come from industry. From 2009–2010 they will be assembling entire microfactory systems. During this period, the bulk of the funding will be from industry. While the objectives of this project are far-reaching, it appears that KIMM is poised to advance state-of-the-art technology to take advantage of new fundamental concepts developed during this project. KIMM is basically developing enabling technology for the microfactory.

KIMM personnel also provided the reasoning behind the Korean government's sponsorship of the microfactory. The following needs were identified as being addressed by the microfactory concept:

- To save energy, space, and resources
- To satisfy the increasing demand of various products to personal taste
- To reduce environmental pollution
- To manufacture highly integrated microproducts
- To support IT (information technology) and NT (nanotechnology)

In their presentation it was indicated that "it is necessary to develop an Intelligent microfactory system for the next generation to support the 5T (IT, NT, biotechnology (BT), environmental technology (ET), space technology (ST)) high-tech industries."

RESEARCH AND DEVELOPMENT ACTIVITIES

There are a number of research areas that are being investigated at KIMM. Most of them target larger-scale manufacturing. The team was informed that the bulk of the microfactory work had only just begun as the project is only a couple of months old. However, there was a good bit of machine tool, microelectronics fabrication and machine design work being conducted that directly supports the microfactory project. This section briefly discusses these areas.

Machine Components

Presently, KIMM is developing the ability to design and fabricate a variety of key components necessary for ultra high-precision machines. Projects developing precision stages, linear slides (aerostatic and hydrostatic) and high-speed magnetic spindles were presented by KIMM personnel. In particular, the magnetic spindle was impressive at a speed of 70,000 rpm.

Information regarding error motion of the spindles and slides was not available. Other components and systems that are currently being designed and fabricated by KIMM include a white light interferometer (WLI) and linear motors. KIMM also has the facilities and experience to conduct life testing and reliability testing on a number of the components that they have developed.

Ultraprecision Machining System

An ultraprecision two-axis machining center was demonstrated at KIMM (Figure C.17), which was developed by three researchers in three years. The system was relatively small in size, a footprint of about 2 m^2. The machine employs linear drives (X, Y lateral) and aerostatic slides that were developed at KIMM. It is controlled using a Delta Tau programmable motion and control (PMAC) PC-based controller and uses single-crystal diamond tools. The machine has a maximum feed rate of 1,300 mm/min. An accuracy value was not given. A resolution of 2 nm was provided. This is the absolute resolution of the glass scales used on the machine in ideal temperature-controlled conditions. However, given the temperature control of $\pm 1°$ C, it is doubtful that the system is currently capable of sub-μm accuracy. There was also no indication that the temperature of the system's environment would be better controlled in the future. However, given the size of the machine, this task would not be particularly difficult. The machine was designed to be able to diamond turn, diamond fly cut and shape.

Laser Machining

One process area that has been identified by KIMM as promising is the use of laser machining. The target application is increasing the density of optical storage media (e.g., DVDs). Currently, the spacing of the information on a DVD is 400 nm, yielding approximately 6 GB of data storage space. The initial target spacing for the next generation of laser machining systems is 100 nm, resulting in a 25 GB storage capacity on a DVD. Part of the research being conducted at KIMM in this area is the testing of a variety of new lasers to generate smaller cutting widths at high speeds. Several laser micromachines were shown for a variety of processes including microhole drilling and grooving. Figure C.18 shows a laser micromachine that is used to make small 3D shafts. Such shafts are targeted as the spindles for next-generation optical storage drives.

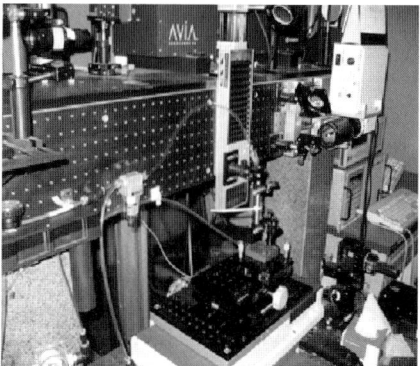

Figure C.17. Ultraprecision
machining sys-
tem.

Figure C.18. Laser micro-
machine.

Future Issues

KIMM is planning to investigate a number of processes including micromachining, nanostamping, injection molding, laser machining and microassembly. They will continue to investigate miniaturizing and improving a variety of machine elements (e.g., spindles, slides and stages) that are core technologies for the microfactory. As their projects mature, they will work with industry to ensure that the technology is transferred to industry.

SUMMARY AND CONCLUSIONS

KIMM appears to be attacking more development problems as opposed to fundamental issues. They are developing the ability to produce the next generation of systems such as white light interferometers, and precision motion and machining capabilities. Clearly, they will be in an excellent position to supply the know-how to produce key components for the next generation of micromanufacturing systems. Furthermore, with their excellent track record of working with industry and technology transfer, they seem poised to provide enabling technology to Korean industry as various microfactory projects develop.

Site:	**Kyocera Corporation**
	6, Takeda Tobadono-cho, Fushimi-ku
	Kyoto, 612-8501, Japan

Date Visited: December 6, 2004

WTEC Attendees: K. Rajurkar (Report author), B. Allen, J. Cao, K. Ehmann, R. Horning, T. Kurfess, E. Uemura

Hosts: Atsuomi Fukuura, Leader, SAW Development Section, R&D Center Keihanna 3-5-3 Hikaridai Seika-cho, Soraku-gun, Kyoto 619-0237, Japan, Tel: +81-774-95-2130, Email: atsuomi.fukuura.nf@kyocera.jp

Naomi Kaihotsu, Manager, U.S.-Europe Team, Head Office, 6 Takeda Tobadono-cho, Fushimi-ku, Kyoto 612-8501, Japan, Tel: +81-756-04-3625, Email: naomi-kaihotsu.nf@kyocera.jp

Hirohiko Katsuda, Department Manager, SAW Device Development Dept., R&D Center Keihanna 3-5-3 Hikaridai Seika-cho, Soraku-gun, Kyoto 619-0237, Japan, Tel: +81-774-95-2124, Email: hirohiko.katsuda.fj@kyocera.jp

Chiaki Matsuyama, Section Manager, R&D Center Keihanna 3-5-3 Hikaridai Seika-cho, Soraku-gun, Kyoto 619-0237, Japan, Tel: +81-774-95-2124, Email: chiaki.matsuyama.nf@kyocera.jp

Shinji Nambu, PhD, General Manager, R&D Center Keihanna 3-5-3 Hikaridai Seika-cho, Soraku-gun, Kyoto 619-0237, Japan, Tel: +81-774-95-2124, Email: shinji.nambu.sd@kyocera.jp

Shigeo Tanahashi, Department Manager, R&D Center Keihanna 3-5-3 Hikaridai Seika-cho, Soraku-gun, Kyoto 619-0237, Japan, Tel: +81-774-95-2124, Email: shigeo.tanahashi.fj@kyocera.jp

BACKGROUND

Kyocera Corporation was founded in 1959 in Kyoto, Japan as a start-up venture by Dr. Kazuo Inamori and seven of his colleagues with a vision of creating a company for the manufacture and sale of innovative, high-quality products based on advanced materials and components. About

58,500 employees work at Kyocera organizations worldwide. Kyocera has operations in Japan, the U.S., China, Europe, Singapore and Brazil. There are 1,100 employees working at the Kyoto Headquarters.

About 400 engineers are working in R&D. Kyocera invests 4% of its revenue in R&D activities.

RESEARCH AND DEVELOPMENT ACTIVITIES

The visit to Kyocera consisted of two parts. First, the group visited Kyocera's headquarters in Kyoto. Mr. C. Matsuyama (section manager) described the vision, scope, products and R&D activities of Kyocera. He and Ms. N. Kaihotsu (manager, U.S.-Europe Team) gave us a tour of their Fine Ceramics History Museum and Showroom where their past and current products are displayed and processes are demonstrated (either with actual physical equipment or by video). Kyocera makes numerous products for several different markets, viz., consumer (wireless phones, cameras, solar energy equipment, ceramic kitchen utensils and stationery), office and professional (digital and printers), and industrial products (fine ceramic components, semiconductor parts, fiber optics components, automotive ceramics, electronic components, industrial lenses, liquid crystal displays (LCDs) and industrial cutting tools). In the second part of the visit, the panel visited the R&D Center at Keihanna (established in 1995) where a number of presentations on Kyocera's current projects were given. No laboratory or manufacturing facilities were visited. What follows is a summary of significant Kyocera products and R&D activities.

Advanced Ceramic Products

Recently developed advanced ceramics called "super ceramics" have unique properties of very high hardness (with a Moh's Scale rating above 9.5 compared to Diamond's Moh's Scale rating of 10). Therefore, the super ceramic cutting tools are increasingly being used in cutting steel and other hard materials in high-speed and high-precision machining operations. Because of their high hardness, these ceramics are also being used to make friction-resistant components in industrial equipment such as paper manufacturing, textile production, and wire drawing.

These ceramics are extremely stable and, therefore, can be machined to dimensional accuracy within a fraction of a micron. Many high-precision industrial machinery, testing and measurement devices and lots of semiconductor equipment require such stability.

The high heat resistance (can function up to 2,500°F) makes these ceramics attractive for applications in automotive engines where the higher performance is a function of higher operating temperature. Kyocera has built and tested ceramic engine prototypes in production cars and has observed better fuel economy, lower pollution and greater engine power. It is estimated that in the future, most automotive engines will consist of up to 50% ceramic material.

In contrast to metals, these ceramics do not corrode, rust and deteriorate over time in extreme environments or chemicals. Therefore, components made of these super ceramics are being used in pumps (including medical pumps) and valves. Their chemical inertness makes these ceramics suitable for implantation into the human body with minimal or no risk of being rejected by the immune system. Using this characteristic, Kyocera has developed long-lasting ceramic components for orthopedic joints and tool replacement systems.

As ceramics are perfect electrical insulators and act like capacitors, their piezoelectric characteristics allow ceramic wafers to generate small electric signals when subjected to acoustic waves or mechanical vibration. These characteristics are essential for further miniaturizing modern electronic products such as cellular phones, pagers, and laptop computers, and future electronics products such as a video telephone that one can wear as a wristwatch and very thin TV sets suitable for hanging on a wall. For example, the smallest ceramic capacitor is 0.2 mm × 0.4 mm in size, with more than 100 layers inside.

Kyocera is also a world leader in producing "photovoltaic" solar cells. Kyocera's solar cells convert solar energy into usable electricity without the problems associated with noise, moving parts or pollution.

Kyocera produces ceramic cutting tools for machining a variety of materials. At the International Manufacturing Technology Show (IMTS) 2004 in September, 2004, they introduced two new chemical vapour deposition (CVD)-coated grades for cutting ductile cast iron. Kyocera also offers a line of microsize cutting tools. They make 2 million tools each year including 1.5 million tools which are smaller than .020 inch in diameter for applications in resistance welding, injector nozzles, extrusion dies, medical/dental components, mold-making and semiconductor manufacturing. Kyocera makes electro-discharge machining (EDM) electrodes of tungsten-copper, chromium-copper, tungsten carbide, graphite and tantalum of 0.001 inch diameter and aspect ratios of 15 and 70 (depending on the electrode material) with a high volume consistency in diameters from 0.005 inch to 0.250 inch, and tolerances of ±50 millionths of an inch with excellent surface finishes.

Research and Development

General Manager Dr. S. Nambu and Department Manager Mr. S. Tanahashi gave presentations, which are summarized in this section.

Kyocera's model of their R&D strategy is conceived in terms of vertical integration of technologies ranging from materials through electronic and optical components and devices to equipment and systems-related R&D activities. Their core technologies include, on the device side, radio communication devices, thin film electronic components, functional single crystals, components for optical fiber communication, optical devices for information equipment and solar cells. On the materials side, their core technologies include high-frequency processes, photo voltaic/semiconductor processes, optics/photonics processes and thin film processes. Underlying these technologies is a substantial analysis and simulation effort.

Examples of devices that push current technological limits include RF SAW filters for mobile communications developed by chip scale packaging (CSP) technology that measure 1.6 × 1.4 × 0.5 mm (see Figure C.19), surface acoustic wave (SAW) duplexers, and components for optical fiber communication. Their push is toward devices that contain 3D structures that, at the same time, double as the device package as well as toward smaller integrated multi-functional devices. For electro-optical devices, tolerances of 0.5 µm are achievable on feature sizes of about 300 µm. Metrology is performed by scanning electron microscopy (SEM) methods. The materials for their devices are generally developed in-house.

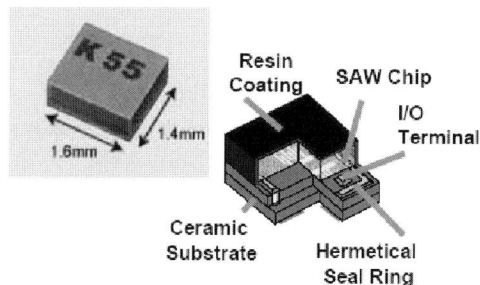

Figure C.19. Band pass filter.

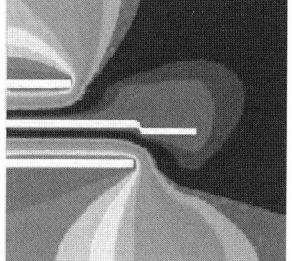

Figure C.20. Electric field analysis.

In the area of computer simulation technologies the major types of analyses performed involve thermal, stress and electric field analyses, principally by finite element methods (FEM). An example of an electric field analysis of a capacitor is shown in Figure C.20. Molecular dynamic (MD) simulations are still considered not sufficiently developed to be useful for predicting macroscopic but are acceptable for predicting microscopic characterizations. Material design programs are still considered too difficult to use. Other divisions at Kyocera are looking into multi-scale simulations, which have been used for manufacturing process prediction, optimization tool, failure analysis, and product performance.

SUMMARY AND CONCLUSIONS

Kyocera manufactures a much-diversified line of products made from ceramics. They are continuously improving ceramics processing and in leading innovative applications in the electronics, medical and automotive fields. Their most profitable products are those for cell phones, since smaller parts generate higher profits. They not only develop solar cells but also use the solar cell panels on the south side of their headquarter building for generating a small portion (1.3%) of their operational energy needs for the building. At the R&D Center Keihanna, 10% of developers are PhDs and most have MS degrees.

REFERENCES

Kyocera. http://global.kyocera.com/ (Accessed October 20, 2005).

Site: **Kyoto University**
                     Sakyo-ku, Kyoto, 606-8501, Japan

Date Visited: December 6, 2004

WTEC Attendees: M. Madou (Report author), D. Bourell, K. Cooper,
                     T. Hodgson, R. Horning, E. Uemura

Hosts: Dr. Atsushi Matsubara, Dept. of Precision Engineer-
                     ing, Tel: +81-75-753-5863,
                     Fax: +81-75-771-57286,
                     Email: matsubara@prec.kyoto-u.ac.jp
                     Dr. Soichi Ibaraki, Department of Precision Engineer-
                     ing, Tel: +81-75-753-52227,
                     Email: ibaraki@prec.kyoto-u.ac.jp
                     Toshiyuki Tsuchiya, Dr. Eng., Tel: +81+75-753-4753,
                     Fax: +81-75-753-5250,
                     Email: tutti@mech.kyoto-u.ac.jp
                     Dr. Isaku Kanno, Dept. Mechanical Engineering,
                     Tel: +81-75-753-3561,
                     Email: kanno@mech.kyoto-u.ac.jp

BACKGROUND

Kyoto Imperial University was founded by Imperial Ordinance on the 18th of June 1897, the second university to be established in Japan. As of 2004, Kyoto University had 10 faculties, 16 graduate schools, 13 research Institutes, and 21 research and educational centers. The faculty staff we met were from the precision engineering and mechanical engineering departments.

RESEARCH AND DEVELOPMENT ACTIVITIES

Dr. Kanno is researching piezo-electric thin film materials, particularly lead zirconate titanate (PZT). He highlighted two different deposition methods for PZT used in his lab. The first one was PZT sputtering from a single powdered PZT sputtering target, the other sputtering from a three-target system. The substrates he uses are magnesium oxide or $SrTiO_3$ to get the right poling of the PZT material. He studies the piezo electric properties of the material with X-rays. The innovation in his work is the use of gradients and multi-layer PZT films. The piezoelectric films are used in microfluidic devices, specifically for micropumps that are put to work in fluidic structures made from a combination of glass and polydimethylislioxane (PDMS). His broader research is fabrication and characterization of functional oxide thin films and their applications in the field of microelectromechanical systems (MEMS). He is trying to fabricate piezoelectric thin films with high piezoelectric properties by control of their nanostructure with modulation of their composition. Furthermore, he investigates the measurement methods of piezoelectricity of thin film materials.

Dr. Tsuchiya's research involves measuring mechanical properties of thin films. He demonstrated for example a tensile strength measuring set-up for polysilicon thin films. He also researches processes for nano formation of gold particles by bringing two flows of reagents together to precipitate the nanoparticles at the mixing interface. To induce the precipitation he pulses a citric acid solution into H_2AuCl_4 solution. This way he is able to create gold particles of about 60 nm in diameter. He has not determined an assembly process for the gold particles yet, although he showed some models for doing so. His research is based on surface micromachining and its sensor applications, and is currently focused on the development of the mechanical property database and the measurement of the fatigue properties for micro- and nanomaterials.

Professor Matsubara has been working on micromachining, and high-precision positioning as a key technology for micromachining. For the implementation of higher accuracy and higher productivity manufacturing technologies, various research topics are actively researched from the viewpoint of mechanics, measurement, design, and control. Dr. Matsubara showed a high-speed computer numerical control (CNC) machine he developed that will work with Rockwell 53 hardened steel with tool speeds up to 5 m/s for high-precision machining. This machine is able to cut and grind, with 2 to 2.5 µm accuracy and 0.1 µm resolution. It uses a vanadium carbide (VC)-coated tool end mill for milling and a cubic boron nitride (CBN) tool for grinding and cutting. It performs laser hardening and has a mechanical probe for measuring. It also has a tool for truing or shaping the cutting tool as it wears away. Tool treatment and machining in the same multi-functional system is desirable for micromachining. The second CNC machine Matsubara showed had been modified to include sensors. One problem he has encountered is that the available sensors are not suitable for integration in high-precision machining due to inappropriate configuration. He needs to develop his own new sensors in order to achieve integration. The resolution of this second machine is in the 5-6 µm range. A third machine he reviewed was a "home-made" one-axis machine that gave resolution of 5 nm. This machine used a ball and screw mechanism for moving the stage. He mentioned that one difficulty he must overcome is the waviness of travel of the stage. He is looking to a linear motor system to help solve this problem - although at a loss of feed force. His next goal is to extend this to a three-axis machine (X, Y, and rotation) but requires funding to continue. He is now using diamond tools but would like to move to CBN tools also on this machine.

Dr. Ibaraki has been researching various topics related to machining and machine tools. His latest research interests are in the motion control of high precision positioning systems, machine tools of novel architecture such as a linear motor-driven machine tool or a parallel kinematics machine tool, computer-aided design (CAD)/computer-aided manufacturing (CAM) for high-productivity machining of dies and molds.

Other discussions:

1. From discussions with the faculty it was made clear that Japan is starting to experience similar loss of manufacturing jobs that occurred in the U.S. in the 1970s. It was remarked that younger researchers from industry are starting to look for academic positions. Part of this is caused by the downsizing of research in the company laboratories.
2. With smaller sized machines it is easier to control environmental factors and maintain accuracies over the smaller dimensions of the work piece. However, this limits the feed size. While desktop fabrication (DTF) is deemed important, no specific applications for it have been shown. The WTEC team thought that the use of DTF by artisans may have merit.

SUMMARY AND CONCLUSIONS

This group is one of several we encountered in Japan and Korea working on better thin film piezoelectric materials. Besides their use in microfluidics (academic) commercial interest is in smaller motors (commercial). Although this country is still way ahead in manufacturing expertise, they are starting to be concerned about manufacturing moving offshore. This group was not particularly impressed with the DTF concept. Dr. Tsuchiya's research in nano formation of gold particles by bringing two flows of reagents together to precipitate the nanoparticles at the mixing interface is original and seems like a promising approach.

Site: **Matsuura Machinery Corporation**
                     Tokyo Jyo-nan
                     1-20-20 Minami-Kamata, Ota-ku
                     Tokyo, 144-0035, Japan

Date Visited: December 6, 2004

WTEC Attendees: Thom Hodgson (Report author), D. Bourell,
                     K. Cooper, G. Hane, M. Madou

Host(s):   Tomio Tomoda, Gen. Mgr., Office of Laser Business
                     Development, Tel: +81-776-56-8125,
                     Fax: +81-776-56-8153,
                     Email: tomoda@matsuura.co.jp
                     Kazuo Miura, Assoc. Gen. Mgr., International Sales
                     Dept.

BACKGROUND

Matsuura Machinery Corporation develops rapid prototyping machines. We met the Matsuura engineers at the Tokyo Jyo-nan Chiiki Chushokigyo Center where the machine is installed. We were able to see their rapid prototyping machine in action, and discuss the attributes of the machine. The machine is still undergoing development, although there are machines already in the marketplace because the current version is fully functional. For the time being, the machines are sold only in Japan. The projected price of the machine is approximately $650,000, which is applicable to the domestic market.

RESEARCH AND DEVELOPMENT ACTIVITIES

A new metal laser sintering/milling hybrid machine, LUMEX 25C, was demonstrated (See Figure 4.29). It combines rapid production and high-speed milling. It uses a powder mixture of iron, copper and nickel. It layers the mixture over the base plate and then sinters it with a 300 watt (500 watt max) CO_2 laser. After 10 layers, the edges of the part are milled. This is a very nice combination of additive and subtractive processes. This process is not necessarily micromanufacturing, but offers tolerances in the 25 micron range.

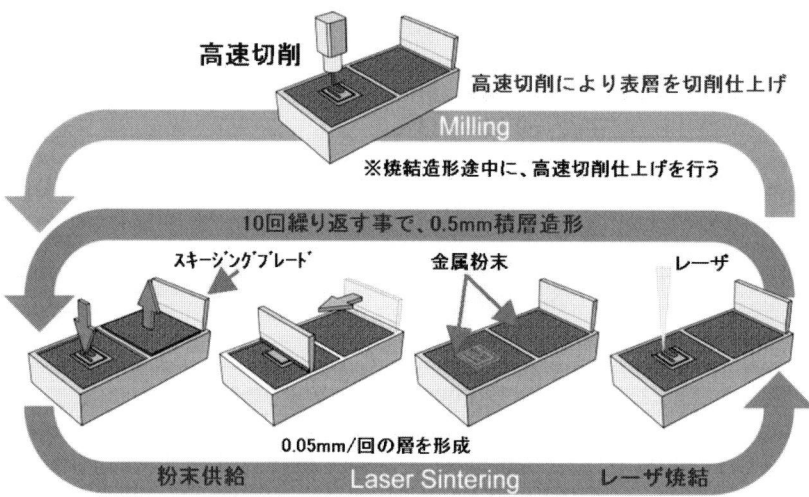

高速切削

高速切削により表層を切削仕上げ

Milling

※焼結造形途中に、高速切削仕上げを行う

10回繰り返す事で、0.5mm積層造形

スキージングブレード   金属粉末   レーザ

0.05mm/回の層を形成

粉末供給   Laser Sintering   レーザ焼結

Figure C.21. Laser sintering and milling.

In furthering the development of this new process, they would like to improve the surface roughness from a present capability of 10~15 microns, to 5 microns and increase the hardness obtainable from 25 RC to 40 RC. This will require alloy development.

Their customers will certainly want to use other materials than the present mixture they now use, especially to make functional or mechanical parts.

A shortcoming of the milling process is that cooling liquids cannot be used within the environment, and the machining process is limited due to heat considerations.

The ability to obtain smooth curved surfaces was remarkable. Internal surfaces normally cannot be milled, but can be with this system to a certain extent.

SUMMARY AND CONCLUSIONS

The LUMEX 25C is a very promising approach to combining additive/subtractive processing. To date, mostly tooling dies with conformal cooling channels have been made for injection molding. Porosity is a problem and the process is relatively slow. Scaling into the micromanufacturing domain is not necessarily easy or evident.

REFERENCES

Matsuura Machinery Corporation. http://www.matsuura.co.jp (Accessed October 20, 2005).

Site:	**Metal Industries Research and Development Centre (MIRDC)** **1001 Kaonan Highway** **Kaohsiung, Taiwan 811** **http://www.mirdc.org/tw/**
Date Visited:	December 13, 2004
WTEC Attendees:	D. Bourell (Report author), K. Ehmann, K. Rajurkar, K. Cooper, G. Hane
Hosts:	Weng-Sing Hwang, PhD, President, Tel: +88-67-351-3121 Ext. 2110, Fax: +88-67-353-6136, Email: wshwang@mail.mirdc.org.tw Antony H.C. Lee, PhD, Senior Project Manager, Metal Processing R&D Department, Tel: +88-67-351-3121 Ext. 2504, Fax: +88-67-353-2758, Email: antony@mail.mirdc.org.tw Kun-Min Huang, Deputy Director, Metal Processing R&D Department, Tel: +88-67-351-3121 Ext. 2460, Fax: +88-67-353-2758, Email: kmhuang@mail.mirdc.org.tw Michael-Fu, PhD, Director, Metal Processing R&D Department, Tel: +88-67-351-3121 Ext. 2500, Fax: +88-67-353-2758, Email: hcfu@mail.mirdc.org.tw Ming-Chin. Tsai, PhD, Engineer, Metal Forming Technology Section, Tel: +88-67-351-3121 Ext. 2549, Fax: +88-67-353-7530, Email: mctsai@mail.mirdc.org.tw S. Huang, C-F. Wu, T-K. Su, K-B. Yo

BACKGROUND

The Metal Industries Research and Development Centre (MIRDC) is a government laboratory in southern Taiwan. Its annual budget is $30 million. There are 465 technical staff members, 4% PhD, 40% graduate and the rest undergraduate. The main branch is in Kaohsiung with two smaller branches in Taipei and Taichung. About 40% of the Centre funding comes as government block funding, 20% government contract funding and 40%

from industry. Industry shoulders 20–30% of the cost for all projects, as is mandated. Approximately 55 projects transfer to industry annually from partnering with about 170 industrial plants. Intellectual property is generally vested in partnerships with actively participating industries, although a licensing mechanism is also in place. The main area of research is secondary processes for metals, both technology and equipment development. Over the last 10–20 years, Taiwan's metals industry has been moving towards micromanufacturing to accommodate the microelectronics industry. The mission of the Centre is to promote growth and upgrading of metals and their related industry in Taiwan. They define micromanufacturing as processes associated with parts that are 1 µm to several mm in size. We visited the micro/mesomechanical manufacturing (M^4) R&D department. It falls organizationally under the metal processing R&D department of the centre.

RESEARCH AND DEVELOPMENT ACTIVITIES

The M^4 R&D department has three main laboratories dealing with micromanufacturing. The first is the microforming lab. It has facilities for microforging, microstamping, microplastic forming (in cooperation with U. Engel of the University of Erlangen/Nuremberg) and hydrostatic extrusion. For the latter, aluminum wire 50 µm in diameter can be extruded successfully from thick bar stock.

The second laboratory was the microjoining lab. Here, plasma welding, laser welding and resistance welding are used for micromanufacturing. One example was welding of 50 µm thick caps on sensor housings. A microresistance welder was used for microwelding iron (Fe)-nickel (Ni)-cobalt (Co) caps onto iron headers of diodes. Another apparatus was used to produce ~1 mm wide plasma welds of 200 µm thick 304 stainless steel sheets to form ~50 mm diameter thin-walled tubes. The tubes were further processed into vacuum bellows.

The third lab was a microprecision surface treatment lab. Capabilities exist for ion implanting, surface coating and physical vapor deposition (PVD). An example was DVD molds coated with diamond-like carbon for wear resistance.

The micro/mesomechanical manufacturing (M^4) R&D department has run several projects of environmental importance. One involved development of lead-free solders, and the other dealt with development of chromium-free coatings.

Future projects in the 2005–07 timeframe are microdevices including the development of a penny motor and laser diodes, and micromolds including a micromold and a micropress.

Significant discussion time with President Weng-Sing Hwang transpired. The main points and observations were:

- *Challenges.* The main challenges facing advancement and success in micromanufacturing are multifaceted and include materials (homogeneity, grain size, purity), processes, die and tool manufacturing, and assembly.

- *Materials.* There is some limited evidence that material strength decreases for some materials as the part size decreases. This is contrary to conventional materials behavior and research is underway to understand this phenomenon better.

- *Processing.* There is a desire to minimize part handling during micromanufacturing since part registration inevitably introduces errors.

- *Simulation.* FEM is extendable to micromanufacturing since it is a continuum mechanics model, but the materials property inputs need to be adjusted accordingly.

- *Microfactories.* There is interest in developing microfactories in Taiwan, but there is not yet funding available to do so.

- *Probability for Success.* Micromanufacturing differs from microelectromechanical systems (MEMS) in that it is industry-motivated and application-driven. The probability for success then is much improved.

- *Design.* There is no design effort in micromanufacturing at MIRDC. This is left to the universities and industry.

- *Design Rules.* Design rules need to be developed for micromanufacturing. These will be part of a module developed at MIRDC for rapid design and manufacturing for industry.

- *Materials.* All materials used in micromanufacturing are conventional materials. No need has arisen to develop new materials particularly suited for micromanufacturing. Innovations in materials must be performance-driven since volumes are too low for cost/economics to drive it.

- *Materials.* Since the volume of micromanufactured parts is so small, very expensive materials like gold are economically justified, which widens the range of usable materials.

- *Applications.* Promising areas of application are mobile electronic technology, such as cell phones, personal digital assistants (PDAs), cameras and computers.

SUMMARY AND CONCLUSIONS

The Metal Industries Research and Development Centre is a government lab involved in significant research and process development in micro-manufacturing, particularly deformation, joining, molding and surface treatment. The emphasis is on secondary processes and equipment to meet the needs of industry.

REFERENCES

Company Profile and Research. Booklet (n.d.).

Micro-Metal Forming Laboratory Overview. Brochure (n.d.) [in Chinese].

Tsai, M. -C. 2003. The parameters in micro metal parts forming. *Forging* (ISSN: 1023-750X; in Chinese with English abstract), 12:1, 18-27.

Tsai, M. -C., Y. -A. Chen, C. -F. Wu, F. -K. Chen. 2004. Size effect in micro-metal forming of copper and brass." *Forging* 3:2, 41-46 [in Chinese with English abstract].

Tsai, M. -C., Y. -A. Chen, C. -F. Wu and F. -K. Chen. 2005. A new model and size effect for micro-metal forming of unalloyed copper and brass. *Journal of Materials Processing Technology* (submitted).

Tsai, M. -C. and C. -F. Wu. 2003. Cu15Zn micro-plasticity and grain size effects. In *Proceeding of the 2003 Annual Conference of the Chinese Society for Materials Science*, PB-029, November 21-23, Tainan, Taiwan.

Zheng, Y. F. and B. M. Huang. 2002. Superelastic and thermally activated TiNi alloys and their applications in dentistry. *Matl Sci Forums* 394:5, 57-60.

Site:	**Mitsubishi Electric Corporation (MEC)**
	1-14, Yada-minami 5-chome, Higashi-ku
	Nagoya 461-8670, Japan
	http://globalmitsubishielectric.com

Date Visited: December 9, 2004

WTEC Attendees: K. Rajurkar (Report author), B. Allen, J. Cao, K. Ehmann, G. Hane, T. Kurfess

Hosts: Koji Akamatsu, Deputy Manager, Manufacturing Systems Planning Dept., Tel: +81-52-712-2232, Fax: +81-52-723-1131,
Email: Akamatsu.Koji@bx.mitsubishielectric.co.jp
Akihiro Goto, Manager, Tel: +81-52-712-2311,
Fax: +81-52-712-3806,
Email: Goto.Akihiro@ap.mitsubishielectric.co.jp
Itsuro Tanaka, Manager, Global Operations Support Group, Tel: +81-52-712-2563,
Fax: +81-52-712-1163,
Email: Tanaka.Itsuro@ak.mitsubishielectric.co.jp

BACKGROUND

Mr. Tanaka presented an overview of Mitsubishi Electric Corporation and Nagoya Works. In 1870, Mr. Y. Iwasaki founded a company called Tsukumo Shokai, which was named Mitsubishi Shokain in 1872. A shipbuilding department established in 1917 led to an independent entity called Mitsubishi Electric in 1921. Currently, there are 42 Mitsubishi companies that are separate entities and operate independently. Some of these companies are: Mitsubishi Electric Corporation (electronics), Mitsubishi Heavy Industries, Ltd. (ships, aircraft, turbines), Mitsubishi Motors Corporation (automobiles), Mitsubishi Corporation (trading), Bank of Tokyo-Mitsubishi (banking), Nikkon Corporation (cameras, optical equipment), and Tokyo Marine & Fire Insurance Co. (casualty, insurance).

Mitsubishi Electric Corporation (MEC) has paid-in capital (i.e., invested funds) of $1,600 million and consolidated net sales of $30,100 million and employs about 99,000 employees worldwide. The business areas of MEC include:

1. Information, Telecommunication and Electric Systems and Device
2. Heavy Machinery
3. Industrial Products and Automation Equipment

4. Consumer and Other Products

Nagoya Works manufactures programmable logic controllers (PLCs), inverters, robots, alternating current (AC) servos, heavy machinery, computer numerical control (CNC) machines, electro-discharge machines (EDM), lasers, and switches. Nagoya Works, established in 1924, has 1,940 employees with a floor space of 220,000m^2. The current key product lines are grouped as automation (69%), mechatronics (15%), and basic machinery (14%) and others (2%).

Mitsubishi's market (Japanese/world) share for PLC is 58/16%; heavy machinery, 35/12%; inverters, 39/10%; servos, 29/10%; CNC, 31/10%; EDM, 35/25%; lasers, 32%/- and robots, 7%/-. Mitsubishi also has facilities outside Japan:

1. Mitsubishi Electric Dalian Industrial Products Co. in China manufactures inverters, EDMs, servos, and CNCs.

2. Oriental Electric Industry Co. in Bangkok, Thailand, manufactures inductor motors.

RESEARCH AND DEVELOPMENT ACTIVITIES

Mr. Tanaka gave us an extensive tour of Mitsubishi Electric products displayed in the show room, as well as on the shop floor. The product display included a series of models of PLCs, inverters, servos, switches, CNCs, and EDMs. The variety of components (such as gears, dies and molds) machined by EDM and laser are also on display. EDM equipment on display, as well as in operation, includes EA (high performance) series, MA (ultra-high accuracy) series, GA (ultra-large) series and automated system (MA 2000 EDM + Robot + Cell System). The wire-EDM systems of the PA series have been designed to operate in a temperature- (about $20\pm1°C$) controlled environment. A stainless steel gear of thickness 1.2 mm, outer diameter 1.2 mm and inner diameter 0.8 mm was machined by EDM in about five minutes.

Examples of die-sinking EDM include meso- and microscale features or components with high precision. Specifically, lens dies (0.5 mm diameter with a glossy surface of 0.5 μm Rmax surface finish), tungsten carbide insert (1.4 micron Rmax), and stepped spur gear (3.0 μm Rmax) are some of the applications of die-sinking EDM equipment. Wire-EDM systems have been used for fine slit (width 100 μm) machining, drilling of fine holes (0.125 mm) in a carbide (0.8 mm thick) optics connection die, high-precision gear machining, high-precision punch machining, and B-axis hexagonal connector core pin machining. An example of a high surface finish (0.8 micron) cavity and related electrode is shown in Figure C.22.

Figure C.22. EDM electrode and high surface finish cavities.

Mitsubishi's Laser Systems (HV series) are able to cut very complex and precision meso/microparts with excellent accuracy and surface finish. The Mitsubishi CNC 700 series features nanocontrol technology (Reduced instruction set computing-central processing unit (RISC-CPU) and high-speed optical servo network) for high-speed, high-precision control and five-axis control. Accuracies on the order of 2.5 μm are possible. Figure 3.2 shows various parts machined by Mitsubishi Laser Systems.

Mitsubishi continues to improve the power supply, CNC functions, adaptive control systems, automation, and network systems to enhance the overall productivity (speed, accuracy, and quality) of their products. The research and development efforts are directed towards designing the machine structure for enabling a stable performance for a long time, adaptive control for easily achieving high-speed and high-accuracy machining, optimum power supply for ultra high-speed, high-accuracy specifications, automated systems with robotic material handling that automatically changes the workpiece, 3D measuring equipment that measures the workpiece dimensions, and a scheduler that carries out scheduled operations. Remote monitoring of the EDM systems is also an important R&D issue.

Dr. A. Goto presented the research and development related to Mitsubishi's recent project on "micro spark coating" (MSC). The micro spark coating technique forms a hard surface or metal cladding on the metallic material using an electric discharge in a dielectric. The tool electrode is made of semi-sintered powder. The discharge energy melts electrode and top surface of the workpiece. The melted tool material transfers to the molten workpiece resulting in a strong bond between the coating and workpiece (part). The electrode gap (100–200 μm) is maintained by a servo system. The coating layer is about 1 mm. This MSC method does not require

any further post-processing operations. A dense layered coating can be obtained with 5–10 A current, 5–10 microseconds on-time and 50–60 microseconds off-time, while a porous layer can be obtained with
20–30 A and similar pulse parameters. As compared to plating, welding and plasma spray techniques, MSC offers benefits of low cost, no need for bench work, small deformation and uniform quality. The MSC process can use ceramics and metals as cladding material. A specific example of the MSC process (discussed and shown) was that of low-pressure turbine (LPT) blade interlock of an aircraft engine. They are commercializing the process targeting the aerospace propulsion (e.g., jet engines) sector. Dr. Goto and his team have also proposed a simple mechanism of MSC process based on the available and accepted EDM erosion mechanism but have not developed a quantitative model for the process.

During lunch, education, training and research collaboration with universities issues were discussed. Mr. Tanaka commented that universities need to get students to learn basic fundamentals of science and engineering. Mitsubishi provides the specific training needed for engineering and manufacturing functions and R&D activities. The duration of such training may vary depending on the level of position (from six months to two years). Mitsubishi has ongoing research programs with the Toyota Technological Institute and Nagoya University and even other industries.

Mitsubishi has an environment management system that aims to monitor the effect of business activities and products on the environment and to prevent contamination before it occurs. Besides following the local and national environmental laws, ordinances and agreements, Mitsubishi has set voluntary standards within a technical and economically feasible range. Specifically, Mitsubishi is promoting energy conservation and use of clean energy to reduce CO_2 emission, reducing waste, reducing the use of restricted chemical substances by using alternatives and appropriate control, and promoting environment-friendly designs in consideration of product life cycles. Mitsubishi's environmental policies are well documented, open to the general public, and all employees are familiarized with it.

SUMMARY AND CONCLUSIONS

Mitsubishi Electric Corporation manufactures information and telecommunications tools, electric systems, heavy machinery, industrial products and automation equipment and consumer products. They make laser and EDM systems for producing complex and precision components. Recently, Mitsubishi has developed a new process called "micro spark coating" which is used in the aircraft industry for coating of a low-pressure turbine blade interlock. Mitsubishi also has an environment management program to monitor the effect of production and business activities on the environment.

REFERENCES

Gato, A., et al. 2004. Development of micro spark coating. In *Proceedings of the 24th International Congress of the Aeronautical Sciences*, August 29-September 3, Yokohama, Japan.
Nagoya Works Environmental Policy, Mitsubishi. Brochure (2004).

Site: **Nagoya University – Center for Cooperative Research in Advanced Science and Technology Furo-cho, Chikusa-ku, Nagoya, 464-8603, Japan http://www.mein.nagoya-u.ac.jp**

Date Visited: December 10, 2004

WTEC Attendees: K. Rajurkar (Report author), B. Allen, J. Cao, K. Ehmann, G. Hane, T. Kurfess

Host: Dr. Toshio Fukuda, Tel: +81-52-789-4478, Fax: +81-52-789-3115, Email: fukuda@mein.nagoya-u.ac.jp

BACKGROUND

The laboratory was originally established in the 1980s and was taken over by the current faculty in 1990. The laboratory under the guidance of Dr. Fukuda is at two different locations. He has about 40 researchers including MS and PhD students and two post-doctoral fellows. He has many international students in his group. The equipment in the laboratory is funded by the Japan Science and Technology (JST). The laboratory projects include a cellular robotic system (CEBOT), a microrobotic system and micromechatronics, nanotechnology, a bio-micromanipulation system, a telesurgery system for intravascular neurosurgery, machine learning of intelligent robotic system and a humanitarian demining robot.

Dr. Fukuda, although very busy, gave us a tour with necessary details in an hour. Later, his students continued the lab tour.

RESEARCH ACTIVITIES

Dr. Fukuda and his team are conducting state-of-the art research in many areas. A brief introduction of some their research is given below.

An *in Vitro* Anatomical Model of Individual Human Cerebral Artery

Using the scanning technique and then rapid prototype method, an anatomical model of the the human cerebral artery has been developed for surgical simulation, preclinical testing, and medical training. They produce a transparent prototype having the same elastic and frictional properties and thin membranous structure as that of a human cerebral artery. The accuracy of a fabricated artery is about 1 micron. A U.S.-made commercial ultraviolet (UV) laser, SoloidScape, is used to fabricate physical models with a layer thickness of 30 microns. A typical such artery model takes about six hours to fabricate.

Microdevice of Catheter Tip for Injection into Myocardium

A system has been developed for catheter-based transfer for regenerative medicine. The suction cup of catheter tip absorbs the pulsative myocardium. A wire is used to open and close the suction cup in the heart

Multi-Locomotion Robot

This robot with a brachiation controller can achieve smooth movements like animals. The robot has 12 degrees of freedom (DOF) and 14 actuators with wire-driven joints. This robot was demonstrated swinging from one rod to another. Dr. Fukuda mentioned that the Discovery Channel runs that demonstration very frequently. The robot has also been featured on BBC TV.

Dancing Yeast

Dr. Fukuda's student showed a video of dancing yeast, in which six yeast balls were controlled via laser beam tapping.

SUMMARY AND CONCLUSIONS

Dr. Fukuda and his team are conducting state-of-the-art manufacturing research applied to many fields including biomedical and robotics. There was not much time to have a follow-up discussion but Dr. Fukuda sited some of his recent papers for further reading and reference. Single-cell manipulation using laser micromanipulation with microtool and single-cell fixation using photo-cross linkable resin in microchip are examples of further reading.

REFERENCES

Research Activities Report, Robotics and Mechatronics. Unpublished report (2003).
Fukuda Laboratory. http://www.mein.nagoya-u.ac.jp/ (Accessed October 20, 2005).

Site:	**Nagoya University – Laboratory of Structure and Morphology Control**
	Furo-cho, Chikusa-ku, Nagoya, 464-8603, Japan
	http://www.mein.nagoya-u.ac.jp

Date Visited: December 10, 2004

WTEC Attendees: K. Rajurkar (Report author), B. Allen, J. Cao, K. Ehmann, G. Hane, T. Kurfess

Hosts: Dr. Eng. Naoyuki Kanetake, Professor, Dept. of Materials Science and Engineering,
Tel: +81-52-789-3359, Fax: +81-52-789-5348,
Email: kanetake@numse.nagoya-u.ac.jp
Dr. Eng. Makoto Kobashi, Associate Professor, Department of Materials Science and Engineering,
Tel: +81-52-789-3356, Fax: +81-52-789-5348,
Email: kobashi@numse.nagoya-u.ac.jp
Mr. Kume (Graduate Student)

BACKGROUND

Professor Kanetake established this laboratory, but currently, Dr. Kobashi and a graduate student are associated with it. After a brief welcome by Dr. Kanetake, the purpose of our site visit was presented and discussed by B. Allen and K. Ehmann.

RESEARCH ACTIVITIES AND DISCUSSION

Dr. Kanetake presented an overall summary of the research activities of his group. He covered activities related to processing of composite materials, synthesis and control of microscopic structure and morphology, processing of porous materials, control of microstructure and properties, and compressive torsion processing.

Dr. Kobashi made a detailed presentation on a reactive infiltration process for metal matrix composites (MMC). The objective of the research was to develop an innovative cost-effective processing technique using pressure-less infiltration. The approach involves a highly wettable combination with an exothermic reaction in powder phase. They fabricated a titanium carbide and magnesium (TiC/Mg) composite and also TiC/ aluminum (Al) using wettable combination. It was concluded that MMC can be manufactured by the spontaneous infiltration process using good wettabilty and heat of reaction.

Dr. Kobashi also presented their recent research on combustion synthesis of porous Ti composites. Porous titanium, due its high melting point and stability in human body, is suitable for bio-implants. The objective of this research was to develop a process to control pore morphology of porous Ti. The reactive synthesis includes steps of raw material, green compact and porous material. The macroscopic view and microscopic cross-section of porous Ti material indicate that the porous structure can be controlled by regulating the bending ratio of the raw material, the porosity of green powder, and the combustion temperature.

Mr. Kume presented their ongoing research on compressive torsion processing which causes severe plastic deformation for obtaining grain refinement of metals. The results of grain refinement for pure aluminum and for Al alloys were shown. Temperature varied from room temperature to $373°K$, $473°K$, and $573°K$. The temperature has been found to affect the resulting grain size.

Later, we visited the laboratory and saw the equipment (F = 50 ton, torque = 1,000 Nm for a specimen 25 mm in diameter and 10 mm in thickness) in operation. The process takes about two minutes to convert the specimen with an initial grain size of 100 micron to a final uniform distribution of 10 micron grains. They have been working on this project for the last three to four years and they have one Japanese patent. They have not yet published their results in journals or in conference proceedings.

SUMMARY AND CONCLUSIONS

The research activities of this laboratory include the processing of composite materials, synthesis and control of microscopic structure and morphology, and compressive torsion processing. The specific projects under investigation are reactive infiltration for metal matrix composites and combustion synthesis of porous Ti composites.

REFERENCES

Kanetake, E. 2004. Presentation on summary of group research activities.
Kobashi, M. 2004. Presentation on reactive infiltration process for MMC.
Kume Y., M. Kobashi, N. Kanetake. 2004. Grain Refinement of Al Alloy by Compressive Torsion Processing. *Materials Forum* 28: 700-704.

Site: **Nagoya University – Department of Micro System Engineering**
                       Furo-cho, Chikusa-ku, Nagoya 464-8603, Japan
                       http://www.bmse.mech.nagoya-u.ac.jp

Data Visited: December 9, 2004

WTEC Attendees: T. Kurfess (Report author), B. Allen, J. Cao,
                       K. Ehmann, G. Hane, K. Rajurkar

Host: Professor Koji Ikuta, Biochemical Micro System En-
                       gineering Laboratory, Department of Micro System
                       Engineering, School of Engineering,
                       Tel: +81-52-789-5024, Fax: +81-52-789-5027,
                       Email: ikuta@mech.nagoya-u.ac.jp
                       Takayuki Matsuno, Research Associate, Tel: +81-52-
                       789-2717, Fax: +81-52-789-5348,
                       Email: matsuno@mein.nagoya-u.ac.jp

BACKGROUND

Professor Ikuta is the Department Chairman for the Department of Micro System Engineering at Nagoya University and he heads the laboratory.
Professor Ikuta presented a number of highly successful microsystem designs, as well as a number of new and innovative processes for fabricating
these designs. Professor Ikuta's lab is approximately 600 m^2, and is staffed
by himself, one associate professor, one research associate, four PhD students and 21 graduate students. His funding level has varied over the past
several years. From 1996–2001, the Japan Science and Technology
Agency funded him at $1 million per year. Recently, he received the pres-
tigious CRAST award (only a few are awarded every year in Japan) for the
next five years. He has received the bulk of his funding from the government. Clearly, his lab is well-funded.

RESEARCH AND DEVELOPMENT ACTIVITIES

A significant number of projects were presented to the team. Current
projects in Professor Ikuta's laboratory include:

1. Biochemical integrated circuit (IC)/large-scale integration (LSI) chip
    and application (Figure C.23).

2. A multi-jointed active endoscope; an active-controlled catheter and a
    remote-controlled microsurgery system.

3. Master-slave microsurgery system with force sensation.

4. Minimally invasive surgery robotics.

5. Safety welfare robotics.

6. Three-dimensional microfabrication made possible by microstereo-lithography (µSLA) (Figure C.24).

7. A microactuator based on a new principle of incorporating "intelligent" materials such as shape memory alloys and piezoelectric materials (Figure C.25).

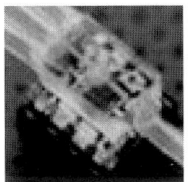

Figure C.23. Opto-sense microreactor

Figure C.24. Microturbine 14 µm diameter

Figure C.25. Cybernetic actuator

Micro-SLA (IH Process)

In 1992 Professor Ikuta's group developed a micro-SLA process that is known as the IH process. This process uses a single-photon polymerization process to develop the photocurable polymer at the focal point of an ultraviolet (UV) beam. Unlike the typical SLA process, the polymer in the integrated harden (IH) process is cured below the polymer/atmosphere interface where the UV beam is focused. A shallow focus is used on the lens (large numerical aperture (NA) lens) such that the beam's energy is small enough not to cure polymer outside of its focal point. A glass slide is also placed above the layer being cured (as opposed to curing the layer at the surface of the polymer). Thus, the layer thickness is not limited to what can be generated by the typical dip and sweep approach of an SLA system; rather, an extremely thin layer can be generated by raising the glass slide small distances above the previously solidified layer. Using this approach, very thin layers can be generated. The IH process resolution in 1992 was 5 µm, and is close to 200 nm at present. Currently, Professor Ikuta's group is targeting process enhancement including a "Mass-IH process" (producing multiple parts simultaneously with multiple beams directed using fiber optics), a "Hybrid IH process" (incorporation of other non-IH microcomponents in the IH structure), and a "Multi-polymer IH process," as well as other innovations.

In 1997, this IH process was upgraded to "Super IH process," which can fabricate freely moving microstructures such as microgears with shafts and stoppers. The enabling technology for this new process is the world's first "deep site pin-point solidification" of a UV-polymer. A 3D microgear with a shaft diameter of 50 μm was introduced on the editor's choice page in "Science." In 1998, "Nano stereolithography with high-speed scanning" was implemented. This process employs a two-photon absorption process of a high-power Ti-Sapphire laser resulting in 100 nm resolution. Using this process, a microgear and pin-set having dimensions of approximately 5 to 15 μm were fabricated in under 10 minutes. This process is capable of generating movable micro- and nanomechanisms as well as simple 3D structures. Examples of such mechanisms include microgears with 15 μm diameters and a nanomanipulator with a three degree of freedom (DOF) motion capability. This process is the key technology employed to fabricate a variety of optical-driven nanomachines.

Laser Trapping to Drive Microsystems

Professor Ikuta demonstrated several very impressive systems that were fabricated using the IH process. Systems such as moving microturbines (15 μm diameter), rotating gears (15 μm diameter) and grippers were demonstrated. All of these devices were driven using laser trapping. Professor Ikuta indicated that laser trapping should provide enough force/torque for most microsystems. Forces are generated in the laser trapping process by the transfer of momentum from light to an object. A similar approach can be used to transfer angular momentum to generate torque. This is how Professor Ikuta drives his microsystems. It should be noted that he specifically designs optimized laser trapping points into his systems. These optimized points were developed empirically.

The application of laser trapping to the various systems was demonstrated. Rotational speeds of 118 RPM were demonstrated for rotational devices. Linearly reciprocating devices were driven at 12 Hz. A three DOF manipulator, on the order of 10 μm in size, was used to manipulate a cell. Nanomanipulators having a thickness of 250 nm were also demonstrated. With these very thin manipulators, the detection of forces at the 5–10 fN levels was demonstrated. Such force levels are typical of those generated in inter-cellular interactions, and are critical for biological applications.

Artificial Cellular Devices

A variety of microfluidic devices called artificial cellular devices (ACD) that are based on the biochemical IC were designed and fabricated by Professor Ikuta's group. Each ACD is disk-shaped and contains a variety of functional, non-SLA, microcomponents such as pumps, valves, concentrators and reactors. These components are inserted into the biochemical IC chips during fabrication. The IH process is then used to encapsulate these components. Furthermore, the ACDs are stackable to enable the construction of complex chemical reaction and analysis units. As some of the ACD components are quite small (at the resolution of the IH process) and others, such as the outer casing, are relatively large (10–20 mm), multiple SLA processes are used to fabricate the ACDs. Thus, the bottom of the outer casing may be initially fabricated in a larger scale SLA system with a broad beam and low resolution. Microcomponents may be inserted. Subsequently paths between the microcomponents—for example, for fluid flow—may be generated using a high-resolution super IH system. Finally, the ACD packaging may be completed in a lower resolution SLA machine. Each of these processes requires the manual loading of the components during their various production stages. Thus, at present the process is very labor intensive. Automation of this process will be necessary before these components can be put into production. While automation will not be a trivial task, it will certainly not be an impossible one.

Surgical Devices

Professor Ikuta's laboratory is also heavily involved in minimally invasive surgery. Many of the systems that are designed and fabricated by his team draw upon the experiences gained in developing microsystems. He demonstrated a number of very impressive devices made by his students. These devices had small actuators and sensors that were fabricated by Professor Ikuta's laboratory. These small actuators were an excellent demonstration of the application of microfabricated technology. All of the devices were manually controlled, but had extremely good motion with excellent accuracy. According to Professor Ikuta, these devices could easily be navigated through very small paths (e.g., blood vessels) for very long distances in the body. As an example of the size of these remote micromanipulation devices, a typical unit is 3 mm in diameter and is controlled by five cables running inside the 3 mm tube. His students machine all components for the device on conventional machines. The components are typically made of stainless steel.

Micromachine Laboratory for Children

Professor Ikuta has also designed and implemented a microsystems interactive experience for children at the National Museum of Emerging Science and Innovation. The experience targets microsystems that are big enough for children to see under a standard microscope. These systems are nominally 100 to 1000 times larger than the minimum size that can be produced by the super IH process. The set-up at the museum is also capable of downloading new designs from outside (e.g., Professor Ikuta's lab) such that it can generate new object shapes for the children to experience. The web site for the museum is http://www.miraikan.jst.go.jp. The museum also hosts a microart contest for children to use the "art-to-part" process. This was funded by a grant from the Japanese government.

SUMMARY AND CONCLUSIONS

Professor Ikuta's group is performing very impressive work in both the process development and the design areas of micromanufacturing. While many of his systems are functional, they appear to be more research and development models rather than production models. This is consistent with the lab's objectives of developing new microsystems. Professor Ikuta indicates that in the near future his team will be working with other industrial groups in an effort to commercialize his work. Clearly, they are an innovative, well-funded and successful team.

REFERENCES

Koji Ikuta, Biochemical Micro System Engineering Laboratory, Department of Micro System Engineering, School of Engineering, Nagoya University.
http://biomicro.ikuta.mech.nagoya-u.ac.jp/~ikuta/index.html (Accessed October 20, 2005).

Miraikan (National Museum of Emerging Science and Innovation).
http://www.miraikan.jst.go.jp (Accessed October 20, 2005).

Summary of Research Project Results under the JSPS FY2000 Research for the Future Program, http://www.jsps.go.jp/
j-rftf/saishu/037-3_e.html (Accessed October 20, 2005).

Site:	**National Cheng Kung University** **No. 1, Ta-Hsueh Road** **Tainan, Taiwan 701**
Date Visited:	December 13, 2004
WTEC Attendees:	D. Bourell (Report author), K. Ehmann, K. Rajurkar, K. Cooper, G. Hane
Hosts:	Chen Kuei Chung, PhD, Assistant Professor, Department of Mechanical Engineering, Tel: +88-66-275-7575 Ext. 62111, Fax: +88-66-235-2973, Email: ckchung@mail.ncku.edu.tw
	Professor Chi Gau, PhD, Institute of Aeronautics and Astronautics, Tel: +88-66-275-7575 Ext. 63684, Fax: +88-66-238-9904, Email: gauc@mail.ncku.edu.tw
	Yung-Chun Lee, PhD, Associate Professor, Department of Mechanical Engineering, Tel: +88-66-275-7575 Ext. 62177, Fax: +88-66-235-2973, Email: yunglee@mail.ncku.edu.tw
	Jehnming Lin, PhD, Associate Professor, Department of Mechanical Engineering, Tel: +88-66-275-7575 Ext. 62183, Fax: +88-66-235-2973, Email: Linjem@mail.ncku.edu.tw
	Professor Kwang-Lung Lin, Director, Department Materials Science and Engineering, Tel: +88-66-275-7575 Ext. 31390, Fax: +88-66-208-0103, Email: matkllin@mail.ncku.edu.tw, Web: http://www.ncku.edu.tw/~nckumems
	Dar-Bin Shieh, DDS, DMSc, Assistant Professor, Tel: +88-06-235-3535 Ext. 5376, Fax: +88-06-2766626, Email: dshieh@mail.ncku.edu.tw
	Professor J.-J. Junz Wang, PhD, Department of Mechanical Engineering, Tel: +88-66-275-7575 Ext. 62189, Fax: +88-66-236-7231, Email: jjwang@mail.ncku.edu.tw

Professor P.D. Lin, PhD, Department of Mechanical
Engineering, Tel: +88-66-275-7575 Ext. 62170,
Fax: +88-66-235-2973,
Email: pdlin@mail.ncku.edu.tw

BACKGROUND

National Cheng Kung University is located in Tainan, Taiwan, and has nine campuses. We visited the Tzu-Chiang Campus, established in 1983. It houses engineering and technology development. There are about 12,000 students on all campuses, about 4,000 of whom are graduate students. The Center for Micro/Nano Technology Research, directed by Dr. K-L. Lin, was officially founded in 2001, although activities within a microelectromechanical systems (MEMS) center were initiated in 1997. Funding for the Center comes from The National Science Council, Ministry of Education and the University. The level of activity is about $3.8 million per year. There are seven permanent staff members in the Center. Seventy percent of all funding is used to purchase equipment. The mission involves outreach to industry, applications and research, focused on access to a large central facility of equipment. Over 120 faculty members and 400 graduate students have used the facility since its founding. This represents both internal and external users. A user fee structure is in place, but fees are very low. A non-exhaustive list of equipment includes a clean room facility, sputter, etch, four-point resistivity, ellipsometer, mask aligner, optical microscopy, spin coating, two atomic force microscopes, nanoindenter, two scanning electron microscopes (SEMs) (one field emission), nanoimprinter, molecular beam epitaxy, micro Raman, and two tunneling electron microscopes (TEM) (one high-resolution transmission electron microscope (HRTEM)).

RESEARCH AND DEVELOPMENT ACTIVITIES

Grinding, Micromachining (J.-J. Wang)

A variety of projects were described briefly. The first involved wafer sawing capability and glass cutting for the optical industry. A diamond cutter was developed with a 3 mm diameter and 2 μm corner radius. This is used in an industrially partnered project to cut glass. Work is underway to understand machine-tool system dynamics both at the macro and micro level. Issues include chatter, damping and stability. Another project underway is the creation and micromilling tools of varying geometry with nominally 100 μm shanks and 50 nm cutting edge radii. The tools have metal shanks and are diamond-coated. Micro electro-chemical machining (μECM) and electro-discharge machining (EDM) are being developed and will be used for construction of the micromilling tools. Some theoretical models of micro end milling are being pursued. The key feature different from macromilling is that the cutting edge radius is large relative to the depth of cut which increases the tendency for plowing.

Laser Precision Engineering Laboratory (J. Lin)

Projects involve formation of nano/micro-sized powder and micromanufacturing. Laser-induced pluming for nanoparticle formation involves dissociation of a target material under the energy exposure of the laser followed by condensation above the target into nano-sized particulate. This project is being funded by both the U.S. Air Force and the National Science Council. Lasers are also used to nanoimprint on metallic substrates. Here stainless steel is coated with a thin layer of nickel. The laser beam traverses through a glass plate and nanoimprints onto the nickel surface. This project is not yet successful. Lasers are also used to cut optical glass. The energy dose makes a 300 μm wide notch in the glass to serve as a crack guide. An interesting observation was the formation of a chip under certain operating parameters. The origin is a complex thermally induced stress state below the surface of the glass. A microparticle generator was described. A carbon dioxide (CO_2) laser ablates a substrate (steel) to produce a fine, 10–20 μm particle mass suspended in the atmosphere. An orthogonal flow of inert gas "blows" the particulate out of the chamber for general use or gathering. In a separate process, work is underway to develop a directed metal laser rapid prototyping system using a 500 μm diameter laser beam to scan a surface and melt an externally directed metal powder stream. A laser cleaning project finds application in the silicon (Si) wafer industry. The laser generates an elastic stress wave that is effective in mechanically removing surface particles. The last project involved microforming of stainless steel tubes. The tubes, 8 mm in diameter and 500 μm thick, were butt-welded using the laser heat source. Finite element analysis for the thermal aspects of the welding has been performed to understand the effects of heat flow in this microjoining process.

Fabrication of Aspheric Microlenses (Y-C. Lee)

Aspheric microlenses have been formed using excimer laser micromachining. The approach is to use a laser-mask assembly to selectively expose the surface of polycarbonate. Selective etching occurs to create the aspheric lens used in optics connector applications. Of particular interest was the design of the mask, which included not only its 2D opening geometry, but also its translation and rotation during the process. The laser was a KrF excimer laser with 248 μm wavelength, operated in pulse mode with 300 mJ/pulse and an energy density of about 2 J/cm^2. The microlens dimensions were on the order of 600 μm diameter and 60–100 μm deep. Accuracy is ±1 μm and the roughness is about 10 nm. It takes about five minutes to create a microlens. The mask design included a concave triangular shape which was rotated and a four-blade fan shape that was planetary rotated. A model was developed that predicted the machining depth and profile based on the geometry of the mask opening and its motion. The comparison of this model to actual experiments was excellent. The optical characteristics of the microlenses were also assessed. The polycarbonate microlens may be used as a pattern for microembossing. First, the lens is coated with 1–2 μm of nickel by electroforming. After polymer removal and backing of the nickel, this hot embossing tool is then used to press the microlens shape into thin sheets of polycarbonate. Finally, analysis and modeling is underway to generate a three-dimensional lens with non-axial symmetry. The working example is an ellipsoidal lens for laser diode coupling applications. Finite element analysis has been completed for creation of the surface height/depth profile, for which surface control to 1 μm was obtained. Experiments will begin shortly.

Nanobiotechnology at NCKU (D-B. Shieh)

Nanoparticle and nanorods are used for biological applications. The nanomaterials are synthesized on-site using wet chemical methods. Projects include using gold particles to image cancer cells by harmonic optical tomography. Another approach is use of iron oxide nanorods (both Fe_2O_3 and Fe_3O_4) for imaging cell components via energy dispersive X-ray (EDX). A treatment approach for cancer was development of nanoparticles consisting of a 6–7 nm iron core coated with gold to produce particles 9–22 nm in diameter. These particles are magnetized and were shown to target and kill cancer cells preferentially.

Micro/Nanodevice Fluidics (C. Gao)

There are a series of projects associated with fluid flow and channel creation for applications in MEMS, microsensors, microactuators, microtubes and advanced CPU chip cooling. Channels are made in an optically clear material using lithographic techniques. The channels are 4 mm by 500 μm and 80 μm deep. Bonding with plate results in closed channels with a bond strength of about 250 psi. Flow characteristics of air and water have been assessed. The heat convection coefficient increases with reduction in channel size, and values as high as 304 W/m^2K have been measured. Other projects include characterization of flow through 10–50 μm micronozzles and nanosensor fabrication. For the latter, carbon nanotubes are grown on a device at 260–450°C. The electrical resistivity is very linear with temperature, allowing the material to function as a nano temperature sensor.

Case Studies of Micro/Nanotechnology (C-K. Chung)

The first case study was monolithic MEMS-based lithographic micro-structures for thermal inkjet (HP) and bubble-jet (Canon) printhead applications. The active material was thermal bubble generated by tantalum-aluminum (Ta-Al) heater. The second project was creation of an ionic polymer-metal composite (IPMC) actuator. The actuator finds potential application in wave guides, biomimetic sensors and artificial muscle. Silver nanoparticles are embedded in Nafion by dissolution and casting. The Nafion is embossed before being electroless plated with silver and electroformed with nickel. The resulting IPMC will bend significantly under applied voltage. For example, a 3 V driving voltage can result in 6.8 mm displacement, 0.222 grams force and more than a 90° bend. The third case study was nanostructured tantalum silicon nitride (Ta-Si-N) films by magnetron sputtering. Two-hundred eighty nm of silicon monoxide (SiO) was generated on a silicon substrate followed by about 900 nm of TaSiN. The hardness was about 15 GPa, and the reported grain size was about 1 nm.

SUMMARY AND CONCLUSIONS

There is a significant and varied amount of micromanufacturing work at the National Cheng Kung University Center for Micro-Nano Technology. The equipment is impressive and most is commercially available. Projects include micromachining, laser precision engineering, laser micromachining, nanobiotechnology, microfluidics and micro-nanotechnology applications. A number of projects have a MEMS flavor, not surprising considering the history of the Center. Since the Center is only three years old, a number of projects are not yet completed but show exciting potential.

REFERENCES

Micro-Nano Technology Research Center, Research/Equipment Description. Brochure (n.d.) [in Chinese].
National Cheng Kung University, Description of Equipment. Pamphlet (n.d.) [in Chinese].
National Cheng Kung University. http://www.ncku.edu.tw/english/ (Acessed October 20, 2005).
Summary of Professor J.- J. Wang's Research Areas. Handout (n.d.).

Site: **National Science Council**
                       20th Floor, Science & Tech Building
                       106, Ho-Ping E. RD., Sec. 2
                       Taipei, 106 Taiwan

Date Visited: December 15, 2004

WTEC Attendees: K. Rajurkar (Report author), D. Bourell, K. Ehmann,
                       G. Hane

Hosts: C.K. Lee, PhD, Director General, Department of En-
                       gineering & Applied Sciences,
                       Tel: +88-62-2737-7524, Fax: +88-62-2737-7673,
                       Email: cklee@nsc.gov.tw
                       Jennifer Hu, Program Director, Department of Interna-
                       tional Cooperation, Tel: +88-62-2737-7560,
                       Fax: +88-62-2737-7607, Email: jenhu@nsc.gov.tw
                       Wei-Chung Wang, Director General, PhD, Depart-
                       ment of International Cooperation, Professor, Na-
                       tional Tsing Hua University,
                       Tel: +88-62-2737-7558,
                       Fax: +88-62-2737-7607,
                       Email: wcwang@nsc.gov.tw

BACKGROUND

The National Science Council (NSC) of the executive branch of Taiwan (called Executive Yuan) was established in 1959. It is the highest government agency responsible to promote the development of science and technology in Taiwan. A Minister, along with three Deputy Ministers, heads the council. The NSC consists of 12 departments/offices and five affiliated organizations. Starting in 1983, the council has established national laboratories and research centers including the National Synchrotron Radiation Research Center, National Laboratory of Animal Breeding and Research Center, National Center for High Performance Computing, National Center for Research on Earthquake Engineering, National Nano Device Laboratory, and the National Space Program Office. The six centers were reorganized in 2003 to improve flexibility and service in two independent nonprofit institutions: the National Synchrotron Radiation Center (NSRC) and the National Laboratory Institute (NLI).

DEPARTMENT OF ENGINEERING AND APPLIED SCIENCE

Dr. Lee briefly introduced the state of the basic and applied research funded by the Department of Engineering and Science (DES). The DES of NSC places equal emphasis on basic and applied research, and it also promotes diversified studies for pursuing theoretical and pioneering research while meeting industrial development needs. The DES also encourages personnel development, publication of academic papers, patent applications, technology transfers, and industrial technology upgrades. Besides academic research programs, the DES is also responsible for national science and technology programs, industry-university cooperative research projects, programs to upgrade industrial technology and enhance human resources, and specialized/forefront/priority research projects. The total budget (annually) for NSC is about $700 million and for DES it is about $150 million. About 81% of the DES budget is allocated for academic research programs, 10% for industry-university cooperative programs, and about 8% for national science and technology programs. The specifics of programs are given below.

Academic Research Projects

Although the NSC actively promotes multi-year and integrated projects, individual performance improvement (PI) projects are also equally encouraged. In 2003, about 9,000 proposals were received and 5,000 were funded. Usually on average, each proposal received about $175,000–$350,000 for one year. Under a special program for starting faculty the funding range is about $65,000 to $320,000. The overhead is only 8% (maximum). The review mechanism involves a preliminary review by a panel (established for three years), which seeks reviews from three peers in the field. The panel then, after considering these reviews, makes appropriate recommendations.

National Science and Technology Programs

The purpose of these programs is to improve Taiwan's competitive advantages and address major socioeconomic issues. Currently the DES is funding programs on telecommunication and system on a chip. In these two projects several other ministries (and related contributions of funds) are involved. Annual funding for such projects may vary from $4–8 million for several years.

Industry-University Cooperative Research Projects

The purpose of this program is to combine industrial needs with academic state-of-the-art research. In 2003, 22 such projects were started with about $3 million from the NSC and $1 million from participating industries.

Program to Upgrade Industrial Technology and Enhance Human Resources

This program aims at the professional development of students and faculty and also at meeting the needs of small- and medium-sized enterprises. About 875 such projects were funded in 2003 with an average funding of $600,000 (NSC:industry about 2:1).

Specialized, Forefront, and Priority Research Projects

The purpose of this program is to enhance the overall quality of academic research and resulting publications. Several research projects were funded in 2002-2003. These include a web technology project, image display project, a fuel cell project and micro- and nanoelectromechanical systems project.

The discussion centered on the above programs. The visit lasted about 75 minutes. No specific programs in micromanufacturing are in place, and currently there is no plan to initiate a related special initiative. Similarly, no special programs for female researchers are in place right now.

SUMMARY AND CONCLUSIONS

The NSC is a Federal government funding agency and operates similarly to the U.S. National Science Foundation (NSF). It supports research and human resource development activities in universities and industry. Besides academic research programs in targeted areas, it also supports industry-university cooperative and priority area projects. Currently at the NSC there is no plan to treat micromanufacturing as apriority area.

REFERENCE

NSC, Department of Engineering and Applied Science. Brochure (2004).

Site: **National Taiwan University**
                         Mechanical Engineering Dept.
                         No. 1 Roosevelt Rd. Sec 4
                         Taipei, Taiwan 106
                         Tel: +88-6-2-2362-1522
                         Fax: +88-6-2-2363-1755

Date Visited: December 15, 2004

WTEC Attendees: K. Rajurkar (Report author) D. Bourell, K. Ehmann,
                         G. Hane

Hosts: Professor Y.S. Liao, PhD, Department of Mechanical
                         Engineering, Tel: +88-62-2363-0231 Ext. 2411,
                         Email: liaoys@ntu.edu.tw
                         Professor Shuo-Hung Chang, PhD, Department of
                         Mechanical Engineering, Director, NEMS Re-
                         search Center, Tel: +88-62-2363-3863,
                         Fax: +88-62-2363-1755,
                         Email: shchang@ccms.ntu.edu.tw
                         Dr. K.C. Fan, Professor, Department of Mechanical
                         Engineering, Tel: +88-62-2362-0032,
                         Fax: +88-62-2364-1186,
                         Email: fan@ccms.ntu.edu.tw

BACKGROUND

The Micro-Nano Technology Research Center was established in 1994 for cultivating talent and basic technology in microelectromechanical systems (MEMS) to promote the country's industrial development. The center is located at the Institute of Applied Mechanics and has a total floor space of 300 m². The center has a yellow room, analysis room, etching room, furnace room, facility room, and an administrative room. The center serves as a core facility for multiple users. All users are trained on industrial safety and environmental protection. The center plays a role in system integration and basic technology research. The center helps in vertical work-sharing, developing professional technologies and carrying out lateral integration through various key industrial development plans, including in the information, communication, semiconductor, automation and biomedicine fields. Within a short period of two years, the center has succeeded in bringing in many universities and industrial companies together to conduct research into microelectromechanical system and also obtained three worldwide patents.

The center also plans many educational training courses for professionals in academics and industry. These include credit courses, basic technology training courses, industrial safety and sanitary training courses and factory training courses. Currently, more than 500 students are enrolled in professional training courses, more than 200 students are certified for general operation of the center's laboratory, and 120 have been qualified to operate all equipment and instrumentation at the center facility.

The center has established academy-industry links for the development of the microelectromechanical industry. Besides on-site technical training (such as at Hsin-Chu Enterprise and Taiwan Si-Wei Electronic Company), the center develops technologies and seeks patents rights together with industries. Some recent such efforts include a back-light baffle and polarized transformer of liquid crystal display (LCD) monitor, surface acoustic wave (SAW) device, temperature and pressure sensors in a mold and an integrated microwave passive device.

Dr. Liao briefly summarized the center emphasis, budget and human resources. The center, established in 1994, has 21 PhD students (part-time), 10 graduate students and several undergraduate students. The center research activities are classified as traditional machining and non-traditional machining.

A. Traditional machining projects include:

1. Machining of difficult-to-cut materials
2. High-speed milling with an emphasis on dry machining
3. Chatter control
4. Diamond cutting
5. Chemical mechanical polishing (CMP)

B. Nontraditional machining projects:

1. Electro-discharge machining (EDM)
2. Rapid prototyping and tooling
3. Laser machining
4. Hybrid machining processes

The center has an annual budget of $0.3 million (50% from NSC and 50% industry) to conduct the above-mentioned research activities and is mainly used for equipment and student support.

RESEARCH ACTIVITIES AND DISCUSSION

Dr. Y.S. Liao presented the details and application examples of a recently developed multifunctional four-axis high-precision micro-CNC machining center (Figure 4.15a). The objective of this project (50% funding from Taiwan National Science Council and 50% Taiwan industry) was to develop a micromachining center capable of performing some standard machining operations to make micro/meso-size 3D complex molds and dies. His research team designed and fabricated the whole system (except the sliding base). The cost of the total system, developed in two years, is $90,000 including the purchased sliding base, which was for $50,000. The machining accuracy of the system is 1 μm. The machining system is capable of performing four functions.

Micromilling

The system is capable of performing low-speed (0–3,000 rpm), middle-speed (1,000–30,000 rpm), and high-speed (2,000–80,000 rpm) milling operations. A micromilling cutter of 0.2 mm diameter is used. The depth of cut varies from 0.1 mm to 0.3 mm with a feed rate of 20 mm/min.

Micro EDM

Microelectrodes are made using wire electro-discharge grinding unit on the system to avoid the problems of alignment. Electrodes as small as 8 μm and high aspect ratio can be developed.

Microwire EDM

A 0.02 mm brass wire is used. Strategies for microwire tension control and wire vibration suppression have been developed and implemented on the unit. A radio control (RC) power generator with a required servo control is used. A novel mechanism designed by the research team makes it possible to perform vertical, horizontal and slanted cutting.

Micro Online Measuring Technology

The technology of measuring the flatness, circularity, concentricity, and other characteristics by microelectrode-produced wire electro-discharge grinding (WEDG) unit has been developed and implemented. Thus, the dimensions and geometric tolerances of the workpiece can be obtained during the process. Additionally, the complex profile can be measured with programming control.

Application Examples

Examples shown (slide as well as actual parts/products) include outer gears (0.6–1.8 mm diameters) without any taper on teeth, micropinion (0.324 mm), and complicated parts, such as the Chinese pagoda (1.25 mm × 1.25 mm × 1.75 mm) shown in Figure 4.15b.

Besides a multifunctional micromachining system, Dr. Liao also briefly described their electrochemical discharge machining (ECDM) project.

Currently, the equipment is at the Industrial Technology Research Institute (ITRI). It is a hybrid process which combines electro-erosion and localized anodic dissolution. An application example of machining 4" wafer for electrophoresis bio-chip (holes < 100 μm) was mentioned. Application examples of excimer laser machining of glass for biosensors (10 μm) and microarray mold on glass were also briefly described.

During the subsequent discussion, he mentioned that they primarily have been working with brass, copper, and stainless steel as work materials. The surface finish resulting from micromilling can be increased with higher rpm. He would also like to replace the WEDG process for making microtools. Currently, the process efficiency is low (range: 5–10%). It was mentioned that Japanese researchers have been using conductive diamond tool with no tool wear.

Figure C.26. Prototype micro-CMM.

We had a very quick (15 minutes) visit to Dr. Fan's metrology laboratory. Dr. Fan showed and described his instrumentation system (Figure C.26). He, along with his collaborators from Hefei University (China), has developed a high-precision micro coordinate measuring machine (CMM) for non-contact 3D measurement in mesoscale system. The measuring range is about 25 × 25 × 10 mm, and the resolution is about 10 nm. Some new design concepts, such as the arch-bridge and the co-planer stage, have been introduced to enhance machine accuracy. The coarse and fine motions of each axis are achieved by the piezo linear motor. A linear diffraction grating interferometer has been developed and is being used for the feedback of linear motion.

SUMMARY AND CONCLUSIONS

Although this site visit was one of the shortest, the research and developed equipment/instrumentation are probably the closest to the scope of this WTEC study. Specifically micro-traditional and micro-nontraditional machining processes research is being conducted on many issues. A multifunctional micromachine developed with limited funds in two years is being used to produce complex 3D micro/mesoparts. A μCMM for non-contact 3D measurement has been developed and tested.

REFERENCES

Fan, K. C., Y. T. Fei, X. F. Yu. 2004. A micro-CMM for non-contact 3D measurement in meso scale. From Proceedings of ICMT2004, November 8-12, Hanoi, Vietnam.

Site: **Olympus Corporate R&D Center**

Date Visited: December 10, 2004

WTEC Attendees: M. Madou (Report author), D. Bourell, K. Cooper,
                             T. Hodgson, G. Hane

Hosts: Shigeya Chimura, Manager, Equipment Technology
                             Department
                             Masahiro Katashiro, Deputy General Manager,
                             MEMS Business Development,
                             Tel: +81-426-91-7261, Fax: +81-426-91-7509,
                             Hiroshi Miyajima, Manager MEMS Business Devel-
                             opment
                             Haruo Ogawa, Division Manager, MEMS Technology
                             Division, Email: h_ogawa@ot.olympus.co.jp

BACKGROUND

Olympus, founded in 1919, and currently with over 27,000 employees, is perhaps the third-largest microelectromechanical systems (MEMS) company in the world (some other very large MEMS efforts are at Samsung in Korea and Bosch in Germany). Sales in the fiscal year ending March 31, 2004, were $5.76 billion. Olympus is working on MEMS applications as well as on tabletop manufacturing (with an emphasis on assembly). Their products range from imaging (digital and film) to medical (endoscopes) to life sciences (microscopes, blood analyzers and genomics) to industrial (microscopes, printers). According to Olympus' Haruo Ogawa, MEMS may help rebuild Japan's power as a manufacturing nation.

RESEARCH AND DEVELOPMENT ACTIVITIES

Olympus was part of the large Desktop Factory (DTF) Consortium, a government-sponsored project (1991–2000), which is still kept alive through less coordinated efforts. Government funding after 2000 shifted back to MEMS.

Olympus is the market leader in endoscopy, the examination of stomachs and colons for suspected cancerous tissue and the treatment or removal of the tissue. The endoscopy technique is being adapted to machine repair, and a smart capsule endoscope is under development. The latter will be swallowed by the patient and make its way to the designated organ, identify the troublesome tissue, and either sever it or deliver a drug to eliminate it. Optical microscanners will be attached to the tips of endoscopes to identify diseased tissue at the molecular and cellular level. Olympus projects this to be on the market by the end of 2008.

Nanobiotech is one of the focus areas of Olympus. A rapidly aging population is the driving force for interest in this, says President Tsuyoshi Kikukawa. That market is estimated at \$18 billion by the end of this decade. In the life science area the company is working on deoxyribose nucleic acid (DNA) (although the development of drop delivery systems for DNA chips was stopped) and protein arrays. Also, some work on sample preparation and free-flow electrophoresis is being carried out.

Internal MEMS efforts at Olympus include optical scanners, atomic force microscope (AFM) cantilevers and microresonators. The MEMS scanners developed by Olympus are magnetic optical scanners for horizontal laser scanning in commercial confocal laser scanning microscopes. Polyimide torsional hinges for an optical scanner have been developed, but the stiffness and Q-factor were found to be insufficient to realize the required resonant frequency and scan angle. A single crystal hinge on the other hand did do the job, and a confocal microscope based on this scanner has since been commercialized (OLS1200).

The Olympus MEMS foundry has a fabrication facility in Nagano (four inch line and expansion plans for six and eight inches) with over a hundred experienced engineers and a design and marketing team at the Tokyo headquarters at Shinjuku-ku. Services started in February, 2002. Today many projects involve either single process or optical MEMS devices. The foundry already has built inkjet printer heads for a customer with hundreds of tiny heads crammed onto the surface of each head. For other companies they have been building MEMS light switches.

At Olympus the microfactory concept is seen as a means to save energy, hand labor, materials and cost. Some lenses handled today at Olympus are 300 μm in diameter, and, at this size, even skilled workers have trouble assembling with high efficiency. Olympus envisions an island type of modality where each worker controls a number of manufacturing/assembly steps on one workstation rather than in a production line. Desktop equipment enables a small local environment, easy to upkeep, where products can be assembled with a joystick without needing special skills. Olympus is building a microfactory that will build and assemble MEMS. For now the key focus is on microassembly. One microassembler is already in use to assemble lens modules for cellular phones and parts for medical endoscopes. Olympus sees the following benefits of a microfactory:

- Improvement in assembly accuracy

- Improvement in productivity per square meter (small footprint)

- Energy and materials savings

- Easier to keep clean

SUMMARY AND CONCLUSIONS

A number of the MEMS foundries in the U.S. went broke in the last decade. Olympus believes they can make foundry services profitable. They estimate the current MEMS market at $4.6 billion and that it will be $460 billion in ten years. Barriers to downscaling are seen in unpredictable scaling effects, limited number of characterization tools at the micro dimensions and the change in materials properties at the microscale.

REFERENCES

Miyajima, H. 2004. Development of a MEMS electromagnetic optical scanner for a commercial laser scanning microscope. *JM³* 3:2, 348-357.

Site:	**RIKEN (The Institute of Physical & Chemical Research)** **Materials Fabrication Laboratory** **2-1, Hirosawa, Wako, Saitama, 351-0198, Japan** **http://www.mfl.ne.jp** **http://www.riken.go.jp**

Date Visited: December 10, 2004

WTEC Attendees: T. Hodgson (Report author), D. Bourell, K. Cooper, R. Horning, M. Madou, E. Uemura

Hosts:

Dr. Eng. Hitoshi Ohmori, Chief Scientist, Director of Materials Fabrication Laboratory,
Tel: +81-48-462-1111, Fax: +81-48-462-4637,
Email: ohmori@mfl.ne.jp

Dr. Weimin Lin, Research Scientist,
Tel: +81-48-462-1111, Ext. 8577,
Fax: +81-48-462-4657, Email: w-lin@riken.jp

Dr. Yutaka Watanabe, Research Scientist,
Tel: +81-48-467-8725, Ext. 8576,
Fax: +81-48-462-4657,
Email: watanabe@nano.gr.jp

Yasuchika Fukaya, Director, The NEXSYS Corp.
(RIKEN Venture), 1-7-13 Kaga Itabashi-ku Tokyo,
173-0003 Japan, Tel: +81-35-943-7966,
Fax: +81-35-943-7977,
Email: nexsys@nexsys.ne.jp

Nobunaga Akiyama, President, TZILLION CCEDSS
CO., LTD., Central Ohtemachi Bldg 803, 1-5-11
Uchikanda Chiyoda-ku Tokyo, 101-0047 Japan,
Tel: +81-33-292-0164, Fax: +81-33-292-0165,
Email: tzillion@tzillion.co.jp

Yoshiko Akiyama, TZILLION CCEDSS CO., LTD.,
Email: yoshiko@tzillion.co.jp

BACKGROUND

RIKEN is a research institute funded by the Japanese government. It is developing a strong outreach effort for commercialization of its research. The projects presented during the visit were mostly machining/grinding projects on macro-sized objects with nano-sized features/accuracy by the

Ohmori group. TZILLION represents products developed by RIKEN. There are about 120 professionals involved in the Ohmori lab.

RESEARCH AND DEVELOPMENT ACTIVITIES

Ohmori et al. developed a three- to six-axis computer numerical control (CNC) multi-functional machine for machining, lapping and polishing, grinding, metrology, and injection molding. One of their fundamental grinding processes, electrolytic in-process dressing (ELID), was developed by Dr. Ohmori. The grinding wheel has diamond grains in a cast iron bond. As the cast iron bond wears, the wheel is naturally redressed. This involves an electrochemical process. The redress probably involves forming iron oxide (rust). Applications range from lenses to silicon to ceramic parts to complex die steel molds. The ELID system has been added to a variety of more traditional machining operations.

The Ohmori group developed a four-axis grinder (with an additional two axes for six-axis control), and an on-machine measuring system. The grinder has 1 nm feed resolution. The cast iron-bonded diamond wheel is 5.0 cm in diameter, and rotates at 20,000 rpm. It grinds lenses, ceramics, and other materials using a water-based cutting fluid. There are limits to the size of the grinding target (a radius of about 30 µm). The smallest grinding tool used to date has a 0.8 mm radius.

The Ohmori group has also developed a neutron beam refractive lens. It has a Fresnel design and is used for cancer treatment applications. A finer mesh (smaller diamond pieces more closely spaced in contact with the part) for the grinding wheel gives a stronger finished part. In other words, putting a smoother surface on the part tends to make it stronger (a smoother part has fewer opportunities to generate cracks/failures under stress). The neighboring surface characteristics around a punched 10 µm hole are improved by a finer mesh of diamonds on the grinding wheel that produced the punch. Grinding with two wheels simultaneously limits deformation of the part and improves grinding speed by 30 times. They can make a square, hexagonal, or octagonal tool in ~10 minutes. In connection with using 5 nm diamond grains, they noted that the bonding material needs to be able to deform to increase efficiency of the grinding process. They also commented that they had developed a non-contact sensor. They have about 10 years experience in this area.

Figure C.27. Mock-up of diamond grinder with ELID (greatly enlarged).

They used microtools (end mill at 30,000 rpm) to cut 30 μm 'S' grooves on a plate. They made micromolds with 50 μm features.

They have a nanoprecision (screw) drive system using hydro-static oil pressure for support that achieves on the order of 2–20 nm control.

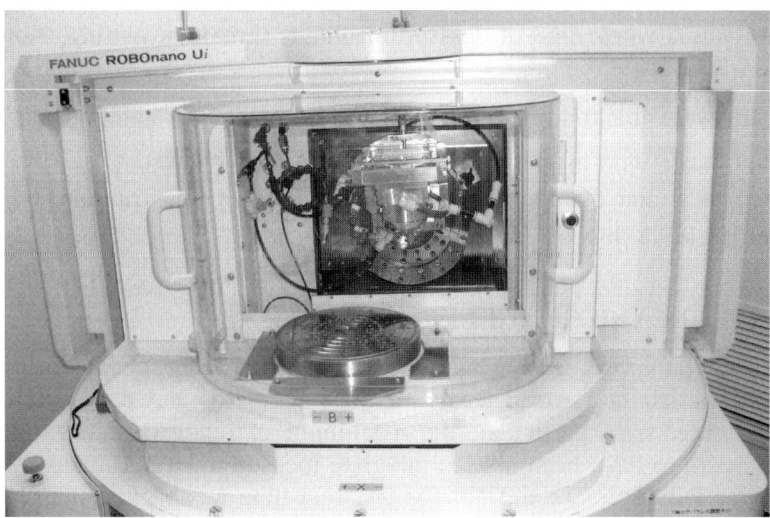

Figure C.28. Nanoprecision five-axis friction-free micro-machining system.

They developed a large ultraprecision machining system to make large mirrors. They have developed metrology to measure and then compensate the mirror after the fact. Their on-machine polishing of mirrors uses a magneto-rheological fluid conditioner and is very effective in getting roughness down to the 1 nm area.

They have built a nano-precision five-axis machining system with a 100,000 rpm air-turbine spindle. It combines machining and grinding. When we saw the actual machine, it was shielded as are most rotating tools. However, they commented that when the tool breaks, it simply drops to the bottom of the machine. The imparted ultra-high rotational speed of the tool apparently keeps it from moving in the lateral direction.

They have done considerable work in making microgrooves on surfaces. They showed profiles of (copper) surfaces with 1,000 nm, 500 nm, and 250 nm grooves. The quality of the surfaces deteriorated significantly as the feature size decreased. They commented that grain size became an issue with 100 nm grooves. The routing of the cutter was an issue in terms of keeping the cutting surface free of chips. They moved the cutter in rectangular patterns in order to scatter the chips.

They have developed simulation and modeling of injection molding using both commercial software and their own software. They were able to show empirical experiments that verified their analysis.

They had developed a desktop system for the manufacture of micro-tools. They were able to show that more precise (smooth) surfaces on the tools resulted in tools that were stronger (required more force to break). There was a whole line of desktop machines (grinding, cutting, molding, measuring).

They were able to grind 50 μm thicknesses on an 8 in. silicon wafer.

They were working on prosthesis finishing (titanium and stainless steel hips, knees, and other joints) in order to achieve low friction and anti-corrosion properties.

SUMMARY AND CONCLUSIONS

RIKEN is a very well (government)-supported laboratory with impressive output. It is a world leader in making micro- and nanofeatures on macro objects, in particular, the machining of highly accurate mirrors. They seem to have been able to integrate measurement into many of their systems. However, we did not see any real-time feedback systems. They are seriously trying to commercialize the machines that they develop through RIKEN "ventures." The world market for most of their machines may be in the dozens. However, the technologies they represent are enablers for producing tools for other mass-market products.

REFERENCES

Itoh, N., H. Ohmori, T. Kasai, T. Karaki-Doy, K. Horio. 1998. Characteristics of ELID surface grinding by fine abrasive metal-resin bonded wheel. *International Journal of the Japan Society for Precision Engineering* 32:4, 273-274.

Liu, C., H. Ohmori, W. Lin, T. Kasai, K. Horio. 1999. Development and characteristics of fixed abrasive polishing utilizing small grinding tools with ELID for aspheric optical surfaces. *Abrasive Technology, Current Development and Application* I SGE, 147-153.

Ohmori Laboratory. http://www.mfl.ne.jp (Last accessed October 20, 2005).

Ohmori, H., N. Ebizuka, S. Morita, Y.Yamagata. 2001. Ultraprecision micro-grinding of germanium immersion grating rlement for mid-infrared super dispersion spectrograph. *Annals of the CIRP* 50, 221-224.

Ohmori, H., K. Katahira, J. Nagata, M. Mizutani, J. Komotori. 2002. Improvement of corrosion resistance in metallic biomaterials using a new electrical grinding technique. *Annals of the CIRP* 51, 491-494.

Shindo, H., H. Ohmori, T. Kasai. 2000. Development of micro-profile ELID-grinding system for stamping tools. From *The 3rd International Symposium on Advances in Abrasive Technology*, October 30-Novemer 2, Hawaii.

Site: **Samsung Electro Mechanics Corp., R&D Center**

Date Visited: December 15, 2004

WTEC Attendees: T. Hodgson (Report author), B. Allen, J. Cao,
                    K. Cooper, T. Kurfess, M. Madou

Host(s):   Dr. Seog Moon Choi, PhD, Principal Engineer,
                        Tel: +82-31-210-3944,
                        Email: sms.choi@samsung.com
                        Kwang-Wook Bae, Chief of CTO Strategy Planning
                        Ghun Halm, PhD, Principal Engineer
                        Sang-Hyun Choi, Assistant Research Engineer

BACKGROUND

Samsung Electro-Mechanical Corporation, founded in 1973, manufactures chips and circuit boards, mobile communications components, computer components and peripherals, audio and video components, and general-purpose components. Sales in 2003 were $2.16 billion. They have eight plants in Asia. An excerpt from President and CEO Ho-Moon Kang's recent message gives some idea of their philosophy and corporate strategy: "… backed by a solid foundation of core digital technologies, world class management, and quality that is second to none, we are well on our way to leading your digital world." They are now in the top five and expect to be in the top three relatively shortly.

RESEARCH AND DEVELOPMENT ACTIVITIES

Our entire stay at Samsung took place in a corporate headquarters meeting room and in the corporate technology display room. We were shown a series of products: seven-color light-emitting diodes (LEDs), PC components and peripherals, a liquid crystal display (LCD) TV digital tuner, an inverter for LCD TV, a tuner for cell phones that receives satellite signals, DVD drive components, flexible polycarbonate (PC) boards for cell phones (SEMIBRID trade name), cell phone chip antennas, active and passive electronic components, multifunction actuators (for speakers & vibrating motors), surface acoustic wave (SAW) filters, small motors (cell phones), and chip signal splitters.

A micromotor for auto-folding a cell phone has a power consumption of 20 milliwatts and is embedded in a case hinge (20 mm in length) made of zinc (Zn) by die-casting.

Samsung has over 100 engineers in microelectromechanical systems (MEMS) R&D; however, they have not had success in MEMS applications but will probably pursue MEMS in future medical products. They are making capacitors smaller but with larger capacitance. They make components for Dell, Macintosh, HP, and other similar firms.

Materials

In their strategic plan, original material technology and system modules will be two foci of the R&D centers. However, we did not have a chance to hear more research topics in these two areas.

Business/Other

Samsung has a corporate strategy of outsourcing mature and labor-intensive products (e.g., classic TV components). They want to keep the high value-added products and rapid-turnover products in-house (some products have a three- to six-month product life, i.e., cell phones, which makes the automation extremely difficult). They clearly are being driven by their ability to develop new product technology and manufacturing technology. They are developing new R&D centers in other countries (Russia, India, Japan, and the U.S.). They are ISO 9000 certified, and are pursuing six-sigma quality.

The Corporate Research Institute was started in 1984. They now have 43 PhDs (13%), 142 graduate students (41%), and 157 undergraduates (46%), for a total of 342 personnel. Their R&D investment was 3.9% of their budget in 2001, 5.3% in 2002, 6.4% in 2003, and will level off at 8.5% next year. Of their PhD employees, they estimated that 30% had foreign educations. They make approximately 70% of their own components, and outsource 30%. Products slated to be outsourced include: PC mice, keyboards, small cooling fans, TV inverters, and conventional TVs.

The present breakdown on output is: semiconductor devices (3%), electronic devices (15%), optics (15%), display and network hardware (17%), printed circuit boards (19%), and digital modules (31%).

When asked about issues of green manufacturing, they noted that China was brown.

They indicated that their goal for the R&D center was to be considered a first class R&D center by 2007 and sited Nokia as their example. Specifically, they planned to be the world's best in two items by 2005, five items by 2007 and ten items by 2010. Their vision is not simply to just improve processes; rather, they must invent and innovate. They are concentrating on the technologies of optics, radio frequency (RF) application specific integrated circuits (ASICs), and materials.

SUMMARY AND CONCLUSIONS

We did not see any manufacturing operations or development laboratories. We were told nothing about their manufacturing or development processes. We were able to gain insights into corporate culture and business strategies. We also were able to understand their vision for how they plan to move into the future. They are outsourcing a number of their products and are focusing on high value-added/short-lifetime products.

The new multi-layer ceramic capacitor is 600 microns in length and 300 microns in width. It consists of 500 layers; each layer is about 1 micron in thickness.

REFERENCES

Samsung Electro-Mechanics Corporation. Brochure (March 2004).

Site: **Sankyo Seiki**
Ina Facility
6100, Uenohara, Ina-shi, Nagano, 396-8511, Japan
http://www.sankyoseiki.co.jp/e/index

Date Visited: December 7, 2004

WTEC Attendees: M. Madou (Report author), D. Bourell, T. Hodgson, K. Cooper, G. Hane

Hosts: Haruhiro Tsuneta, Senior Director of Engineering,
Tel: +81-265-78-5069, Fax: +81-265-78-5584,
Email: haruhiro.tsuneta@sankyoseiki.co.jp
Yuichi Okazaki, from National Institutes of Advanced
Industrial Science and Technology (AIST)

BACKGROUND

Sankyo Seiki is a major manufacturer of a wide variety of microproducts and robots. Microproducts range from music boxes to card readers. The main interest we had in visiting Sankyo Seiki was to learn more about their leading role in the Desktop Factory (DTF) project. The DTF project is a major Japanese effort to regain (revive) the manufacturing prowess of Japan, especially in Nagano prefecture (The Switzerland of Japan). The project started in the early 1990s when manufacturing jobs started to move to China. Sankyo Seiki is today selling the first DTFs.

RESEARCH AND DEVELOPMENT ACTIVITIES

The purpose of a DTF is to enable low-cost, very flexible manufacturing on a desktop. The general idea is that only electrical power is required and that direct input (DI), cleanroom environment (class 100 to 10), pressurized air, and other elements are provided within the confines of the machine. In other words, DTF does not root equipment to the ground. Since the clean environment is in the machine, it can be operated anywhere. This translates into very fast start-up of manufacturing anywhere. We were at first surprised that DTF proponents would envision saving jobs in Japan since shipping the factory would become so much easier. Their response is that DTFs will become increasingly automated and require very few human operators so that exporting these machines does not make sense since no labor costs are saved (free market approach). Another popular scenario sees desktop manufacturing becoming as popular as desktop publishing (socially responsible approach). In the latter case manufacturing becomes much less centralized, and many more people can get involved. In either case the impact on society will be dramatic.

The DTF machines developed at Sankyo Seiki have modular units that carry out different functions. Modules seen in operation include: oven baking, glue application, cleaning, assembly of two parts, direct input (DI) unit, small high-efficiency particulate air (HEPA) filter for atmosphere conditioning, drilling and ID turning. Some of these modules were shown to work together. The DI unit was the largest unit of a series of connected DTFs shown in Figure C.29.

The connection between the modules is with an automatic carrier vehicle (size of a hand). There are 38 patents that have been applied for (domestic and international) around this concept. The cleaner module was already available for outsiders by the end of September 2003. Eighteen enterprises have composed the DTF project and are doing research activities in the Nagano area. And some research organizations and universities, such as the National Institutes of Advanced Industrial Science and Technology (AIST), are participating in this activity.

Important components in the different modules are: particle monitors, robot hands (instead of arms), vacuum pumps, compressor, linear motors, direct drive motors, auto-guided pallets, and de-ionized (DI) water supply. By the end of next year integrated systems for assembly will be available. Machining systems will be a third-generation effort and come later. The DTF is connected to a PC environment and an Ethernet link.

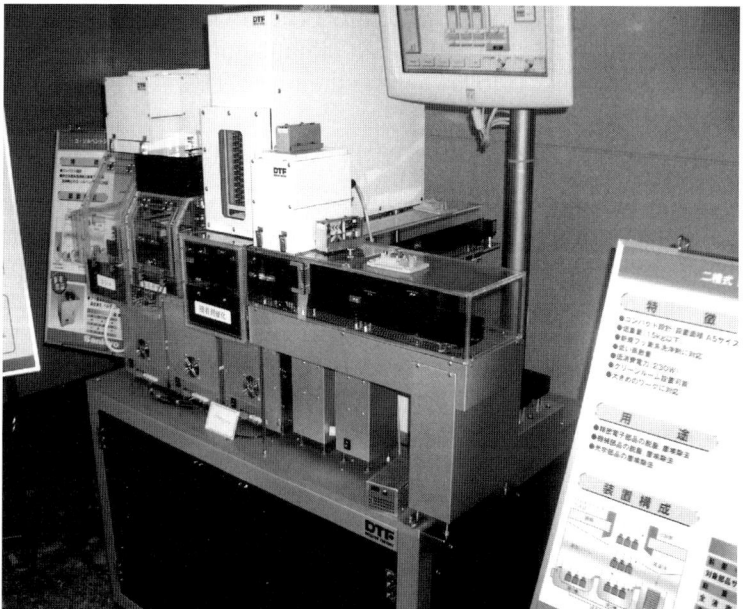

Figure C.29. A series of desktop factory units at Sankyo Seiki.

SUMMARY AND CONCLUSIONS

This is a very impressive effort by a company that has been involved in manufacturing precision components since 1946, very well-planned and executed. It is hard to see, though, how such a flexible and portable manufacturing system will keep manufacturing jobs in Nagano. The extension to micromanufacturing is not easy at this point, since parts are typically larger than 1 mm. Current positioning technology hampers extension into the microscale.

Site: **Sansyu Finetool Co., Ltd.**
                             4-1-55 Hieda-Cho
                             Takahama-City, Aichi, 444-1321, Japan
                             http://www.sansyu.co.jp

Date Visited: December 10, 2004

WTEC Attendees: K. Ehmann (Report author), B. Allen, J. Cao,
                    T. Kurfess, K. Rajurkar, E. Uemura

Host: Akiji Kamiya (Chairman), Tel: +81-566-53-1140,
                             Fax: +81-566-53-4956
                             Dr. Eng. Masayuki Suzuki (President),
                             Tel: +81-566-52-2511, Fax: +81-566-52-6540,
                             Email: Msuzuki@sansyu.co.jp
                             Ricardo Salas, Tel: +81-566-53-1781,
                             Fax: +81-566-52-6540,
                             Email: makuta.jp@sansyu.co.jp

Notes: Masayuki Suzuki became president of Sansyu Engi-
                             neering Service Ltd. from January 2005 (Tel: +81-
                             566-53-1407, Fax: +81-566-53-4956). Current
                             president of Sansyu Finetool Co. is Munetaka
                             Kamiya (Tel: +81-566-53-1781,
                             Fax: +81-566-52-6540).

BACKGROUND

Sansyu was established formally in 1954 as a manufacturer of nee-
dles/hands for wristwatches with gradual expansion to presswork for different watch components. Although today this is not the company's mainstream activity, Sansyu remains involved in the manufacture of precision
blanking dies and microstampings. About 35 years ago, the company began to transition to injection molding die production. Today, the focus is
on precision injection molding technologies. In the 1990s the company
started to expand to overseas facilities. Now it consists of eight facilities in
Japan, Hong Kong, Indonesia and the U.S. In Japan, it employs about 400
people while the total number of employees worldwide is about 800. The
breakdown is about 50/50 between female and male employees (30/70 in
Japan).

Sansyu's core expertise is in micromolded single and multiple materials
and insert-molded precision plastic components, micro metal stampings
and assemblies, injection molds, stamping dies and automation equipment

for microcomponents. In a number of these fields they can be considered technology leaders.

RESEARCH AND DEVELOPMENT ACTIVITIES

The group was greeted and briefed by President Dr. Suzuki and Mr. Salas (Japan Branch of Makuta Technics, Inc., one of the Sansyu companies headquartered in Columbus, Ohio). Mr. Salas gave a presentation about the company and its core technological competencies. Subsequently, the company's showroom, mold-making facility and the injection molding facility were toured. During lunch, the group had the opportunity to conduct a very substantive conversation with Chairman Mr. Kamiya, (the son of the company's founder). A synopsis of the main observations follows.

Injection Molding

Sansyu's principal technological strengths rest in the over-molding of dissimilar materials and in the use of multi-slide cored molding of very small-sized components with tight tolerances. They were the first in a number of instances in exceeding the existing technological limits of the day. Some examples of achievable component designs and features are shown in Figure 3.10. Figure 3.10a shows a polyoxymethylene (POM) encoder disk, previously manufactured by etching that has 80 μm slits injection molded with a tolerance of about 10 μm. Figure 3.10b shows a miniature precision gear made of the same material with a module 0.1, OD 2.6 ±0.03 mm with 24 teeth. Figure 3.10c shows an array of small holes in a polybutylene terephtalate (PBT) part while Figure 3.10d shows fins in a polycarbonate PC part, and Figure 3.10e shows one of their smallest components made of POM with a volume less than 1 mm^3.

Figure 3.11 depicts examples of parts that require a complex system of moving cores in the mold to facilitate automatic ejection. The components shown are made of PBT. Other noteworthy features include components with 20 μm holes, components with wall thicknesses on the order of 80 μm, and concentric feature tolerances less than 7 μm. Typical materials used are polymethylmethacrylate (PMMA), acrylonitrile butadiene styrene (ABS), polycarbonate (PC), thermo plastic elastomer (TPE), POM and other engineering plastics.

Quality control and metrology for the types of components shown above is generally performed by optical microscopy-based methods.

An interesting statement made was that commercially available software is rarely used in attempting to predict the performance of operations for parts like the ones shown above since it would predict that the operation was not feasible. The explanation given was that the models in the commercial software generally neglect boundary-layer effects that are not significant at the macroscale, yet they cannot be neglected and are dominant at the microscale.

The injection molding facility is a state-of-the-art facility which mainly consists of 18 and 30 ton injection molding machines that are fed with raw material from corridors that house the machine's hoppers to avoid contamination of the products and machines on the floor.

Precision Mold Machining

Sansyu uses unigraphics (UG) as their core computer-aided design/manufacturing (CAD/CAM) system. The design and manufacture of molds is automated to the largest possible extent. In spite of this, final manual fine-tuning is a necessity. The training of an apprentice for mold assembly takes about three months. Mold manufacturing relies on customary methods that are honed to their extremes by in-house know-how and expertise. For over-molding, conventionally very large molds were used. Sansyu has developed a system, in collaboration with Sumitomo, of double-shot molding in which the mold slides from one to another station (Sansyu's IP).

Many of the precision tools needed for mold-making are developed in-house. A typical example includes the manufacture of $\phi 20$ μm tungsten carbide micromilling cutters done manually by skilled operators on a suitably modified tool grinder with diamond tools. Machining conditions used with these tools include 1 μm depth of cut at 32,000 rpm, which is considered insufficient. Tool clamping is performed by shrink fits, and spindle runout is estimated to be about 2–3 μm. Tool geometry measurements rely on optical methods, viz., video cameras.

Other than the whole array of cutting operations used in the manufacture of molds, Sansyu also uses wire EDM with wire diameters as low as 20 μm. As stated above, they use conventional technologies and challenge their boundaries. Sansyu does not use microelectromechanical systems (MEMS) and lithography-based techniques for mold-making.

Discussion

The discussions with Chairman Kamiya have brought out a number of interesting points in regard to the direction of their industry and future requirements. These observations are:

1. The demand on the accuracy of microinjection molded components is steadily increasing. Typical tolerances expected by the customers today can be as low as ±3 μm. Demands already exist for ±1 μm tolerances. Today, trial and error is the only way to meet such demands. The trend toward further tightening of the tolerances is certain.

2. A push toward ever-smaller injection-molded parts is also an existing trend.

3. There is a need for small, single shot machines in the 500-1,000 kg range that would utilize small molds. The rationale being:

   a. Single cavity molds would assure better consistency than multi-cavity molds.

   b. Small single cavity molds are more practical (in a multi-cavity mold if one cavity fails the whole process needs to be stopped).

   c. Small machines occupy less space and are more efficient. They could be used in parallel.

4. Design engineers and process engineers work together on each new product; however, there are no specific design rules for microprecision parts.

5. Sansyu is currently working with two universities on R&D projects; however, they would like to see more universities having open-house activities so that they can establish more ties.

SUMMARY AND CONCLUSIONS

Sansyu is a leading manufacturer of microinjection molded components as well as a supplier of molds for such components. In addition, they have expertise in other precision micromanufacturing technologies, namely microblanking and assembly. A notable feature of their technological edge lies in their ability to exploit and push the limits of existing technologies to meet the technical requirements of their customers. They see an ever-growing demand for their products and competencies epitomized by ever-decreasing part sizes and part features with tightening tolerances.

REFERENCES

Sansyu Micro Precise Injection Molding. http://www.sansyu.com/index.htm (Accessed October 20, 2005).

Site: **Seiko Instruments Inc. (SII)**
                       563 Takatsuka-Shinden, Matsudo-shi,
                       Chiba 270-2222, Japan
                       http://www.sii.co.jp

Date Visited: December 7, 2004

WTEC Attendees: K. Ehmann (Report author), B. Allen, J. Cao, G. Hane,
                    T. Kurfess, K. Rajurkar

Hosts: Manabu Oumi, PhD, Senior Researcher, R&D Divi-
                       sion, Micro & Nano Technology Center,
                       Tel: +81-47-391-2131, Fax: +81-47-392-2026,
                       Email: manabu.oumi@sii.co.jp
                       Takashi Niwa, Senior Researcher, Process Develop-
                       ment Group, Micro & Nano Technology Center,
                       Tel: +81-47-392-7880, Fax: +81-47-392-2026,
                       Email: takashi.niwa@sii.co.jp
                       Osamu Matsuzawa, Process Development Group, Mi-
                       cro & Nano Technology Center,
                       Tel: +81-47-392-7880, Fax: +81-47-392-2026,
                       Email: osamu.matsuzawa@sii.co.jp
                       Toshimitsu Morooka, Senior Researcher, Process De-
                       velopment Group, Micro & Nano Technology Cen-
                       ter, Tel: +81-47-392-7880, Fax: +81-47-392-2026,
                       Email: toshimitsu.morooka@sii.co.jp

BACKGROUND

The Seiko Group consists of three independent companies: Seiko Corporation, SII and Seiko Epson established in 1881, 1937 and 1942 originally under the names K. Hattori & Co., Ltd., Daini Seikosha Co., Ltd.,
and Daiwa Kogyo, Ltd., respectively. These three independent companies
share the cumulative expertise gained in the watch business. Today major
products include consumer products (including watches, electronic dic-
tionaries), mechanical components (for watches, hard disks, thermal printers, and ultrasonic motors), electronic components (complementary metal
oxide semiconductor (CMOS) integrated circuits (ICs), quartz crystals,
liquid crystal display (LCD) modules, microbatteries, optical fiber connectors), semiconductor manufacturing equipment (analyzers, focused ion
beam (FIB) apparatus, photomasking repair equipment, probe microscopes), network devices (restaurant ordering systems, time authentication
services, large format inkjet printers), and others. SII specifically has an-

nual sales of ¥150,400 (216,400-consolidated) million and 4,300 (9,400-consolidated) employees and has a presence in Asia, Europe and America. The Process Development Group visited by the panel focuses on 3D microfabrication by developing microelectromechanical systems (MEMS) process technologies, bonding, and packaging.

RESEARCH AND DEVELOPMENT ACTIVITIES

The visit lasted three hours with the first two hours devoted to presentations by four senior research staff members, lead by Dr. Oumi of SII, and comprehensive discussions. The presentations were followed by an approximately one hour tour of the clean-room and FIB facilities.

The focus of the R&D group the panel has visited can be broadly classified into MEMS-related technologies, specifically micromechatronics and nanotechnology. Micromanufacturing, as defined in the context of the present study, e.g., watch, hard disk drive, printer manufacturing, is performed at a different location (Morioka Seiko) in the far north of Japan.

The subject matters discussed are now highlighted.

MEMS and Micromanufacturing (T. Niwa)

A comparative assessment of MEMS and traditional methods was given, emphasizing that MEMS methods are cheaper for the production of parts in small quantities. In this realm SII is striving to achieve 1 mm thick molds by deep reactive ion etching (DRIE) methods. The focus of the presentation was, however, on a new nanofabrication technology the group is developing for the manufacture of ϕ100 nm nano apertures (f100 nm). The process consists of two steps: 1) preparation of a tip with a 20–50 nm apex radius by etching, and 2) by puncturing aluminum (Al) film by the tip under pressure (as depicted in Figure C.30). Currently, the process is being performed manually with an approximately 50% yield. Exact measurements of the achieved dimensions and repeatability are not known yet, partly because exact measurements of the geometry are difficult. Tip geometry and form were confirmed by scanning electron microscope (SEM). The quality of the apertures is checked indirectly by evaluating their optical performance. The primary target applications areas are optical storage heads and near-field optical microscopy.

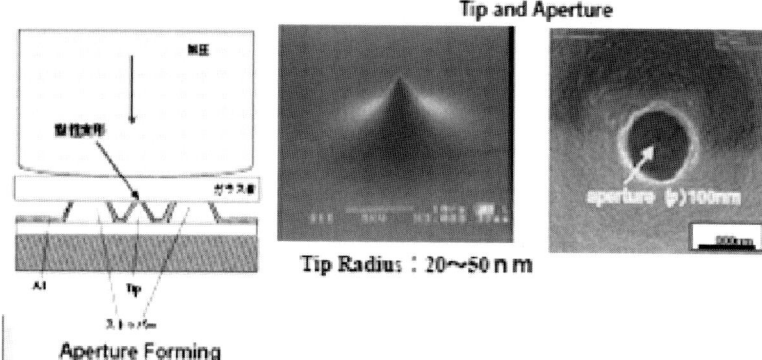

Figure C.30. Nano-aperture forming.

The second application discussed was the fabrication of an optical switch assembly. The focus was on the DRIE process used for the fabrication of the mirror and of the actuator. Most of the discussion was directed toward an understanding of the measurements performed for the verification of functional characteristics of the mirror including metrology and dynamic responses. The principal approach was the use of laser measurements for the verification of the angular orientation of the mirror and the measurement of optical intensity. Vibration testing was performed by measuring optical losses in response to impact. FEM analysis was applied in the design stages of the device.

Microcantilever (O. Matsuzawa)

Four different cantilever structure-based sensing devices were introduced (self-sensitive cantilever for displacement, Nano4probe for electrical resistivity, self-sensitive dual cantilever for multi-resolution displacement, and the thermal cantilever for temperature measurement). The self-sensitive cantilever which is commercialized and used in dynamic force microscope (DFM) and atomic force microscope (AFM) applications is described here in more detail. The major innovation is the replacement of the customary optical probe displacement measurement arrangement that requires a laser and associated optics with strain measurements at the base of the cantilever by a piezo-resistive element connected in a Wheatstone bridge configuration along with a reference resistance (see Figure C.31 below). The principal advantages are reduced overall size, the ability to be operated in the dark field, and ease of use and setup. Standard samples are used to perform calibration. The size of the cantilever is approximately 50 μm (w) × 120 μm (l) × 5 μm (t), with frequency response of 300 kHz.

The Nano4probe, fabricated by FIB, uses the four-point resistivity measurement method implemented by four narrowly pitched (200 nm) electrodes. The dual cantilever displacement probe uses a bimorph to switch between the large and small area scanning probes, while the thermal probe uses a heat pile-based idea to eliminate heat flow between the sensing tip and the thermocouple to obtain accurate measurements at the tip of the probe.

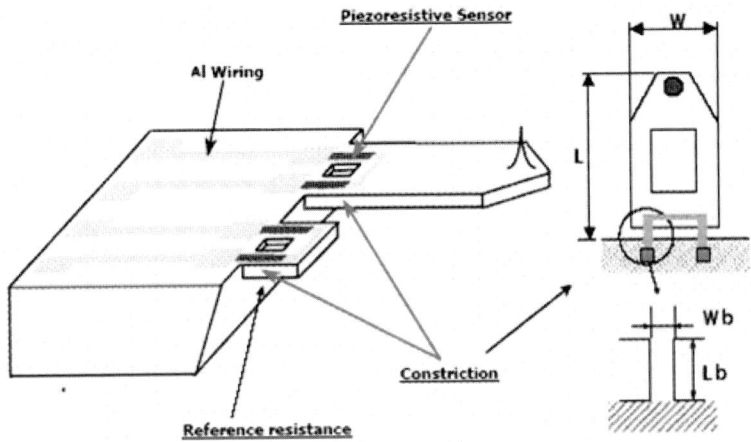

Figure C.31. Structure of self-sensitive cantilever.

Transition Edge Sensor (T. Morooka)

This is a high sensitivity X-ray detector that uses MEMS technology and superconductivity. The developed detector has a 10 eV resolution at 5.9 keV as compared to the conventional silicon (Si) lithium (Li) detectors with a 130 eV resolution. The design of the detector lends itself to mass production by anisotropic Si wet etching (front side only) in four inch wafers. Detector size is approximately 300×300 μm. Applications include microanalysis for the semiconductor industry and materials science as well as applications in astronomy and astrophysics.

Laboratory Tour

Class 1000 and 100 facilities were visited. There are about 50 total employees. The capabilities of the laboratory include: various etching systems, spin coating, general sputtering, reactive ion etching, Canon coating systems and contact type mask aligners (both sides), various optical microscopes, photo exposure microscopes, and SEM. The FIB facility's major capability is built around a SMI-2200 system. The newest FIB facility at Oyama (SMI model 3000) produced a demo wine glass about 2.75 μm in height and 100 nm in wall thickness that corresponds to the resolution of the machine.

SUMMARY AND CONCLUSIONS

The principal activities of the R&D group center around the development of microfabricated devices by MEMS methods. These activities include the conception of new devices, their analysis, prototype development and testing as well as the consideration of methods that would facilitate their mass production. A number of innovative devices were discussed with a substantial number of questions raised in the discussions directed towards metrology and functional verification issues. A tour of the clean room and FIB fabrication facilities has also been conducted.

REFERENCES

Seiko Instruments Incorporated. http://www.sii.co.jp/corp/eg/index.html (Accessed October 20, 2005).

Site:	**Seoul National University**
	San 56-1, Sillim-Dong
	Kwanak-gu, Seoul, 151-742, Korea

Date Visited: December 13, 2004

WTEC Attendees: M. Madou (Report author) B. Allen, T. Kurfess, J. Cao, T. Hodgson

Hosts: Min-Koo Han (Dean, College of Engineering),
                Tel: +82-2-880-7001, Fax: +82-2-875-0335,
                Email: mkh@snu.ac.kr
                Professor Chong Nam Chu, PhD (Mechanical and
                Aerospace Engineering), Tel: +82-2-880-7136,
                Fax: +82-2-875-2674, Email: cnchu@snu.ac.kr
                Professor Dong-il ("Dan") Cho, (Electrical Engineer-
                ing, Head, Microsystem Technology Center),
                Tel: +82-2-880-8371, Fax: +82-2-877-9304,
                Email: dicho@asri.snu.ac.kr

BACKGROUND

After the end of the Japanese occupation of Korea at the end of World War II, the Seoul National University (SNU) got off the ground in 1946. They moved onto the current campus in 1980. About 80% of the PhDs here come from the U.S. R&D directions are also following more of the U.S. government directions than in Japan. The College of Engineering has six departments with 305 faculty members (six of which are female). They work from an $80 million/year budget, a small part of which comes from the U.S. Defense Advanced Research Projects Agency (DARPA) and the National Institutes of Health (NIH). About 50% of their courses are in English. Their publication records in very respectable international journals often exceed those for MIT, Stanford, Georgia Tech, and other prestigious U.S. institutions.

RESEARCH AND DEVELOPMENT ACTIVITIES

The University supports a multi-user MEMS processes (MUMPS)-type (see Microelectronics Center of North Carolina (MCNC) in the U.S.) multi-project chip process in their Semiconductor Research Center. Some projects carried out here include micromirror arrays similar to the Texas Instruments (TI) micromirrors as well as microaccelerometers and catheters to probe the content of biological cells. Professor Chu runs a micro-

EDM research program in the machining center. For larger dimensions DI water is used as a cutting medium and kerosene is used for smaller dimensions. Tool wear is less in DI water. One application is the making of masks for organic thin film transistors (OLED). A mask with aspect ratio of 20 was made. This was accomplished by combining EDM and electrochemical machining (ECM). In ECM, experiments were conducted varying the pulse frequency, and much better results (cutting closer to the electrode shape) were obtained by using faster and faster pulse rates. Simulation of the ECM process has been started and is geared towards predicting the shape of the tool in order to obtain the desired final workpiece shape. Today new tool shapes are still mainly determined experimentally. In particular, it was noted that thermal effects could be detrimental to the workpiece final shape. Dr. Chu does not see much merit in further miniaturization of the micro-EDM machine. There was some early work on electrochemical deposition in his group, but it was deemed to be uncontrollable and the work stopped.

The entrepreneurial Professor Cho sits with his students in an open office space. He is working in MEMS and nanobiotechnology. He improved the Cornell single-crystal reactive etching and metallization (SCREAM) process by substituting the isotropic dry etch with a wet etch. This results in much smoother walls. He claims no silicon on insulator (SOI)/SCREAM products made it to the market because of surface roughness. His improved micromachining process is much more manufacturable than SCREAM or SOI processes. Most of his current work is in inertial sensors. He started a commercial venture to commercialize his three-axis gyroscopes. Professor Cho is negotiating a U.S. army contract on his inertial sensors. He is also involved in capsule endoscope work (swellable endoscope), which he calls a microbiopsy tool. For the endoscope work and cell culturing on polydimethylislioxane (PDMS) surfaces Professor Cho collaborates with SNU medical doctors. He is also involved with radio frequency (RF)-MEMS and a micromachined probing card.

SUMMARY AND CONCLUSIONS

The Korean intellectual elite is more closely tied to the U.S. intellectual elite than to the elite in Japan. This makes the research topics in Korea in general quite similar to the ones in the U.S. They very clearly measure themselves against the best U.S. universities. Professor Cho also embodies an entrepreneurial approach you are not so likely to find in Japan.

Site: **University of Electro Communications**
                           1-5-1 Chofugaoka, Chofu
                           Tokyo, 182-8585, Japan

Date Visited: December 6, 2004

WTEC Attendees: T. Hodgson (Report author), D. Bourell, K. Cooper,
                           G. Hane, M. Madou

Host(s):   Professor Hisayuki Aoyama, Dept. of Mech. Eng. &
                           Intelligent Systems, Tel: +81-424-43-5426,
                           Fax: +81-424-84-3327

BACKGROUND

The University of Electro Communications is a technical university. We visited Professor Hisayuki Aoyama's laboratory in the Department of Mechanical Engineering & Intelligent Systems. We were able to tour what appeared to be his entire laboratory. We saw a number of demonstrations of the microrobotic equipment his students have developed. This is an incredible learning laboratory where students learn by doing.

RESEARCH AND DEVELOPMENT ACTIVITIES

Professor Aoyama is using off-the-shelf (from Akihabara, the "electronic city" section of Tokyo) components to build microrobotic factory elements. He involves young graduate-level students to build all of his equipment. He motivates his students by making the process fun for them. It is clear that he enjoys a special relationship with his students.

Desktop Factory

The overall goal is the production of microparts at low cost and high flexibility. Microrobots can drill and indent, and yet another robot can put a pin in the drilled hole. Finally, the manufactured device can be brought to a microscope for inspection. The microrobots are piezo-driven and global control is by an overhead charged-couple device (CCD) camera, and local coordination is achieved with a pair of position-sensitive detectors (PSD) and light-emitting diode (LED) on the small robots. The robots can move to a specified point with less than 60 μm of positioning error.

Robot Mating for Artificial Insemination

This was demonstrated by the (artificial) insemination and fertilization of a fish egg in a "female" robot by a "male" robot.

Microdrop Mixturing

Another demonstration project was a system for mixing very small quantities of liquids. The target drop was approximately 1 mm in diameter. A small amount of another liquid was added robotically (from liquid held in a tube by capillary action). The liquids were then mixed using surface tension vibration (100–300 hertz). The system has potential in certain drug delivery applications.

Micromanipulator Driven by Sony PlayStation Controller

This system featured magnified overhead and 45 degree views of the target. The PlayStation controller was used to manipulate two end effectors that were in turn used to manipulate small parts. This system could easily be adapted to any number of microassembly applications.

Microindentation

In this case, three eddy current sensors were used exclusively for positioning. A vibrating diamond-tip stylus was used to microindent the surface of a target part.

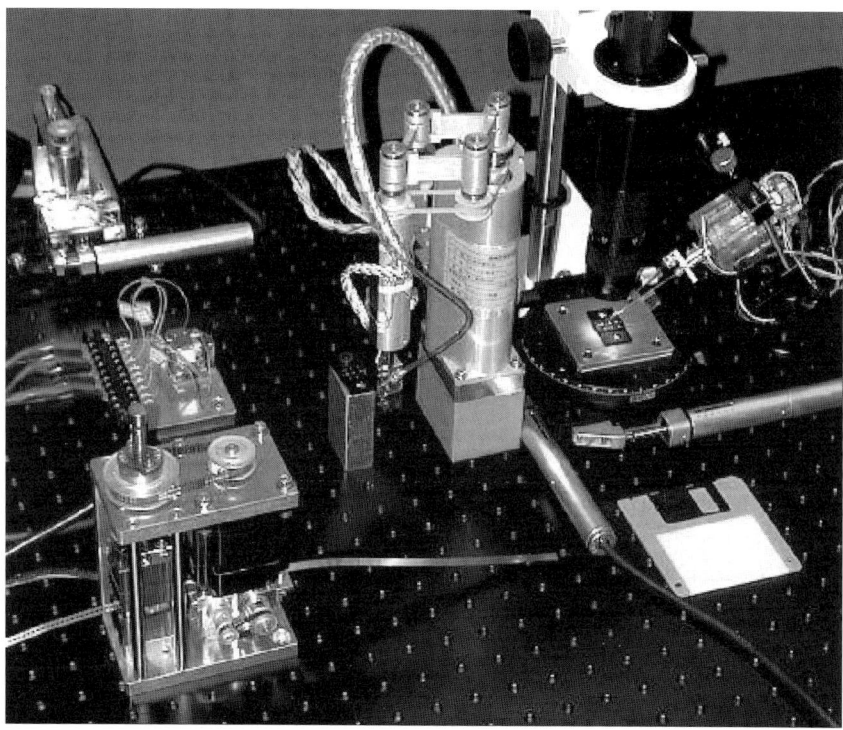

Figure C.32. Desktop factory.

SUMMARY AND CONCLUSIONS

Professor Aoyama runs an extremely innovative laboratory where the students learn by building automated systems that are fun. All of the systems were built from locally available parts. It appeared that any of the systems we saw could be built for less than $1,000, exclusive of the computers that were used to control the motions. It is clear that his students are the ones that will go out to industry to build the systems that will make micromanufacturing actually work.

REFERENCES

UEC Aoyama Lab for Desktop Micro Robots Factory. http://www.aolab.mce.uec.ac.jp (Accessed October 20, 2005).

Site: **University of Tokyo**
                             7-3-1 Hongo, Bunkyo-ku, Tokyo 113-8656, Japan
                             http://www.nml.t.u-tokyo.ac.jp

Date Visited: December 6, 2004

WTEC Attendees: D. Bourell (Report author), M. Madou, T. Hodgson,
                             K. Cooper, G. Hane

Hosts: Professor Mamoru Mitsuishi, Department of Engi-
                             neering Synthesis, Tel: +81-3-5841-6355,
                             Fax: +81-3-3818-0835,
                             Email: mamoru@nml.t.u-tokyo.ac.jp,
                             Web: http://www.nml.t.u-tokyo.ac.jp/~mamoru
                             Associate Professor Shin'ichi Warisawa, Department
                             of Engineering Synthesis, Tel: +81-3-5841-6014,
                             Email: warisawa@nml.t.u-tokyo.ac.jp,
                             Web: http://www.nml.t.u-tokyo.ac.jp/~warisawa
                             Takeshi Ooi, Research Associate, Nakao Laboratory,
                             Department of Engineering Synthesis,
                             Tel: +81-3-5841-6361, Fax: +81-3-5800-6997,
                             Email: ooi@hnl.t.u-tokyo.ac.jp
                             T. Higuchi, Professor, Precision Machinery Engineer-
                             ing, Tel: +81-3-5841-6449, Fax: +81-3-5800-6968,
                             Email: higuchi@intellect.pe.u-tokyo.ac.jp

BACKGROUND

The University of Tokyo has two areas of micromanufacturing. One is
in the department of engineering synthesis (Professors Mamoru Mitsuishi,
Shin'ichi Warisawa, Takeshi Ooi) and the other is in the precision machinery engineering area (T. Higuchi).

RESEARCH AND DEVELOPMENT ACTIVITIES

Department of Engineering Synthesis (Professors Mitsuishi and Wari-
sawa)

The primary function of the lab is micromachining. The first of three
machining projects was micromachining of soda glass. An oscillating diamond machine tool ($\pm 0.63°$, 125 Hz) is able to produce relatively deep
grooves, 3–4 µm, which are 150-200 µm wide. Without oscillation, only 1–

1.5 μm depths are obtainable. The potential applications are biomedicine and optics.

The second project is micromachining of lithium niobate glass using an ultraviolet (UV) laser (355 nm) for optical waveguide applications. Smooth grooves up to 10 μm in depth and 300 μm wide were produced with a spot size of 9.2 μm. Depth is controlled by changing the laser energy. Width is controlled by altering the focal length/spot size. The primary advantage of this process is the ability to process fragile materials like glass.

The third project was surface finishing of a 22.2 mm diameter, spherical cobalt chromium molybdenum (CoCrMo) alloy using computer numerical control (CNC) water-jet machining. The femur head to be machined was initially 12 nm Ra with about 300 nm roundness. The femur head was mounted in a fixture and immersed in water containing 4% 1 μm diamond particulate. A water-jet nozzle was immersed into the slurry and the femur head was polished at jetting pressures of 200–300 MPa (megapascal). The finished part had 7.2 nm Ra and 136 nm roundness.

Hatamura and Kakao Laboratory

Two projects were presented. The first was a computer/strain gage-controlled three-axis table with a 200,000 rpm tool used to create 100–500 μm parts in poly-methyl methacrylate. One or more fine carbide tools mounted on air spindles were used. The second project was nanomanufacturing. Here a preparatory controlled atmosphere chamber was used to fast atom beam (FAB) machine thin-layered silicon and polyimide. The beam was created by accelerating CHF_3 by electric ionization, acceleration and neutralization. The FAB was projected onto a micromask produced using a proprietary process. The target material was then selectively sputtered. One part production was silicon "microtorii." The shrine shape was machined in ~10–20 μm thick silicon using FAB machining. The long dimension was 100 μm. The part was then moved using 15 μm diameter electrically charged probes. Two square holes were FAB machined and loaded with solder paste prior to positioning on the microtorii vertically into the holes. The solder was then laser reflowed to generate the final part.

Another demonstration of micromachining was in the creation of a polyimide cabin. The cabin dimensions were 500 μm by 400 μm by 500 μm tall with walls/roof each about 80 μm thick. The walls were machined by laser cutting and were positioned using the electrically charged probes. The roof was also laser cut, ridge-line creased and folded prior to positioning atop the structure. Two scanning electron microscopy (SEM) imaging systems were available for visualization of the process.

Precision Engineering (Professor T. Higuchi)

We discussed barriers and challenges developed from the August 2004 workshop in Washington, D.C. Professor Higuchi noted that these are not new and that he had described these in a keynote lecture over ten years ago. His approach to micromanufacturing is that the part demands define the manufacturing process. In Japan, assembly and optics (e.g., lenses) dominate micromanufacturing. He described development of a four-axis precision machining system produced by Toshiba Machine Company. The machine footprint is 1.12 m by 0.79 m. Control to 1 nm is now available. The machine has been used to finish nominally 30 mm curved lenses, mirrors and immersion gratings with microscale surface features/patterns. A pattern example was periodically repeating 3 μm grooves that were 0.5 μm thick. A copper optical lens mold was also produced in this fashion. Another demonstration was micromachining of forceps and other parts produced in 303 stainless steel using a diamond wedge cutter. Feature size less than 10 μm was produced.

A more recent project was "pulling" of paraffin-domed cylinders on 15 μm centers. A laser locally melts the material where the cylinder is to be built. Forces (which were undisclosed) are applied to the melt under the laser beam which results in rising the melt along the laser beam axis. The cylinder aspect ratio was approximately 5:10.

SUMMARY AND CONCLUSIONS

The micromachining efforts at the University of Tokyo are well-established, varied and significant. This includes production of surfaces with microscale features and nanoscaled finish, creation of parts under 500 μm in size in polymers, metals and glass, and machining of optics grooves less than 200 μm wide and up to 3.5 μm deep.

REFERENCES

Mitsuishi, M., S. Warisawa, K. Yamada. Ductile mode cutting with torsional vibration for glass material. Information sheet (n.d.).

Nakao, M., K. Tsuchiya, T. Ooi. Common design rules empirically derived from prototyping microfactories for various functional requirements. Brochure (n.d.).

Warisawa, S., M. Mitsuishi, A. Nakanishi, I. Kohno, Micro machining of a glass material by means of a UV laser. Information sheet (n.d.).

Warisawa, S., M. Mitsuishi, H. Sawano. Precision finishing in nanometer order for the femoral head of the artificial hip joint by means of abrasive waterjet. Information sheet (n.d.).

Site: I. I. S., The University of Tokyo
                      Center for International Research on MicroMecha-
                          tronics (CIRMM)
                      4-6-1 Komaba, Meguro-ku,
                      Tokyo 153-8505, Japan

Date Visited: December 7, 2004

WTEC Attendees: K. Ehmann (Report author), B. Allen, J. Cao, G. Hane,
                      T. Kurfess, K. Rajurkar

Hosts: Professor Dr. T. Masuzawa, Center for International
                      Research on MicroMechatronics, 4-6-1 Komaba,
                      Meguro-ku, Tokyo 153-8505, Japan
                      Tel: +81-3-5452-6163, Fax: +81-3-5452-6164,
                      Email: masuzawa@iis.u-tokyo.ac.jp

BACKGROUND

The University of Tokyo currently consists of 10 faculties, 11 institutes, 14 graduate schools and a number of shared facilities. Professor Masuzawa is a member of the Center for International Research on Micro Mechatronics (CIRMM) that falls under the Institute of Industrial Science (I.I.S.).

RESEARCH AND DEVELOPMENT ACTIVITIES

Overview

The visit to Professor Masuzawa's laboratory lasted for about two hours and consisted of two parts: a presentation by Professor Masuzawa giving a state-of-the-art overview of micromanufacturing processes interspersed with specific accomplishments of his laboratory, and, a visit to his research laboratory with presentations and discussion of ongoing projects.

The focus of Professor Masuzawa's activities is on micro-electro-discharge machining (µEDM) processes and their modeling and control and the development of techniques and methods for the characterization of machined microscaled features and surfaces. He has done pioneering work in these areas that has led to a number of commercial successes. One of his most notable accomplishments is the development of the micro-EDM process and machine that incorporates a wire electro-discharge grinding (WEDG) unit for the *in situ* preparation of the electrodes. This development has allowed for a dramatic increase in accuracy and micro-EDM capabilities. This concept has been commercialized by Panasonic. He has also conceived the concept of uniform electrode wear in micro EDM and the micro-EDM lathe that allows for the creation of intricate external and internal features (Figure 3.7).

Assessment of Micromanufacturing

Table C.1 summarizes Professor Masuzawa's assessment of a number of micromanufacturing processes focusing on their advantages and disadvantages, current capabilities, and notable accomplishments in his laboratory.

Table C.1
Professor Masuzawa's assessment of a number of micromanufacturing processes

Process	Advantages/ Disadvantages	Masuzawa
Microcutting	High precision; Wide range of materials; Tool and workpiece deflections at the lower limits of dimension	
Microgrinding	High accuracy and finish; All materials; Limited shapes; Residual stresses	
Wire EDM	High precision—about 0.5 µm; 20 µm wire capability available; Limited geometry; Heat affected zone of about 0.2 µm; Material limitation	
Micro EDM	Similar to above; Electrode wear	5.0 µm microholes
WEDG	Similar to microgrinding; Practically no tool wear; No grinding force	4.5 µm micropin; ~150 µm long; Micro end mill;
Micro Ultrasonic Machining	Can machine brittle materials; 3D; Precision; High tool wear; Needs precise tools (WEDG can be used)	Precision increased by vibrating the workpiece instead of tool; 5.0 µm microholes
Laser Beam Machining Nd: YAG	High productivity; Heat-affected zone	
Electro Chemical Machining	Deep holes; Smooth surfaces; Limited materials; Limited precision	
Blanking	High Productivity; Complex tools; Low aspect ratio; Burr height	40×100 µm microholes Blank thickness t=10 µm; Diameter ϕ=20 µm; Blank material—stainless steel; Tool material—Tungsten carbide; Burr height =1 µm
Molding/Casting	High productivity; Complex tools; Material limitations	

Discussion

Professor Masuzawa's laboratory has gained state-of-the-art expertise in a wide range of micromanufacturing processes. The accumulated experiences expressed during the discussions will be summarized here.

The achievement of high accuracy is limited by material uniformity and, in particular, impurities and defects. Grain size effects are a problem not only from a pragmatic but also from the modeling standpoint. At present it is also difficult to obtain materials with a suitable structure and quality (fine grain and uniform) for micromanufacturing.

In-process sensing poses a considerable practical problem (e.g., sensing of forces in ultrasonic motor (USM) to prevent breakage, current detection in micro EDM).

Many processes leave unintended consequences on the machined surface. Methods need to be developed for their minimization or avoidance. For example, removal of the heat-affected zone after the WEDG process. The measurement of this zone is a problem by itself.

To overcome current limitations in precision, smaller machines are a must, in particular from the thermal deformation standpoint. Also, multi-functional machines are needed to eliminate the repositioning and re-clamping of parts since handling is a problem. This conclusion can also be substantiated by the fact that long-standing experiences by Swiss watch-makers have forced them to develop multi-functional machines for some of their operations. In many instances performing the tool manufacture and part processing on the same machine is the only solution (e.g., micro-punching process). All these problems dictate that the design philosophies and concepts that drive the design of machines for micromanufacturing operations need to change. Functional integration and new thinking will be required for precision machine design. This is further accentuated by the fact that components of downsized machines will also require micromanu-facturing operations.

Future trends toward product miniaturization and functional integration will continue and will require the availability of mass-production capabilities (e.g., injection molding, blanking, deformation). In this context, micro-factories have a promising future since they offer the technological capability for economical mass-production of high-quality products.

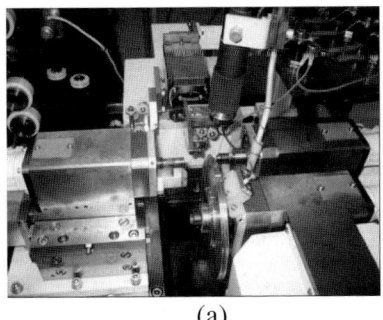
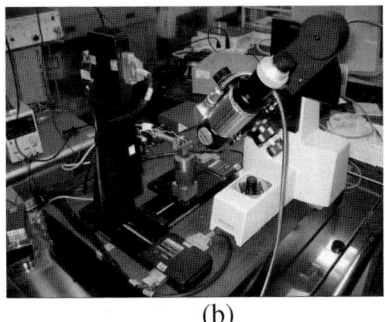

(a)   (b)

Figure C.34. a) EDM-lathe, b) Profiling high-aspect ratioholes.

Laboratory Tour

During the laboratory visit ongoing projects were discussed. Some of the projects demonstrated include: the original prototype of the micro-EDM machine with the wire electro-discharge grinding unit, the EDM-lathe (see Figure C.34a), a metrology system for the profiling of microfeatures, in particular high-aspect ratio holes (see Figure C.34b), an excimer laser beam machining (LBM) apparatus for performing the hole area modulation (HAM) process (which uses a 2D mask but etches out a 3D lens), and others. The laboratory is extremely well-equipped for microscale processing and metrology tasks. Currently, three students and a postdoctoral researcher constitute the team.

SUMMARY AND CONCLUSIONS

In spite of its short duration, this visit was extremely productive since the group was given the opportunity to discuss issues with one of the pioneers in the field of micromanufacturing. The insights gained, based on Professor Masuzawa's experiences, will undoubtedly be an invaluable resource in writing the final report. Perhaps the most consequential message from the visit is that micromanufacturing, as defined in the context of this study, will continue to grow and will demand new advances on a broad front ranging from the provision of better materials, understanding of processes, all the way to new ways in designing manufacturing equipment that will have to be downsized and multi-functional to meet microproduct demands.

REFERENCES

CIRMM Members: Professor Masuzawa, Takahisa. http://www.cirmm.iis.u-tokyo.ac.jp/Members/membershtml/
members_masuz.htm (Accessed October 20, 2005).
Masuzawa, T. and H. K. Toenshoff. 1997. Three-dimensional micromachining by machine tools. *Annals of CIRP* 16:2, 621-628.

Site:	**Yonsei University** **134 Sinchon-dong, Seodaemun-gu** **Seoul 120-749, Korea**

Data Visited: December 13, 2004

WTEC Attendees: T. Kurfess (Report author), B. Allen, J. Cao,
K. Cooper, T. Hodgson, M. Madou

Hosts:

School of Mechanical Engineering

Professor Dae-Eun Kim, Nanotribology, ME,
  Email: kimde@yonsei.ac.kr
Professor Sang Jo Lee, Manufacturing, ME,
  Email: sjlee@yonsei.ac.kr
Professor Jae Won Hahn, Laser Photonics, Physics,
  Email: jaewhahn@yonsei.ac.kr
Professor Hyun Seok Yang, Control, ME,
  Email: hsyang@yonsei.ac.kr
Professor Shinill Kang, Nanomolding, ME,
  Email: snlkang@yonsei.ac.kr
Professor Yong Jun Kim, MEMS, EE,
  Email: yjk@yonsei.ac.kr
Professor Byung-Kwon Min, Manufacturing System,
  ME, Email: bkmin@yonsei.ac.kr
Professor Hyo-Il Jung, Bioengineering, Biology,
  Email: uridle7@yonsei.ac.kr

Graduate Program in Information Storage Engineering

Professor No-Cheol Park, Optomechatronics, ME,
  Email: pnch@yonsei.ac.kr
Professor Young-Joo Kim, Nanofabrication, Mat. Sci.,
  Email: yjkim40@yonsei.ac.kr

BACKGROUND

Yonsei University was established in 1885. It has two major campuses, Shinchon and Wonju. The team visited the Shinchon campus. It is a private university consisting of graduate and undergraduate programs. The undergraduate program is comprised of 19 colleges with 89 majors having 36,115 students (approximately 26,000 of these are at the Shinchon campus). The graduate program is comprised of the same 19 colleges with 89 majors and 12,286 graduate students. There are 130 research centers and

1,530 full time faculty members. The annual research budget for Yonsei is approximately $130 million.

The College of Engineering within Yonsei University was established in 1950, and has 10 majors within seven schools. There are 175 faculty members, 3,679 undergraduate students, 1,755 graduate students and 24 research centers within the College of Engineering. The annual research budget for the College of Engineering is $39 million. Twenty-seven faculty members are in the School of Mechanical Engineering. Of these, 24 have doctorate degrees from the United States, two from the United Kingdom and one from Korea.

RESEARCH AND DEVELOPMENT ACTIVITIES

Yonsei University has four primary mission objectives with respect to micromanufacturing. These objectives are:

- To develop novel micro/nanomanufacturing processes.

- To achieve 10 nm feature size for various materials.

- To understand atomic-scale phenomena in processing.

- To educate students in micro/nano-manufacturing processes.

To achieve these objectives Yonsei University has developed a set of strategies that maximize interdisciplinary cooperation within and outside of Yonsei. Furthermore, close collaboration with industrial partners is critical for their team. In particular, they are looking toward their industrial partners to provide them with applications to drive the Yonsei team's development work. The Yonsei team currently participates heavily in the national initiatives in nano technology, and encourages its faculty to collaborate internationally as well. Finally, there is a strong push to develop graduate and undergraduate curricula targeted at training engineers for further research, development and employment in the micromanufacturing sector.

There are three major sponsored projects or centers currently underway at Yonsei. These include Micro-Nano Hole and Line Manufacturing Research Center (Professor S. J. Lee, Director), sponsored by the Ministry of Commerce, Industry, and Energy (MOCIE) with a budget (2004) of $1 million (75% MOCIE, 25% Yonsei). This project began in 2003 and is scheduled to run through 2008. The Center for Information Storage Device (Professor Y. P. Park, Director) is sponsored by the Korea Science and Engineering Foundation (KOSEF), with a budget (2004) of $2.3 million (40% KOSEF, 10% University, 50% Industry). This Center is equivalent to a National Science Foundation (NSF) Engineering Research Center (ERC). This project began in 1997 and is scheduled to run through 2005. The third major center targets Electro-chemical Machining for Microfactory (Professor S. J. Lee, Director). It is also sponsored by MOCIE with a budget (2004) of $300,000 (100% MOCIE), and will operate from 2005-2010. In addition to these three centers, there are two National Research Labs at Yonsei. They are the Nano Photonics Lab (Professor J. W. Han, Director) and the Nano Molding and Micro Optics Lab (Professor S. I. Kang, Director), they are both sponsored by the Ministry of Science and Technology (MOST) and have a budget (2004) of $200,000 per lab (100% MOST). These labs were established in 2004 and will be supported through 2008. A variety of smaller individual research projects are also linked to the micromanufacturing area, including, but not limited to, work addressing microrobotics, systems and control, smart material, and bioengineering applications. These projects are sponsored by a variety of entities including MOCIE, MOST, KOSEF as well as industry.

The research projects at Yonsei have been divided into four major categories. The first one is optomechatronics for micro/nanoscale information devices. Projects in this area include: microlens fabrication optical pickups, near field optical array head and micro glass molding. A primary application for this category is the development of the next generation optical storage devices using blue laser diodes (405 nm wavelength). The target size for this optical device is under 2.2 mm × 6.6 mm with a thickness of 1.9 mm and a weight of 30 mg. In particular, these projects are targeting the fabrication of components and subsystems of the overall optical drive. For example, Figure C.35 shows a very small optical pickup designed and fabricated by Yonsei and Samsung Company. The pickup is used for integrated system of micro/nano optical components in small size optical data storage for mobile applications. (In general, this application appears to be quite popular at a variety of research institutions in Korea. The target application is for consumer electronics such as cell phones and personal digital assistants (PDAs).)

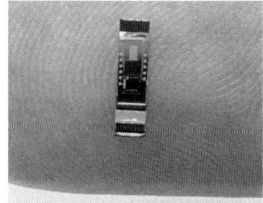
Figure C.35. Integrated optical pickup.

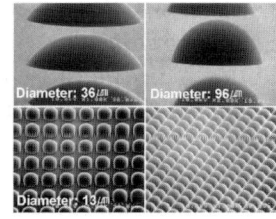
Figure C.36. Stamped lens array.

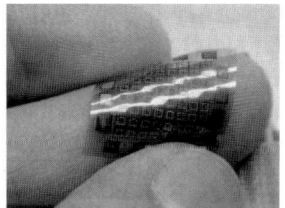
Figure C.37. Flexible bio-sensor.

The second major research category is micro/nanomachining processes. Projects in the area include: mechano-chemical micromachining / scanning probe lithography, ultraviolet (UV) molding, micro electro-discharge machining (µEDM), 3D process development and simulation, magneto-rheological (MR) polishing for micromolding of lenses, fluidic channels and radio frequency (RF) device development, and work on the microfactory initiative that is supported by the Korean government. Again, a major application of this area is in the high density optical storage sector. In this case both elements of the storage drive are fabricated (e.g., lenses), as well as the ultra high-density storage medium itself (e.g., optical read-only memory (ROM) and optical rewritable media). Several techniques were presented including molding parts using a nickel stamp with a self-assembled monolayer. Figure C.36 shows microlenses and microlens arrays that were generated as part of this research. The array is for use in coupling a vertical cavity surface emitting laser (VCSEL) array to fiber optics. In terms of micro EDM, the smallest electrode has a diameter of 20 μm. Research issues include how to make a precise small hole and how to achieve the smoothness of a trajectory. They indicated that going for a smaller structure of an EDM machine seems inevitable to achieve the above goals. In the area of micromolding, they collaborated with industrial partner Sumitomo to specially design a machine for micro/nanomolding. They redesigned the scroll to minimize its size while maximizing its pressure, used microelectromechanical systems (MEMS) temperature sensors and reduced the time needed to heat up the material to 100°C down to 15 seconds. They also worked with Philips in the Netherlands to develop processes for making 1 μm lenses and 1 μm holes. For this work, they received part of the intellectual property (IP) rights.

The third major category is application of laser and ion beam technologies for microfabrication. Projects in this area include work in nanomold fabrication using a focused ion beam (FIB), fabrication of patterned media using immersion holography lithography, and laser micromachining and monitoring. Several examples of holographic patters were presented. Typical pattern geometric parameters included pitch sizes of 200 nm, line widths of 100 nm and aspect ratios of 1.18.

The fourth major research category is Bio-MEMS applications, including work on bio-sensors and flexible electronics. In particular, the focus of this work targets sensors that are flexible in nature permitting them to operate easily in a biological environment. Examples of wireless sensors such as a blood pressure module or a tactile sensor for use in noninvasive medical procedures were presented. Figure C.37 shows a flexible tactile sensor for use in such a procedure.

EDUCATION

Following the research lab tour, Professor S. J. Lee showed the WTEC team the Yonsei Student Venture Center (YSVC). The university has invested $5 million into the center, which hosts 10 student teams at any given period to encourage students to innovate. Performance is evaluated every six months, and two teams will be rotated out of the center following the performance review. One successful story was generated by a sophomore who developed a security system, which generated $30 million in income.

SUMMARY AND CONCLUSIONS

The research being conducted at Yonsei University is extremely impressive. The approach to micromanufacturing is well-integrated with industrial partners, as well as the long- and short-term plans of the Korean government. Furthermore, it appears as if the faculty at Yonsei plan on a long-term institutionalization of micromanufacturing through the development of graduate and undergraduate courses to train future generations of engineers to support this technical sector.

REFERENCES

Choi, W., J. Lee, W. -B. Kim, B. -K. Min, S. Kang, S. -J. Lee. 2004. Design and fabrication of tungsten carbide mould with micro patterns imprinted by micro lithography. *J Micromech Microeng* 14, 1519-1525.

Goto, K., Y. -J. Kim, T. Kirigaya, Y. Masuda. 2004. Near-field evanescent wave enhancement with nanometer-sized metal grating and microlens array in parallel optical recording head. *Japanese Journal of Applied Physics* 43: 8B, 5814-5818.

Kim, S. -M. and S. Kang. 2003. Replication qualities and optical properties of UV-moulded microlens arrays. *J Phys D: Appl Phys* 36, 2451-2456.

Kim, W. B., B. -K. Min, S. J. Lee. 2004. Development of a padless ultraprecision polishing method using electrorheological fluid. *Journal of Materials Processing Technology* 155-156, 1293-1299.

Kim, Y., N. Lee, Y. -J. Kim, S. Kang. 2004. Fabrication of metallic nano-stamper and replication of nano-patterned substrate for patterned media. *Nanotechnology* 15, 901-906.

Sohn, J. -S., M. -B. Lee, W. -C. Kim, E. -H. Cho, T. -W. Kim, C. -Y. Yoon, N. -C. Park, Y. -P. Park. 2005. Design and fabrication of diffractive optical elements by use of gray-scale photolithography. *Applied Optics* 44:4, 506-511.

Sung, I. -H. and D. -E. Kim. 2004. Molecular dynamics simulation study of the nano-wear characteristics of alkanethiol self-assembled monolayers. *Appl Phys A* 81:1, 109-114.

Sung, I. -H., D. -E. Kim. 2005. Nanoscale patterning by mechano-chemical scanning probe lithography. *Applied Surface Science* 239, 209-221.

APPENDIX D: SITE REPORTS—EUROPE

Site: **BASELWORLD**
                    Basel, Switzerland

Data Visited: April 5, 2005

WTEC Attendees: T. Kurfess (Report author), H. Ali, K. Cooper,
                    K. Ehmann, T. Hodgson

Hosts: None

BACKGROUND

BASELWORLD is the leading trade show for the watch and jewelry industry. Approximately 2,200 exhibitors are present at the event. BASELWORLD occupies over 160,000 m² of exhibition area on several floors. Many world-famous names in the watch and jewelry industry choose to show their products exclusively at BASELWORLD. Over 85,000 retailers and wholesalers attend this show annually in the spring.

RESEARCH AND DEVELOPMENT ACTIVITIES

Metal Cutting Machines

Several different types of machining systems were demonstrated at the fair, including multi-axis machining centers such as the one shown in Figure D.1 and computer numerical control (CNC) turning centers, as shown in Figure D.2. The precision of the machine shown in Figure D.1 is 5 μm. The cost of the three-axis version of the machine is $140,000. The five-axis version of this machine is $190,000. The turning center shown in Figure D.2 is produced by Pfiffner and costs approximately $150,000 and has six outside diameter (OD) tools and five inside diameter (ID) tools as well as an active tailstock. It has an accuracy of 3 μm. These are typical examples of the machine tools that were shown at the show. Nominally, these machines are targeted at machining components that can be used in watches and jewelry. The largest components are on the order of 20–30 mm (rings, watch faces and cases) to small components such as watch gears and watch winding shafts having diameters of 100 μm.

Figure D.1. Five-axis machining center.

Figure D.2. Turning center.

Additive Type Systems

A wide variety of forming systems was demonstrated including the one shown in Figure D.3. This particular system is used for producing jewelry and is typically used on software metals such as gold and silver. However, it does have the capability to generate parts in steel. Mario di Maio Corporation in Italy makes the press. This particular company has a wide variety of machines for use in small component/jewelry fabrication.

Figure D.4 shows one of several rapid prototyping machines that are used to produce wax models of various jewelry pieces. This system is from Solidscape. It has a resolution of 25M addressable drops per square inch (25.4 mm^2), a build envelope of X=12 in. (30.48 cm), Y=6 in. (15.24 cm), Z=6 in. (15.24 cm) with a build layer of 0.0005 in. (0.013 mm) to 0.0030 in. (0.076 mm). The achievable accuracy is ±0.001 in. (±0.025 mm) per inch in X, Y and Z dimensions. The system can generate a surface finish between 32–63 µin (RMS) with a minimum feature size of 0.010 in. (0.254 mm)

Figure D.3. Press for
small-scale parts.

Figure D.4. Wax RP system.

Inspection Systems

A wide variety of machine tool and inspection companies were represented at the fair. The only systems shown for inspection were typical optical inspection systems such as the one shown in Figure D.5, produced by Vision Engineering. The system utilizes a microscope for 2D measurement of features. It has a variety of settings and options that permit a variety of inspection levels, from manual to on-screen metrology. It is similar in nature to a variety of systems that are available on the market. The electrical testing units such as the one shown in Figure D.6 demonstrate some nice designs for microcomponent handling and testing microsystems' electronics.

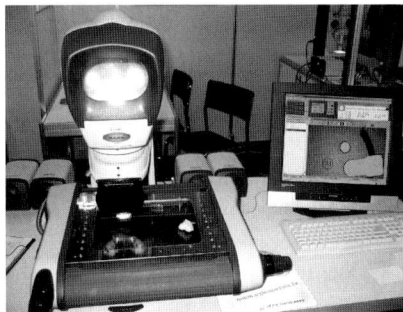 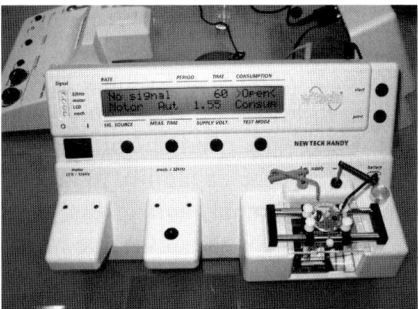

Figure D.5. Optical inspection system.   Figure D.6. Electrical inspection unit.

SUMMARY AND CONCLUSIONS

A number of microproduction and metrology systems were presented at the fair. While these machines produce components that can be considered microcomponents, the systems are still relatively large. The most precise systems are as large as a small machine tool (approximately 3×2 m^2 footprint). Several table top machine tools were on display at the show, but they did not have the accuracy required for machining microcomponents. It is also suspected that these small tabletop units lack the rigidity required for precision machining of microcomponents. With respect to prototyping and forming systems, they can be used to fabricate small components at reasonable process. As with the machine tools, these do not have the necessary resolution. Finally, some inspection systems were demonstrated; however, the systems were 2D in nature. They did not represent any new technology.

REFERENCES

BASELWORLD, http://www.baselworld.com/li/cc/ss/lang/eng/ (Accessed October 20, 2005).

Bumotec, http://www.bumotec.ch/ (Accessed October 20, 2005).

Carl Benzinger Präzisionmaschinen GmbH, http://www.benzinger.de/ (Accessed October 20, 2005).

K.R. Pfiffner, http://www.pfiffner.com/ (Accessed October 20, 2005).

Mario Di Maio, http://www.mariodimaio.com/english/home.html (Accessed October 20, 2005).

Rosilio Machines Outils, http://www.rosilio.fr/ (Accessed October 20, 2005).

Solidscape, http://www.solid-scape.com/index.html (Accessed October 20, 2005).

Vision Engineering, http://www.visioneng.com/ (Accessed October 20, 2005).

Willemin-Macodel SA, http://www.willemin-macodel.com/ (Accessed October 20, 2005).

Site: **Robert Bosch, GmbH**
                             FV/PLM5
                             Post Fach 30 02 40
                             70442 Stuttgart, Germany

Data Visited: April 5, 2005

WTEC Attendees: T. Kurfess (Report author), H. Ali, K. Cooper,
                    K. Ehmann, T. Hodgson

Hosts: Mr. Wilhelm Hopf, Director, Corporate Research and
                             Development, Tel: +49711 811-8637,
                             Fax: +49711 811-2601 30,
                             Email: wilhelm.hopf@de.bosch.com
                             Dr. Martin Schoepf, Corporate Research and Devel-
                             opment, Tel: +49711 811-33689,
                             Fax: +49711 811-2830,
                             Email: martin.schoepf@de.bosch.com
                             Mr. Wolfgang Hickauf, Manager, Corporate Research
                             and Development, Tel: +49711 811-8633,
                             Fax: +49711 811-2830,
                             Email: wolfgang.hickauf@de.bosch.com
                             Mr. Joachim Hofer, Teamlieter, Injector, Tel: +49712
                             135-1848, Fax: +497121 811-263913,
                             Email: joachim.hofer@de.bosch.com
                             Mr. Karsten Vormann, Corporate Research and De-
                             velopment, Tel: +49711 811-1436,
                             Fax: +49711 811-2830,
                             Email: karsten.vormann@de.bosch.com

BACKGROUND

This panel visited the Robert Bosch Fuel Injector unit of the Diesel Systems Group. The plant, located in Rommelsbach, Germany, was purchased
by Bosch in 1964, and has made diesel injection components since 1998.
There are a total of 82 administrative and engineering personnel at the
plant and 561 personnel involved in production. They produce diesel injec-
tors for engines having 3 to 10 cylinders for sizes ranging from 0.8 liters
through 5 liters.

Approximately 8% of Bosch's worldwide sales are spent on R&D for a
total of €2.7 billion. This investment in research and development is re-
flected in an increase in the number of patents that they have received in
the past year. The automotive group accounts for 65% of Bosch's sales.

Bosch employs a total of approximately 231,600 personnel worldwide, and had a capital expenditure of €2 billion in 2003.

RESEARCH AND DEVELOPMENT ACTIVITIES

Bosch produces a significant number of small components with standard size machine tools. They have had significant success in continuous quality improvement and control. Within this particular plant, they produce the same diesel injection pumps year-round, allowing them to focus on repeatability.

Controls

The panelists were shown a variety of controls in the Bosch facility; however, in the future, all machine tools will use Bosch controllers. In particular, the next generation of Bosch controllers, the IndraMotion MTX, is the direct result of Bosch purchasing Indramat and will be used exclusively. These new controllers will be open architecture in nature, and allow Bosh to push the limits of their production processes. The production engineers at this facility indicated that they would have "inside" information regarding the new Bosch controllers that would allow them to push the limits of their machines. This information will most likely be available to only Bosch facilities and may be a distinct advantage for Bosch.

Machine Tools

Bosch has a variety of machining centers, turning centers and grinding machines. They state that their most important process is grinding. They use a variety of grinding operations including OD plunge, bore and through-feed centerless. Some of their operations use cubic boron nitride (CBN) wheels. The average age of their machines is 8–10 years. Once a machine is worn out, it is sent to another part of Bosch that does not require the same performance as this particular plant. The old machines are then replaced with new ones. It is interesting to note that the relative young age of the machine tools in this facility permits it to produce at higher quality levels and more efficiently then facilities employing older machines. It was also noted that purchasing a new machine is slightly more expensive than rebuilding an old machine. However, a strategic decision has been made at this plant to keep the age of the production line equipment relatively young. This is not the case at Bosch's facility in Michigan. Consequently, the plant is in a better position to provide product specified to an increased quality level, when that level is specified in the future. In essence, they have made a capital investment in the future. Plants with older machines will not be able to move to the next generation of quality requirements.

Testing

The facility visited produces 1.5 million pumps per year, running 24 hours per day for five to six days per week. All pumps are tested for four separate functions. Testing for the final assembly takes seven minutes per pump. There is also a fair amount of fixed gaging, in particular air gages in the facility. All of these tests are highly automated and do not need to be flexible or reconfigurable due to the fact that the same components are produced by the plant year-round.

Future Issues

Bosch indicates that they need to focus on process modeling to better understand how to control their processes. They are currently developing cutting process models and plan to develop models for cutting, tool wear, burr formation, chip formation and force models in machining. One of their long-range goals is to be able to accurately predict the ramp-up of a new production process. They are looking towards keeping their production lines modern, and employing open architecture controls (IndraMotion MTX).

SUMMARY AND CONCLUSIONS

Bosch is a forward-thinking company. They are investing in the future with respect to both their production capabilities as well as their understanding of their production processes. Many of the components they make can be considered microparts, or at least have microfeatures. They would like to see new smaller scale machine tools that have the same capabilities as the larger ones that they currently employ. They feel that such machines would help them realize significant cost savings. They are looking at some possible applications of smaller scale machines tools; however, it appears that they are not investing too heavily in this technology area at present.

REFERENCES

Bosch Group. http://www.bosch.com (Accessed October 20, 2005).
Bosch Rexroth AG. http://www.boschrexroth.com (Accessed October 20, 2005).

Site: **École Polytechnique Fédérale de Lausanne (EPFL)**
                        Institut de Production & Robotique (IPR)
                        Ch. Des Machines – ME A3
                        CH – 1015 Lausanne, Switzerland

Date Visited: April 6, 2005

WTEC Attendees: T. Hodgson (Report author), K. Ehmann, T. Kurfess,
                        K. Cooper, H. Ali

Hosts: Professor Hannes Bleuler, Tel: +41 21 69 35927,
                            Email: hannes.bleuler@epfl.ch
                        Professor Jacques Giovanola
                        Dr. Jean-Marc Breguet

BACKGROUND

École Polytechnique Fédérale De Lausanne (EPFL) is a Swiss Federal Institute of Technology and one of two major technical universities that we visited in Switzerland. Professor Bleuler of the Micro-Engineering Department showed us around the various projects housed in the Institut de Production et Robotique (IPR). The Micro-Engineering Department has existed for 20 years and has 20 faculty members. It is the largest engineering department at EPFL.

RESEARCH AND DEVELOPMENT ACTIVITIES

Miscellaneous Activities

Initially, we were shown several projects. The first was a force feedback simulation device for the teaching laparoscopic surgery. The device, which has been commercialized by Xitact (http://www.xitact.com), featured some very nice graphics, and was very effective as a teaching tool. They showed us the results of their development of a magnetic bearing to replace the conventional bearing in hard drives (supported by a major manufacturer of hard drives). The project was successful but cost too much and used too much power. We understood that the company uses it for testing components. Using similar technology, they developed an inertial sensor, tunable to specific frequencies. It is good for low-frequency earthquake sensing. We were shown the Delta robot (also commercialized) which is the fastest in the world (40 G's). In many cases it is so fast that it is gripper limited.

High Precision Robotics Group (HPR)

There are about 10 people in the group. It has activities centered in microbotics. They have built a series of actuators that are useful as building blocks for laboratory systems within a scanning electron microscope (European project ROBOSEM). They feature piezo actuators built on the slip stick principle, with a velocity of 5 mm/sec., a range of 8–20 mm, an actuator resolution of 5 nm, and a sensor resolution of 100 nm. Three vertical actuators can be combined to tilt end effectors, tools, platforms, and other components. When combined with an XY actuator, you get five degrees of freedom (DOF). As an application they developed a selective carbon deposition system that writes a 100 nm wide line on polished copper. It is highly repeatable. They are also using it to manipulate carbon nanotubes. A material testing system does nanoscratching and nanoindenting of surfaces.

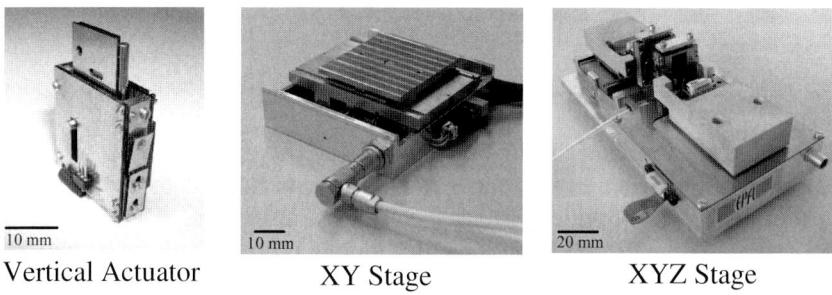

Vertical Actuator   XY Stage   XYZ Stage

Figure D.7. IPR building block actuators.

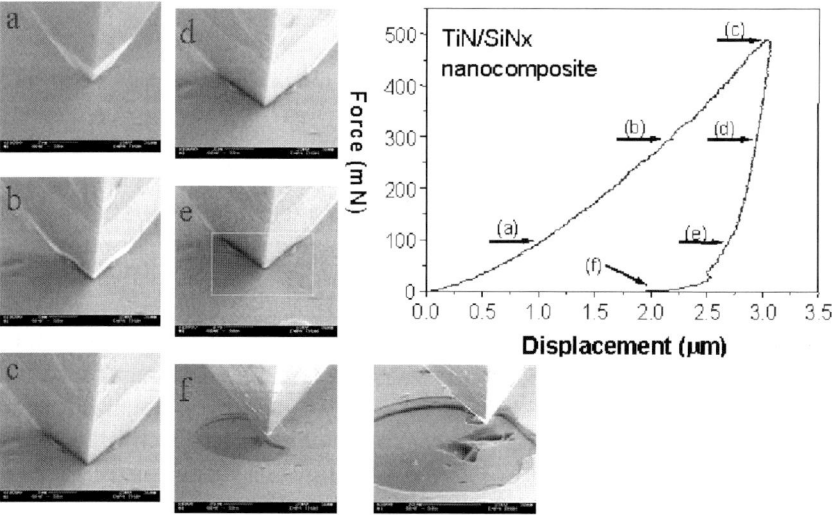

Figure D.8. Example of material characterization.

Figure D.9. Cooperating cm^3 cooperating microrobots.

They have also built miniaturized cooperative robots: essentially a cluster of cm^3 cooperating microrobots consisting of different modules with actuation, energy (battery), electronics and tools. They move using piezo slip stick technology. They have experimented with swarms of really small, stupid robots doing relatively complicated tasks. They have developed monolithic push-pull piezo actuators that are very simple and adapted to mass production, but have limited force capability. Speed is 1–2 mm/sec. A patent has been applied for.

Spark-assisted Chemical Engraving

Electrochemical discharges were first used in 1968 to drill microholes in glass. They developed a procedure for spark-assisted chemical engraving. The engraving is done in an electrolyte on glass. They have also drilled holes as small as 150 µm. Engraving proceeds at the rate of 50 µm/sec. Applications are in the fields of microfluidic systems, encapsulated accelerometers and pressure sensors. They have developed a mathematical model of the process.

Low-temperature Co-fired Ceramics (LTCC)

LTCC allows the integration of electronics since the relatively low temperature of the sintering (800–900°C) doesn't damage the electronics. It is good for high-temperature applications and is stable in high humidity applications (it is somewhat brittle though). There is 13–15% shrinkage during the sintering process. After that the dimensional stability is good.

High-speed Micromilling Spindles

Micromilling has suffered from a lack of capable spindles. Professor Giovanola is designing a high-speed spindle using air turbines (air turbines do have their own problems from the internal pressures generated). His first goal is to achieve 300,000 rpm with 1 micron accuracy (now undergoing tests). His second goal is to achieve 1 million rpm. There are potential problems including high tool wear. He is working on analytical models to predict performance of cutting tools at the very high rpm's proposed. He is concerned with how metal grain size will affect the cutting process versus macroscopic cutting. He has sapphire tools being made for this project.

Electrostatic Glass Motor

By taking advantage of the relatively long relaxation times for charges in glass, charges can be moved around the glass rotor using non-contact charge induction and inducing motion. The motion is much smoother than the electromagnetic or piezo-electric alternatives. Potential applications appear to be very small disk drives, and very precise positioning of optical elements such as mirrors and lenses.

SUMMARY AND CONCLUSIONS

The IPR is an outstanding example of a quality academic organization involved in a wide range of developments in micromanufacturing. Swiss professors are well-funded internally relative to the U.S. professors, and they are taking good advantage of the opportunity afforded them. They have been successful in commercializing a number of their developments.

REFERENCES

Institut de Production et Robotique. http://ipr.epfl.ch (Accessed October 20, 2005).
Xitact. http://www.xitact.com (Accessed October 20, 2005).

Site: **Fraunhofer Institute – Manufacturing Engineering and Automation (IPA)**
Institut Produktionstechnik und Automatisierung
Department of Cleanroom Manufacturing
Nobelstrasse 12
70569 Stuttgart, Germany
http://www.ipa.fraunhofer.de

Date Visited: April 5, 2005

WTEC Attendees: K. Ehmann (Report author), H. Ali, K. Cooper,
T. Hodgson, T. Kurfess

Host(s):   Udo Gommel, Tel: +49-7-11-9701633,
Fax: +49-7-11-9701007,
Email: gommel@ipa.fraunhofer.de
Thomas Lenz, Tel: +49-7-11-9701138,
Fax: +49-7-11-9701007,
Email: thomas.lenz@ipa.fraunhofer.de
Ralf Muckenhirn, Tel: +49-7-11-9701217,
Fax: +49-7-11-9701010,
Email: muckenhirn@ipa.fraunhofer.de

BACKGROUND

The Department of Cleanroom Manufacturing was founded in 1984. The Department, under the leadership of Dr. Johann Dorner, focuses on providing support to industry on issues related to miniaturized and contamination-sensitive manufacturing. Their expertise encompasses questions concerning methods, equipment and manufacturing technologies and consulting in cleanroom manufacturing. It currently employs 30 engineers/scientists and 25 scientific assistants. Their infrastructure consists of a 200 m² cleanroom surface area (ISO Class 1 to 6), airflow laboratory, test laboratory for media/materials, microassembly laboratory, and simulation laboratory. The Department's total budget of €5 million (in 2004) breaks down as follows: basic financing 10%, publicly funded projects 32%, and industrial projects 58%.

The principal specific activities of the department are: semiconductor manufacturing and microelectronics, microsystem technology, information and communications technology, solar technology, bio-engineering, medical technology and pharmaceutics, automotive industry, and device and equipment technology. In all these areas their expertise covers the whole spectrum from product conception to realization.

RESEARCH AND DEVELOPMENT ACTIVITIES

The visit lasted about 2.5 hours and consisted of several presentations (the introductory by Mr. Udo Gommel) and a brief visit to the laboratory.

Match-X Modular System (Mr. Thomas Lenz)

The Match-X system is a collection of standardized modules for the constitution of microsystems consisting of electrical, optical and mechanical modules whose interconnections are governed by a standardized interface. The VDMA standard sheet 66305 concisely describes the structure of the interfaces that enable the flexible integration of different components into microsystems. The now commercially available construction kit modules (http://www.match-x.org) can be supplemented by user-developed modules. The kits enable the assembly of the most diverse set of products. By combining and varying components, manufacturers can design and manufacture customer-specific systems without having to develop entirely new products. Example modules include: CPUs, A/D converters, signal conditioners, actuators, pumps, and labs on a chip. (See Figure D.10).

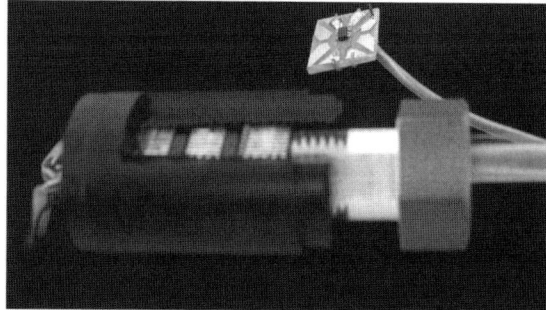

Figure D.10. Match-X components and integrated system.

Mechanical Microgrippers (Mr. Thomas Lenz)

The Group is involved in the development of different types of microgrippers that use different principles of actuation (viz., piezo-electric,

SMA, electrostatic) for small displacements (< 1 mm) and small gripping forces (< 1 N). The latest attempt is the development of a hydraulic micro-gripper (See concept in Figure D.11). So far only concepts have been ana-lyzed, and two sealing methods are being considered (Gap ring and O-ring seal). The anticipated advantages include higher gripping forces and the possibility of independently controlling the gripper's fingers.

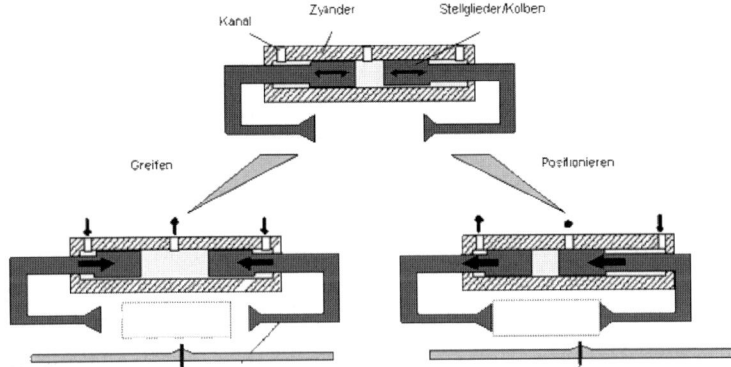

Figure D.11. Concept of the hydraulic microgripper.

Manufacturing of the Future (Mr. Ralf Muckenhirn)

Under this project heading, an integrated manufacturing concept was developed for the assembly of hard disk drives for IBM. The concept was successfully demonstrated but not implanted. However, the technologies developed are finding their ways into other applications. The overall sys-tem philosophy consists of the integration of the logistics, processing, equipment hardware and software. The starting point is design segmenta-tion that facilitates the definition of the required manufacturing steps/modules that result in a process recipe.

The developed hardware consisted of modules (boxes) W1.3 m × L1.2 m × H2 m. Each of the modules contains a cleanroom (volume) with its own filtration (class 100), an assembly robot, an I/O area, docking doors, media-couplers and alignment brackets. Modules communicate through a central handling system. Trays of parts, containing inlays for accommodat-ing parts of different shapes, are shuttled between modules. Modules are configured into various processing clusters (see Figure D.12 for an exam-ple). Modules can be disconnected during operation and connected to other clusters allowing for a dynamic reconfigurability of a particular cluster. Each module may contain specific processing components dictated by its basic function. The module controllers are also modularized and are auto-matically reconfigured. The price of a module was estimated to be in the range of $150,000–$200,000. About 40 modules would have been required for the fully automated assembly of hard disk drives.

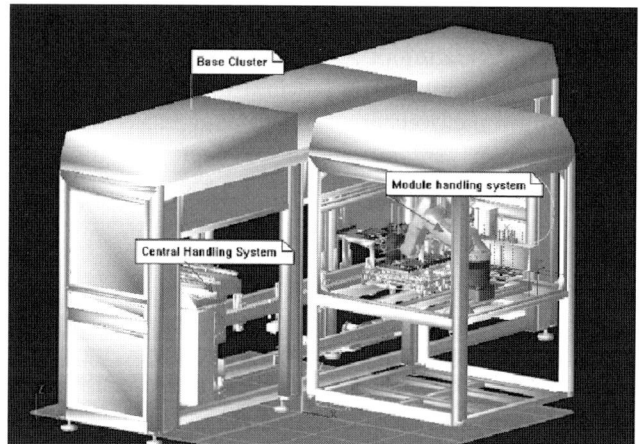

Figure D.12. Assembly cluster constituted from modules.

Miniaturized Reconfigurable Microassembly System (Mr. Udo Gommel)

The results of the "MiniProd" project that was sponsored by industry and Federal Ministry for Education and Research (BMBF) were presented. The aim of the project was to develop a miniaturized highly flexible and integrated microassembly system commensurate to product size. A "plug-and-produce" principle was envisioned in which the user configures the system from individual functional modules. The modular philosophy has been applied to the hardware, software and control architecture of the system.

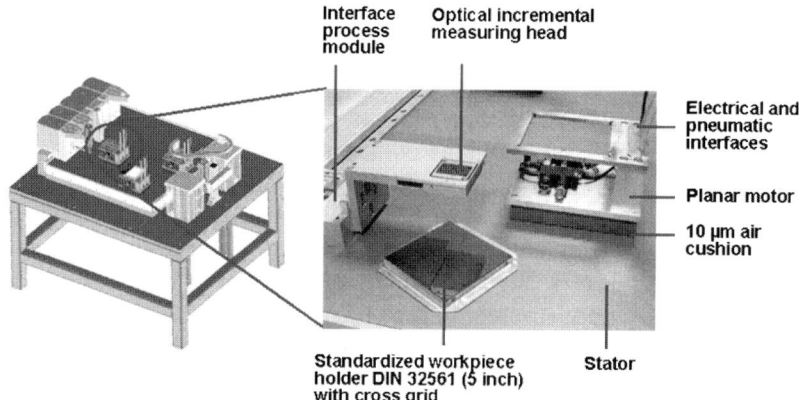

Figure D.13. The MiniProd system.

A conceptual representation of the system is shown in Figure D.13. It uses a planar motor for accurately positioning and transporting components between processing modules (e.g., assembly and bonding). The absolute positioning accuracy is 2 μm without and 0.15 μm with temperature compensation. One of the project sponsors is in the process of developing a commercially available product based on this concept. The project was completed in 2004 and demonstrated on the assembly of a laser diode.

SUMMARY AND CONCLUSIONS

A number of systems and concepts for micromanufacture were presented. One of the principles that permeates the work in IPA and that is closely related to the objectives of the current study is the use of principles of modularization and standardization of hardware and software interfaces to achieve a high degree of flexibility in adapting to different production demands. It is also important to note that some of the developed principles are finding their way into industrial practice and commercial products.

REFERENCES

Fraunhofer IPA. http://www.ipa.fraunhofer.de (Accessed October 20, 2005).

Gaugel, T., M. Bengel, D. Malthan, J. Schließer, J. Kegeler, G. Munz. 2004. Miniaturized Reconfigurable Micro Assembly System. In proceedings of *International Precision Assembly Seminar IPAS 2004*. February 11-13, Bad Hofgastein, Austria.

Match-X. http://www.match-x.org (Accessed October 20, 2005).

Reinst- und Mikroproducktion. http://www.mikroproduktion.de (Accessed October 20, 2005).

Site: **Fraunhofer Institute – Production Systems and De-
                                    sign Technology (IPK)
                                    Institut Produktionsanlagen und Konstruktion-
                                      stechnik (IPK)
                                    Berlin University of Technology Institute for Ma-
                                      chine Tools and Factory Management (IWF)
                                    Technische Universität Berlin Institut für
                                      Werkzeugmaschinen und Fabrikbetrieb (IWF)
                                    Pascalstraße 8-9
                                    10587 Berlin, Germany**

Date Visited: April 7, 2005

WTEC Attendees: M. Madou (Report author), M. Culpepper, J. Cao,
                          D. Bourell, S. Ankem, G. Hazelrigg

Hosts: Dirk Oberschmidt, Dipl.-Ing.: Group Manager, Micro
                            Production Technology, Tel: +49-30-6392-5106,
                            Fax: +49-30-6392-3962,
                            Email: dirk.oberschmidt@ipk.fraunhofer.de
                          Ulrich Doll, Research Engineer: Head of Department,
                            Production Systems/Manufacturing Technology,
                            Tel: +49-30-39006-147,
                            Email: ulrichdoll@ipk.fraunhofer.de
                          Sascha Piltz, Research Engineer, Technical University
                            Berlin, Department of Machine Tools and Manu-
                            facturing Technology, Tel: +49-30-6392-5105,
                            Fax: +49-30-6392-3962,
                            Email: piltz@iwf.tu-berlin.de
                          Kai Schauer

BACKGROUND

Fraunhofer Institute for Production Systems and Design Technology
(Fraunhofer Institut für Produktionsanlagen und Konstruktionstechnik—
IPK) is one of six Fraunhofer Institutes located in Berlin. IPK's mission is
to turn scientific research into practical applications by overseeing industrial processes from the research and development stage through to production. IPK has a staff of 446 scientists, service staff, and students. Its 2004
budget was €24 million.

IPK shares the Production Technology Center (Produktionstechnischen
Zentrum—PTZ) campus in Berlin with the Berlin University of Technol-

ogy Institute for Machine Tools and Factory Management (Technische Universität Berlin Institut für Werkzeugmaschinen und Fabrikbetrieb—IWF). Organizationally, most departments at the IWF are mirrored in the Faculty for Process Sciences (Fakultät Prozesswissenschaften—FGI) at TU Berlin. Both IWF and IPK are members of the European Society for Precision Engineering and Nanotechnology (EUSPEN).

Visiting panelsists were invited to review the activities of four departments. The meeting took place at the Center for Microsystem Technology (Zentrum für Mikrosystemtechnik—ZEMI) in Berlin-Adlershof, of which IPK and IWF are founding members. The Center, which supports collaborative programs for small and medium enterprises (SME) in the microsystem technology and biotechnology fields, received €15.85 million in funding from the European Union in 2004, primarily for equipment.

RESEARCH AND DEVELOPMENT ACTIVITIES

In addition to the development of conventional manufacturing technologies, IPK and IWF have also recently begun conducting research and development of microproduction processes. As mentioned above, their primary objectives in this area are to assist small- and medium-size firms to develop manufacturing techniques and systems and apply them to cost-efficient manufacturing process chains. IPK and IWF support and assist firms in realizing their microproducts from the idea stage through prototyping to the market stage.

Specifically, IPK and IWF support the following:

- Research, development and application of manufacturing techniques, including: ultraprecision turning, milling and grinding; high-precision milling and drilling; micro wire-, die sinking-, drilling- and path-electrical discharge machining; and generative processes such as microinjection molding and rapid prototyping

- Development, integration and application of machine elements to adapt manufacturing equipment to the requirements of micromanufacturing

- Development of micromilling tools, mold tools for microstamping, and tools for micromolding and massive forming

- Order-specific application of manufacturing techniques for the production of high-performance material components (e.g., high-temperature heat exchanger elements made from Co-based alloys), microelements for medical applications (e.g., minimally invasive surgical instruments made from Ti-based alloys), optical elements (e.g., prismatic lens arrays made from nonferrous metals), and microstructured forming tools for mass production of microsystems (e.g., microinjection molding tools)

- The combining of manufacturing techniques with complete process chains

In cooperation with IPK, the IWF recently established a satellite facility at the Center for Medical Physics and Technology (Zentrum für medizinische Physik und Technik—ZMPT) at Berlin-Adlershof. The facility includes 175 square meters of fully air-conditioned manufacturing cells and labs. Currently, the working group at the ZMPT facility consists of eight scientists, 15 student assistants, and one service employee.

Figure D.14. Microproduction Laboratory.

The laboratory features micro wire and die-sinking EDM equipment, high- and ultraprecision micromillers and injection molders, and optical metrology equipment. This equipment permits research and development activities in micro EDM, ultraprecision machining, high-precision micromilling, and microinjection molding for evaluation purposes.

Ultraprecision Machining

The facility features a Moore Nanotechnology System Nanotech® 350 FG five-axis-UP-Machining Center. The machining center is capable of travel of up to 350 mm in the X axis, 300 mm in Y, and 150 mm in Z with a motion accuracy of 0.3 μm over the full travel range. The programming resolution is 8.6 nm, with a work spindle speed range of 50 to 6,000/min.

Maximum speed C is 2,000/min, and maximum speed B is 50/min. Motion accuracy is less than 50 nm (axial and radial), and the high-speed spindle speed range is 25,000/min to 100,000/min. Accuracy of the machining achieved is 0.1 μm.

Sperical and aspherical optics are a major focus of the lab, as is the machining of nonferrous metals and ceramics. Research topics include: machining technologies for turning, milling, and grinding complex free-form surfaces; machining of microoptical prototypes from nonferrous metals and polymers with Ra < 2 nm; the development of technologies for the machining of steel and advanced ceramics; and testing of new cutting tools for ultraprecision machining.

Figure D.15. UP-machine tool and machining example.

High-precision Micromilling

The main primary emphasis here is on the machining of steel. Miniaturized end mills are made from cemented carbides. Because new design concepts are needed for the end milling of very small parts, the group is working with tool suppliers to deliver appropriate new tools. Currently, spindle run-out is too large; spindles with higher speed capacity are needed. Research topics in this area include: the development of process technologies and strategies for high-precision milling using miniaturized end mills; the development and optimization of machine tools and machine tool components; testing and development of innovative miniaturized milling tools; and benchmarking of micro and precision cutting tools.

Micro Electrical Discharge Machining (μEDM)

The emphasis here is on the development of new process variants (e.g., microcontouring operations), technologies, process strategies, and machine tool components that can be adapted to the demands of micromanufacturing. Currently, structural dimensions down to 20 μm can be machined with an accuracy better than 2 µm and an obtainable surface roughness of Ra < 0.1 µm. The facility has the capability to machine materials of high mechanical strength and hardness, such as hardened steel, tungsten-carbide, and electrically conductive ceramics. New electrode materials (e.g., tungsten-copper and cemented carbide) and new electrode concepts have also been tested.

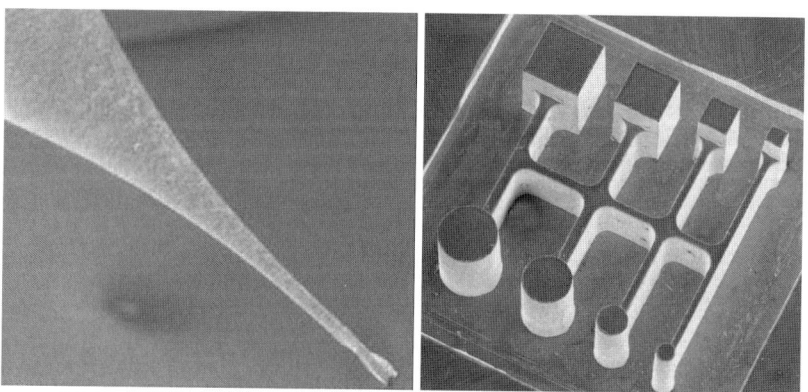

Figure D.16. High-precision micromilling: new developed mill with diameter 0.1 mm and machining example in hardened stainless steel.

In wire EDM, for example, instead of using abrasive and cost-expansive tungsten wires, they use a brass-coated steel wire with a tensile strength of more than 2,000 N/mm². IPK and IWF have developed new machining concepts using ultraprecision milling to fabricate microstructured form electrodes. Research topics here include: optimization of machining parameters and process strategies for micro die-sinking and microwire EDM; fundamentals of material removal; surface and sub-surface generation; research on new electrode materials and concepts (e.g., form and wire electrodes); development of new process variants (e.g., micro ED-milling, vibration-assisted micro ED-drilling, and ED-grinding); the development of new machine tool systems; structuring of micromechanical components; and micro die and mold-making applications.

1. Microinjection molding:

A machine tool is used for evaluation purposes. Microstructures are replicated on macroscopic components (m2M). Research topics include: design, simulation, and manufacture of micromolds; feasibility studies for the mass production of polymer microparts (i.e., new materials and replication limits); the production of micromechanical and microoptical prototypes; and the replication of microstructures on macroscopic components (m2M).

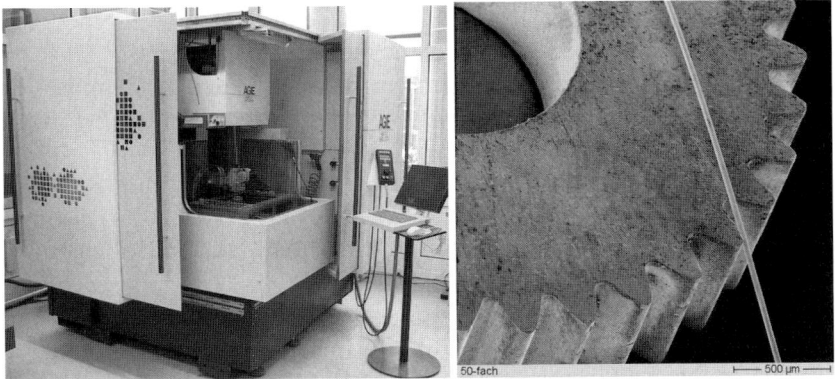

Figure D.17. Microwire EDM: machine tool Agiecut Evolution 2 SFF and microwire-eroded gear with wire electrode, diameter 0.03 mm.

2. Micrometrology

Optical methods are used to measure the geometry and surface roughness of micromechanical parts. Two different systems are used. A Zeiss LSM 5 Pascal confocal laser-scanning microscope is used to measure the geometric features of components with steep flanks, with lateral resolutions of 0.2 μm and axial resolutions of 40 nm. ZygoLot NV5010A white light interferometer capable of a resolution of 20 Å is also used.

Figure D.18. Confocal laser scanning microscope and example.

The working group at ZMPT has already successfully realized several research projects for the developement of manufacturing elements, tools and technologies. For example, the group configured and technically qualified the DFG-SPP Micro-Mechanical Production Technique machining center, in which microwire- and die-sinking ED has been integrated. An essential aspect of this effort was the preparation of technology tables that are either integrated into the control or offered separately for adaptation to different applications.

The group also scrutinized and then developed a process model for SPP's Micromachining of Metal Composite Materials project. A DFG project named Modelling of Dimensional Effects in Scaled Machining Processes required the development of systems for simulating the micromachining of heterogeneous materials.

Together with industrial users, the IWF has developed miniature milling, drilling and threading tools for the Innovative Technologies for the Manufacturing with and Application of Precision Machining Tools in Micro Techniques program.

At IWF and also within the BMBF-funded network-project called Micro-Structured and Processed High-Performance Materials for the Warm Forming of Special Glasses, several micro-ED alternatives for the structuring of dies came under scrutiny and were evaluated. The group is also currently carrying out another DFG project for the micro-ED machining of rotating parts.

Additionally, IPK participates in the Masmicro network project, a European consortium of 36 partners with a budget of €20 million over four years. Masmicro is interested in the development of the mass production of microparts through forming techniques. For Masmicro, IPK develops technologies for the production of microstructured forming tools.

Other current research projects include ultraprecision machining, micromeasurement techniques, laser material processing, the application of micromachining for medical technologies, and the manufacture of MEMS prototypes using ultraprecision manufacturing techniques. In cooperation with several German companies, IPK is developing a high-precision machine tool that is capable of micromilling, microlaser processing and machine tool integrated optical measuring.

Besides technology and machine tool development, a crucial aspect of IPK's and IWF's activities is the transformation of research results into market-ready applications as a service to potential industrial users. IWF and IPK have realized a series of bilateral research projects in which they consulted with and advised industrial companies on ways to introduce micromanufacturing systems and processes.

The hosts also noted that FGI in Stuttgart carries out the assembly and inspection of machines with very small footprints (called "micro fabs").

SUMMARY AND CONCLUSIONS

As with all other European sites that panelists visited, the IPK/IWF group is exceedingly well-equipped. They are also able to fine-tune and work on their techniques and processes for extended periods of time. As a result, they have accumulated a tremendous amount of practical experience and knowledge. The shared facility appears to be very well-run, and their success in spinning-off companies can be gleaned from noting the high percentage (12%) of former employees who have gone on to start their own companies. While progress might be incremental, it is also continuous and guaranteed.

Site: **Fraunhofer Institute – Production and Laser Technology (IPT)/(ILT)**
Institut Produktionstechnologie (IPT)/Institut Lasertechnik (ILT)
Steinbachstr. 17
52074 Aachen, Germany

WTEC Attendees: M. Culpepper (Report author), D. Bourell

Hosts: Dr. Arnold Gilner, Head of the Department of Micro Technology (ILT), Tel: +49-241-8906-148, Fax: +49-241-8906-121, Email: Arnold.gilner@ilt.fraunhofer.de
Frank Niehaus (IPT), Dipl.-Ing., Tel: +49-241-8904-155, Fax: +49-241-8904-6155, Email: frank.niehaus@ipt.fraunhofer.de
Jörgen von Bodenhausen (IPT), Research Scientist, , Tel: +49-241-8904-233, Fax: +49-241-8904-6233, Email: joergen.von.bodenhausen@ipt.fraunhofer.de

BACKGROUND

The mission of the Fraunhofer Institute for Production and Laser Technology is to transfer new production and laser technologies from research to industrial practice. The Fraunhofer Institutes derive approximately 50% of their funding from public funding and 50% from private/industry funding. In combination, the Institutes have approximately 400 employees, with roughly 70% permanent staff and 30% students working on their diploma thesis. These Fraunhofer Institutes appear to be unique in that they actively seek to provide space within their campuses in which researchers from collaborating companies may set up an office and/or conduct research. There is a clear difference in the degree of success between projects in which a company takes advantage of this opportunity vs. projects in which this does not occur. The services which the Institutes provide consist of consulting, feasibility studies, basic research in production/laser technology, technology transfer, system integration and quality assurance. Both Institutes are affiliated with a local university, RWTH Aachen. The total annual operational budgets are €11.8 million for IPT and €16.4 million for ILT (IPT 2003 Annual report, 2003, ILT 2003 Annual report, 2003).

RESEARCH AND DEVELOPMENT ACTIVITIES

Overview

The mission of the Fraunhofer Institutes for Production and Laser Technologies is to transform technologies from their respective areas to the point at which they become attractive for commercial use. The Institutes are conducting a wide array of basic and applied research in microtechnology. The main highlights of the visit were micromilling (process and tool development at IPT), the formation of optically reflective microscale features on macroscale molding/replication tools (IPT), laser ablation processes used to form microscale molds (ILT) and laser welding processes used to join metals and dissimilar polymers (ILT).

Micromilling (IPT)

A substantial effort is focused upon the development of tooling and cutting process parameters for micromilling. Their approach to micromilling utilizes traditional macroscale precision machine tools to drive and position microscale cutting tools. Figure D.19 (left) shows the process of micromilling a mold. The milling was completed using the precision milling machine (approximately 5 feet tall) shown on the right of Figure D.19.

Figure D.19. Micromilling process (left) with HSPC precision mill (right).

Figures D.20 and D.21 show scanning electron microscope (SEM) pictures of 300 and 50 micron diameter end mills which IPT has developed in collaboration with a local cutting tool supplier. The tools are fabricated using electro-discharge machining (EDM). The design of the tools is based upon experience and macroscale cutting fundamentals. Process parameters for the milling are being investigated with an experimental approach. Results to date are promising, though limited to the formation of geometry. Quantification of cutting forces has yet to be realized.

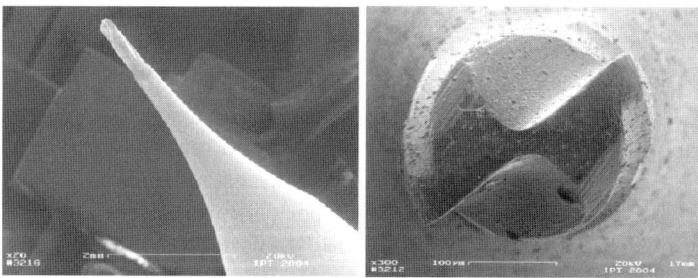

Figure D.20. End mill, 300 micron diameter tool.

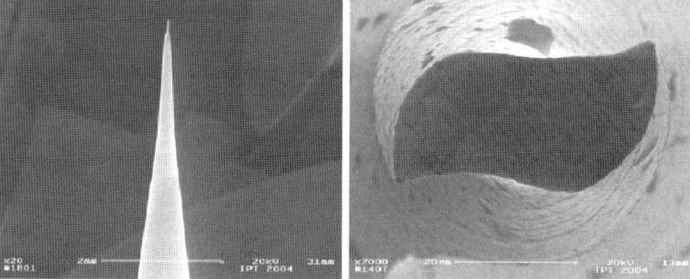

Figure D.21. End mill, 50 micron diameter tool.

There is an effort to utilize hard coatings to prevent tool wear. A major challenge for the IPT team is to be able to control the coating thickness so as not to adversely affect the geometry of the tool. Specifically, the features of importance (e.g. cutting edge) may only be a few microns in size. It is difficult for their vendors to supply coatings that have a thickness that is small compared to these feature sizes. Preliminary results show that the coating may be applied with a thickness of 2 microns. They are working to reduce this to a fraction of a micron. The coatings appear to have a positive effect on wear resistance as shown in Figure D.22.

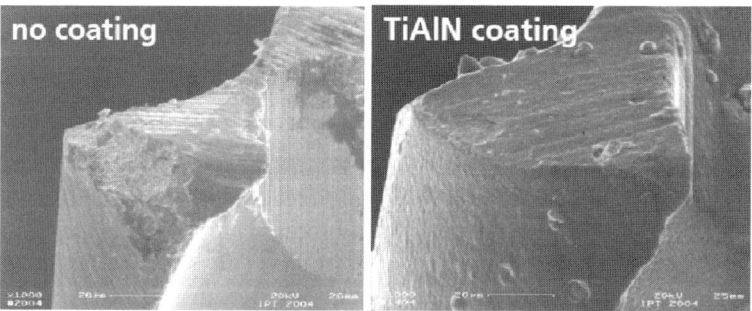

Figure D.22. Comparison of the wear on micro end mills without (left) and with (right) hard coatings.

Using these tools, IPT has been able to create the positive and negative rectangular geometries shown in Figs. D.23 and D.24. The work has thus far been able to generate controlled features (size and position) for structures over 100 microns in minimum feature size. Two major challenges for process development are the detection of tool failure and accurate measurement of cutting forces. A portion of the future effort is directed toward solving these problems.

| 300µm Tool | 400µm Tool | 500µm Tool |
| Magafor 9500-H | Magafor 9509-H | Magafor 9509-H |

| ⇒ depth of slot = 1000 µm | ⇒ depth of slot = 2000 µm | ⇒ depth of slot = 2500 µm |
| (40mm long) | (40mm long) | (40mm long) |

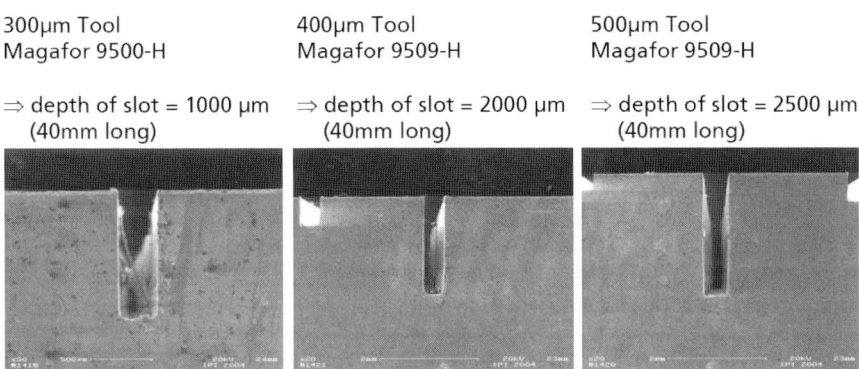

Figure D.23. Slots cut by micromilling processes.

Figure D.24: Rectangular bosses fabricated via micromilling processes.

Grinding of Microscale Parts and Features (IPT)

We also observed microgrinding processes. Figure D.25 shows the types of devices which IPT has fabricated using a grinding process. These processes utilized traditional, macroscale machine tools to position and turn grinding wheels. In some instances (see Figure D.25 bottom left), the grinding tools possessed microscale features. The work in this area is primarily experimental. With these grinding processes, it is possible to obtain feature size/placement accuracy on the order of microns.

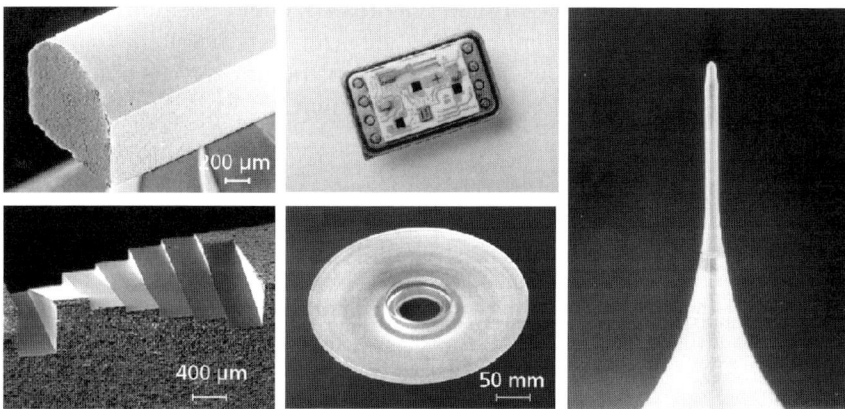

Figure D.25. Grinding of micro- and mesoscale parts and features.

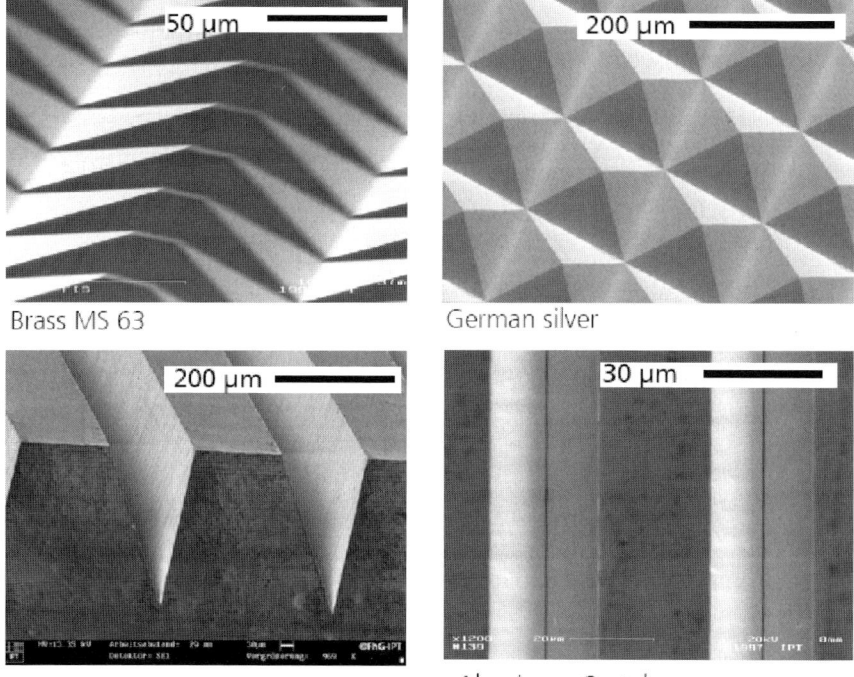

Figure D.26. Micromilled features on macroscale (tens to hundreds of centimeters size) objects/surfaces.

Turning of Meso/Microstructure on Macroscale Parts I (IPT)

IPT has leveraged their expertise in precision engineering, precision machine design, and precision metal cutting to fabricate microscale features on macroscale parts. Figure D.26 shows several examples of the more complex surfaces which they have fabricated. This fabrication is conducted using traditional macroscale equipment and a combination of micro- and macroscale tooling. A significant driver of this process/technology is optically reflective surfaces. For instance, this research is, in part, focused upon making molds for high-visibility street signs. The top-right picture in Figure D.26 shows the surface texture of a mold which when replicated via polymer molding, results in a highly reflective sign surface. The effort at IPT has been focused on the development of the next generation of tooling which can be used to form/mold these signs.

IPT has also successfully formed a variety of complex optical surfaces via traditional turning equipment which has been augmented with a fast tool servo (FTS). A fast tool servo is a device which enables one to position the cutting point of a tool along the axis of lathe rotation. The positioning of the tool is synchronized with the rotation of the part so that non-axisymmetric shapes may be cut. This may occur at a speed that is comparable to traditional turning (hundreds to thousands of rpm). IPT has developed several custom FTSs for various applications, three examples of which are shown in Figure D.27.

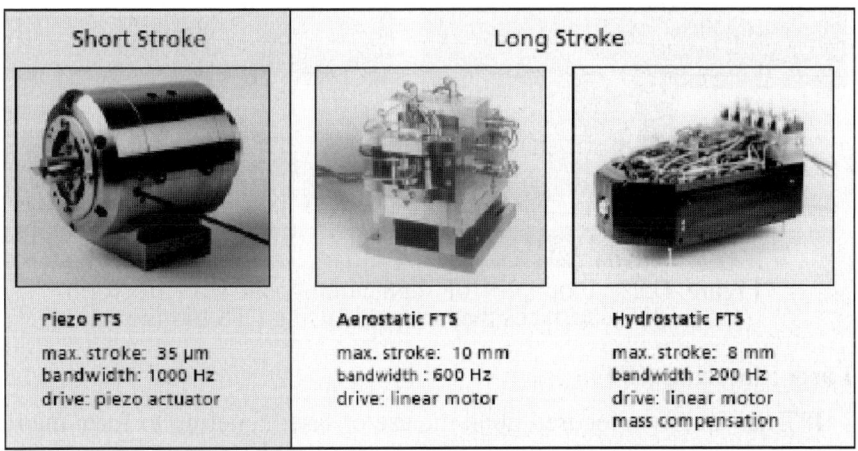

Figure D.27. Fast tool servo mechanisms developed by IPT. The left and middle FTSs are approximately 12 x 6 x 8 inches. The rightmost FTS is approximately 24 x 12 x 8 inches.

IPT is working on applying this technology to fabricate micro- and mesoscale optically reflective surfaces on macroscale parts. Figure D.28

shows an example of a macroscale part which is fixtured to a precision lathe (left) and the resulting surface topography (right) achieved via FTS turning.

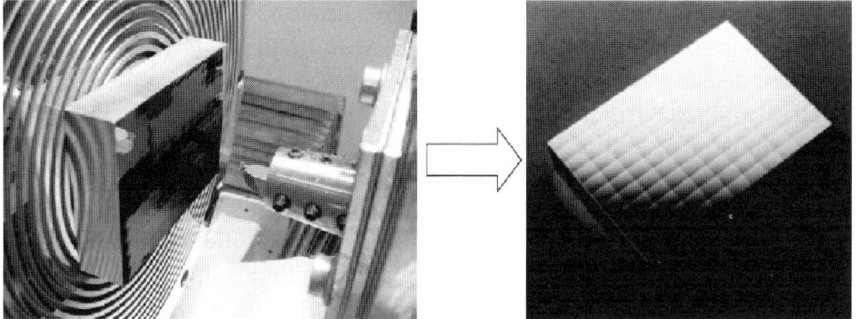

Figure D.28. Rectangular, optically reflective surface topography formed by turning with a fast tool servo.

Figure D.29 shows a few types of molds that were made using this process, a replicated optical part (polymer), and an example of how these optics may be used to reflect and shape light to form signs.

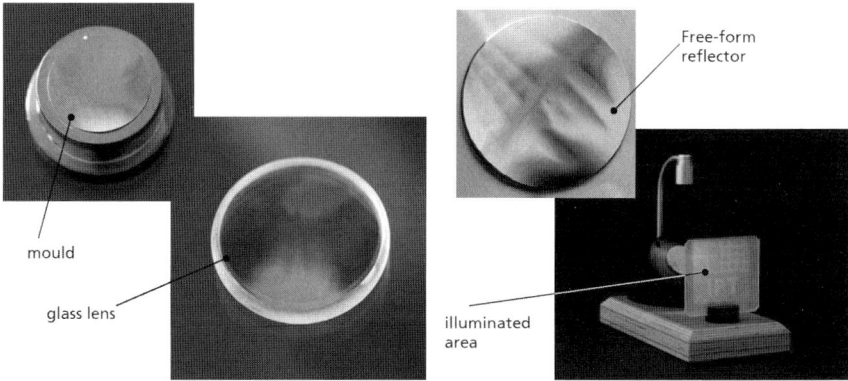

Figure D.29. Examples of replication tools with freeform surfaces that are made using FTS turning.

Laser Sintering (IPT)

IPT has an effort focused upon the use of laser sintering to form microscale structures. Although the work is in a nascent stage, some promising results have been obtained. Figure D.30 shows several geometries which have been fabricated via this process. There are limitations on making sharp corners, due mainly to the radius of the laser spot (several microns). It appears that there is a collaborative effort with ILT Aachen to advance this research.

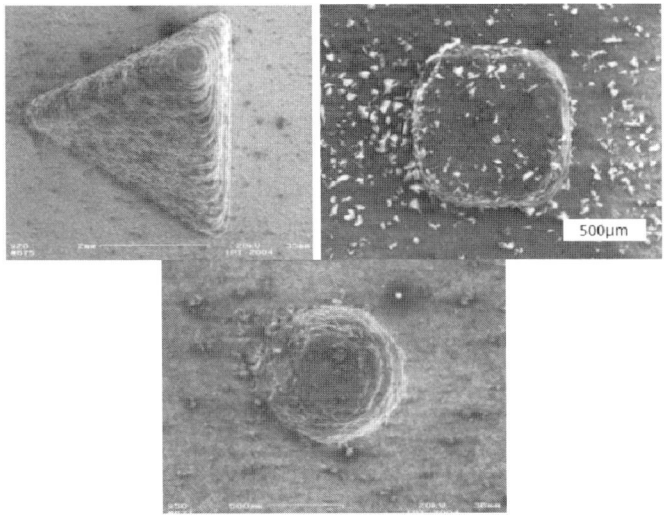

Figure D.30. Prototype features fabricated by laser sintering, a pyramid at left, a cube in the center, and a hemisphere on the right.

Laser welding (ILT)

ILT has developed a diode laser technology that enables them to precision weld microscale metallic parts. Figure D.31 shows an example of a mesoscale gear that has been microwelded upon a shaft. This weld is surprisingly only a few hundred microns deep. A variety of continuous and pulsed welding processes are under investigation. This technology is primarily targeted at applications in the automotive, electronic and watch production fields.

Figure D.31. Example of a welded microgear (gear diameter ~ 1 mm).

ILT has developed a novel process for laser welding of sub-surface interfaces between microscale polymer parts. The process is illustrated in Figure D.32. The wavelength of the laser radiation and the material properties of the top polymer component are matched such that the top component is transparent to the laser radiation. The bottom polymer component is either comprised wholly of material that absorbs the laser radiation or is impregnated with particulates that absorb the specific wavelength of the laser. As such, the energy from the laser radiation heats the materials at the sub-surface interface above their melting points.

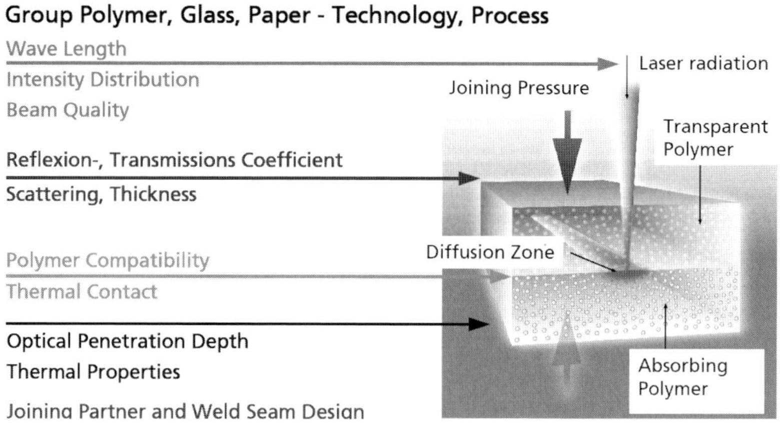

Group Polymer, Glass, Paper - Technology, Process

Wave Length

Intensity Distribution

Beam Quality

Reflexion-, Transmissions Coefficient

Scattering, Thickness

Polymer Compatibility

Thermal Contact

Optical Penetration Depth

Thermal Properties

Joining Partner and Weld Seam Design

Laser radiation

Joining Pressure

Transparent Polymer

Diffusion Zone

Absorbing Polymer

Figure D.32. Process used to weld polymers at a subsurface interface.

Figure D.33 shows two examples of polymer parts that have been welded using ILT processes. A major focus of the work is placed upon modeling the thermo-mechanical behavior at the joined interfaces. This work aims to minimize (1) the amount of energy which is required to form an adequate bond, (2) the thermal energy which spreads into the adjacent material(s), and (3) the depth to which thermal energy penetrates the joined components.

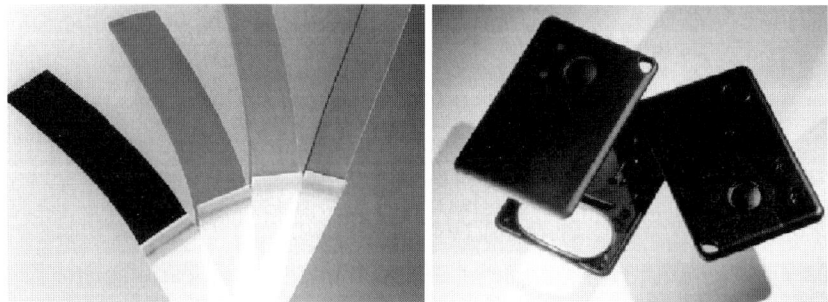

Figure D.33. Welded polymer assemblies; dissimilar materials (left) and keyless go card covers (right).

Laser ablation (ILT)

ILT has leveraged their knowledge of laser technology to fabricate meso- and microscale molding tools by laser ablation. Figure D.34 shows examples of the different types of materials and tool geometries that have been processed via laser ablation.

Steel   Ceramics   Polymers

Figure D.34. Materials and shapes formed via laser ablation process.

The accuracy and surface finish that may be obtained via this process depend somewhat upon the type of material. Rough numbers of several microns accuracy and a few microns surface finish are generally descriptive of the process capability.

Precision Positioning via Laser Induced Deformation (ILT)

Laser energy has been used for several years to induce position changes via permanent deformation of electronic and microoptical package leads. This process is for the most part experience/experiment-based and requires several "tweaking" shots with the laser to obtain a desired alignment. ILT has developed a process in which they are able to design compliant mechanisms or flexures which deform deterministically when heated or deformed by laser. The localized deformation process is illustrated on the left of Figure D.35, and two examples of deformable alignment mechanisms are located on the right of Figure D.35.

Figure D.35. Process used to permanently deform flexible structures for precision alignment of optical and microscale components. Laser heats metal (left) and induces a permanent deformation in a compliant mechanism/flexure. The two right-most images show two types of deformable structures which are several mm in size and may be used to achieve micron-level alignment.

The benefit in this approach is that one may ascertain the location at which to induce the local material deformation and the amount of energy required to induce the desired deformation(s). Deterministic, multi-axis alignments may be obtained in this way. Multi-axis alignment is nearly impossible using non-deterministic, iterative approaches.

SUMMARY AND CONCLUSIONS

Both institutes have an impressive portfolio of research in microtechnology-related processing. IPT's expertise in the design and use of precision equipment is a strength which they have leveraged to form novel and useful microscale features/parts. They have successfully used a mix of (a) traditional macroscale machine tools and (b) custom-made, macroscale hardware. They have an impressive array of precision machine tools and precision measurement equipment for prototype and experimental work. IPT has a strong experimental program focused upon determining the process parameters, tool designs and equipment characteristics that are required for microscale metal-removal processes. Their efforts are producing many proof-of-concept prototypes and generating observations that will form a base of data for future analytic work.

ILT's expertise in laser technologies (laser design, laser physics, laser processing) has been used to produce new/improved laser fabrication processes. The efforts at ILT are more broadly focused, covering applications ranging from bonding, new laser designs (e.g., diode lasers), precision alignment processes and tooling design/fabrication. Their work, a mix of experimental and analytic modeling, is likewise producing many proof-of-concept prototypes and observations for model validation. Their technology has been used in several products, most notably the joining of the casings for "go cards." Unfortunately, time constraints prevented a tour of their facility, and therefore firsthand observations of their equipment and testing facilities are not available.

Overall, the scientific approaches (experimental or analytic) used by the Institutes and the quality of engineering work is among the best seen to date. The permanent staff is very talented, and the students we interacted with were comparable to the best with which the panel has ever dealt. The students (From Aachen University) were particularly impressive as there did not seem to be a technical/engineering question that they could not answer. The link with Aachen University is clearly a competitive advantage for both Institutes.

REFERENCES

IPT 2003 Annual report, published by the Fraunhofer Institute for Production Technology, Steinbachstraße 17, 52074 Aachen, Germany, http://www.ipt.fraunhofer.de/cms.php?id=1411 (Accessed October 20, 2005).

ILT 2003 Annual report, published by the Fraunhofer Institute for Laser Technology, Steinbachstraße 15, 52074 Aachen, Germany, http://www.ilt.fraunhofer.de/ilt/pdf/eng/JB2003-eng.pdf (Accessed October 20, 2005).

Site: **Fraunhofer Institute – Reliability and Microintegration (IZM)**
                     Institut Zuverlässigkeit und Mikrointegration (IZM)
                     Gustav-Meyer-Alee 25, 13355 Berlin, Germany
                     Tel: 49-30-46403 100, Fax: 49-30-46403-111
                     E-mail: info@izm.fraunhofer.de
                     URL: http://www.izm.fraunhofer.de

Date of Visit: April 6, 2005

WTEC Attendees: K. Rajurkar (Report Author), S. Ankem, D. Bourell, J. Cao, M. Culpepper, G. Hazelrigg

Hosts: Dr. Oswin Ehrmann,
                        Email: ehrmann@izm.fraunhofer.de
                     H. Potter, Email: potter@izm.fraunhofer.de
                     H. Oppermann,
                        Email: oppermann@izm.fraunhofer.de
                     J. Sommer, Email: sommer@izm.fraunhofer.de
                        Tel: +49-30-46403-100, Fax: +49-30-46403-111
                     Dr. M. Wiemer, Email: wiemer@che.izm.fhg.de,
                        Chemnitz, Reichenahiner Strabe 88, 09126,
                        Germany, Tel: +49-30-37-15397-474,
                        Fax: +49-37-15397-310

BACKGROUND

The IZM Institute specializes in microelectronics and microsystems integration packaging research and development. It provides a research and technological knowledge base to industries in the areas of photonic packaging, radio frequency (RF) and wireless, 3D systems integration, microelectromechanical systems (MEMS) packaging, microreliability and life estimation, wafer-level packaging and thermal management.

The Institute has about 250 employees, including scientists and engineers. Additionally, it also hires students from regional and other universities to assist on the research projects. The total budget of the Institute is about $40 million with 20% from the German government and remaining 80% from the industry through R&D projects. The institute is organized into various departments including mechanical reliability and micromaterials, environmental engineering, chip interconnect technologies, photonic and power system assembly, microdevices and equipment, micromechatronics systems and micromechanics, actuators and fluidics. Some of the recent research projects are described below.

RESAERCH ACTIVITIES

Hetrointegration

The project deals with the new generation of microsystems embedded with sensors and actuators for future objects of daily life outfitted with information processing abilities. This research focuses on developing a hybrid technology to assemble single components and maintain the benefits of monolithic integration in terms of miniaturization and reliability and flexibility. This method of integrating different components with electronic, optical, bioelectronic and micromechanical functionality at wafer level is known as "hetrointegration" The needed design of the interface between human and electronic systems is another important issue that is being addressed.

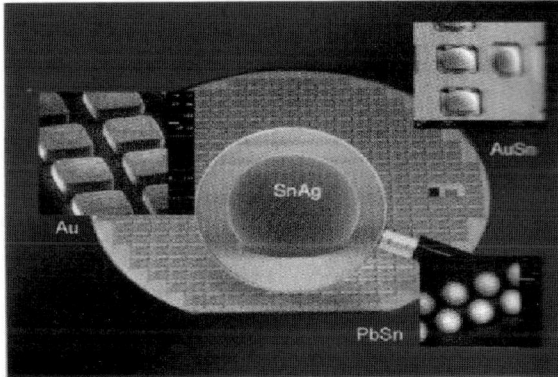

Figure D.36. Wafer-level packaging.

Wafer-level Packaging (WLP)

The WLP process involves packaging that is performed at wafer level. The program considers wafer bumping (using electroplating, electroless deposition, and stencil printing. redistribution technologies (such as processes using photodefinable dielectrics)), wafer-level inspection and wafer dicing. Currently, this group is working on lead-free bumping, development of processes on 300 mm wafers and wafer-level test.

3D System Integration

The IZM works on various 3D integration techniques such as stacking of packages, stacking of integrated circuits (ICs), and vertical system integration (VSI). The program goal is to establish a competence center in the field of 3D system integration with industrial partners.

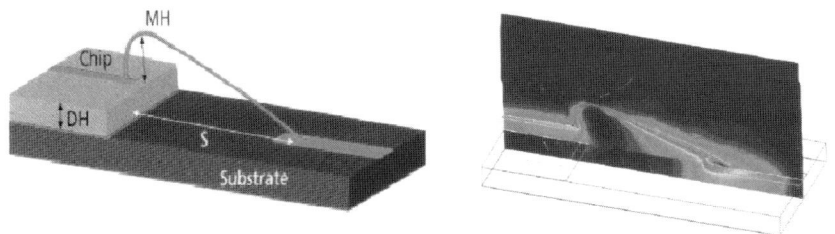

Figure D.37. EM simulation, magnitude of E-field of upper wire geometry.

RF and Wireless

The IZM provides customized, complete RF and wireless system development, i.e. methodology, board, assembly, prototyping and characterization and testing.

Photonic Packaging

In this program single packages, modules or subsystems consist of at least one optoelectronic device or optical interconnector. The research and development activities include wafer and single-chip bumping, fluxless chip assembly, active and passive fiber alignment and laser fusing of fibers on microoptical components. Additionally, issues related to electrical optical circuit board (EOCB) and radiation sources and detectors are also addressed.

Thermal Management

The issues of heat spreading and removal pose design challenges as the power and power density of microcomponents increase. Additionally, the thermomechanical reliability is also an important issue that needs to be addressed. This IZM program provides solutions for efficient cooling of miniaturized electronic components from the silicon and system level. The group performs thermo-fluidic simulation, thermal and mechanical material characterization, infrared (IR)-thermography, thermo-electric coupling and thermal strain measurements of micro and nanostructures (Figure D.38).

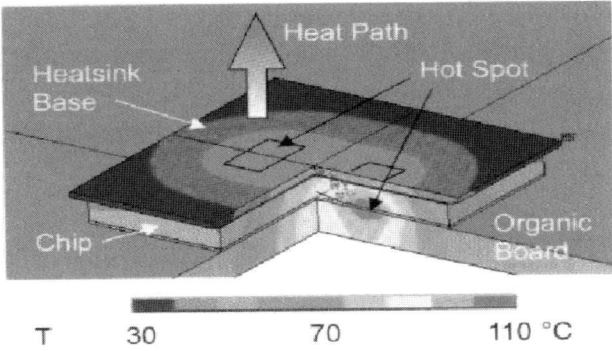

Figure D.38. Thermal simulation of reverse-side cooled die
mounted in flip-chip technology with two hot
spots.

Mechanical Reliability and Micromaterials

This department in the Fraunhofer Institute, focusing on micromaterials
and micropackaging technologies, is one of the leading groups in thermal
and thermomechanical analysis of microcomponents and microsystems in
Europe. Besides its other activities, the department organizes many international conferences. Since the summer of 2000, the department has been
running the Micro Materials Center Berlin (MMCB). This is one of three
German centers with a special competence in micromaterials and micro-
technologies; the German Ministry of Education and Research (BMBF)
funds it. The center is very attractive to partners from industry, and, within
the past four years, it realized research projects with industry with an ap-
proximate value of €6 million in the field of materials research for applications in microtechnology. The department also continues to coordinate the
IZM Micro Reliability Research Program. During the ASME Annual
Meeting in Orlando/Florida, the Electrical and Electronic Packaging Division of ASME honored Professor Bernd Michel, head of the department
with the Packaging Award for "Outstanding Contributions to the Field of
Engineering Mechanics to Electronic Packaging." Bernd Michel is the first
European scientist to receive this award.

The primary research emphasis of this group is to study the thermo-
mechanical reliability of microcomponents in high-tech systems with special attention on thermo-mechanical simulation combined with advanced
experimental methods. Combination of numerical simulation with experimental results is one of the focuses. Some important research results include:

1. Reliability of automotive electronics (solder joints, sensors)

2. Quality assurance and lifetime estimation of microsystems

3. Stress evaluation of packaging assemblies
4. Evaluation of thermal "misfit" problems in solder regions
5. Nanomechanics and nanoanalytics
6. Crack and fracture avoidance and evaluation in microelectronics and IT

Applications

Main research applications include mechanical reliability of advanced packaging (chip scale package (CSP), flip chip, ball grid array (BGA)), chip cards, air bag sensors, microactuators, PCB deformation and reliability problems and evaluation of solder contacts (Figure D.39). Micro DAC (Deformation Analyses by means of Correlation methods), a powerful tool for deformation analysis, is applied for accurate measurement of deformation fields in critical regions. New concepts of fracture electronics have been developed and studied. Other services include finite element method (FEM) simulation, experimental stress and strain analyses and defectoscopy.

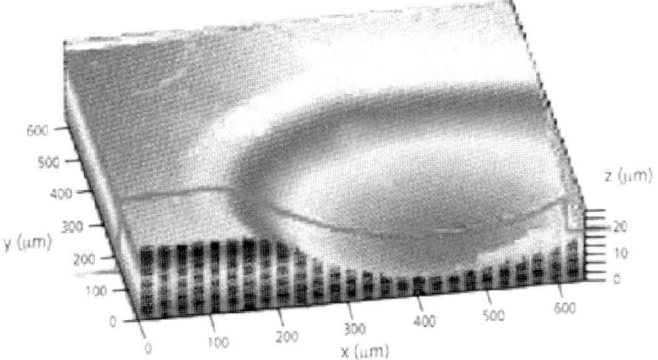

Figure D.39. Geometry parameterization by embossing at a solder ball.

Thermo-Mechanical Analysis

The department conducts active research on the thermo-mechanical properties of micromaterials. The emphasis is on mechanical and thermal stress evaluation, material characterization, mechanical reliability and design optimality, fatigue and creep behavior, fracture electronics and material mechanics. Powerful simulation tools such as finite element modeling are used in the estimation of both deterministic and stochastic reliability. Micromaterials are characterized by estimating material parameters such as coefficient of thermal expansion (CTE), Young's modulus, Poisson's ratio, thermal conductivity, and other factors. Crack and fracture analysis and subsequent crack avoidance have been investigated in advanced packaging technologies. Material mechanics of advanced and electronic packaging technologies are also studied in the facility. Some of the packaging technologies studied are flip-chip, ball grid arrays, chip scale packages, and chip cards. Thin films, hybrid materials and components like adhesives, encapsulants, metals, ceramics and polymers are included in the study.

Microfabrication Techniques

The department has numerous research activities on microfabrication techniques. An innovative microsystem production process based on the concept of division of labor was developed and tested. This process is intended primarily for application in small- and medium-scale industries. The new process enables low- to medium-range production of microsystem components with highest product flexibility. The production process was also tested using a microsystem and the results of the project were collected in a handbook. The department has state-of-the-art experimental techniques available which include laser methods combined with image processing, residual stress evaluation by X-ray stress analysis, defectoscopy and *in situ* measurement techniques for packaging microfabrication modules.

SUMMARY AND CONCLUSIONS

The Fraunhofer IZM-Reliability and Microintegration Institute is one of the leading research organizations in the field of reliability and microintegration of microelectronics and microsystem packaging. The institute provides knowledge for new developments in information and communication (e.g., mobile phones) and automotive technologies (e.g., airbag sensors), and it covers a wide spectrum of research areas ranging from photonic packaging to RF to thermal management.

REFERENCES

Fraunhofer Institute for Reliability and Microintegration (IZM). http://www.izm.fhg.de (Accessed October 20, 2005).

Fraunhofer IZM, Annual Report, 2003 and 2004.

Ramm, P., A. Klumpp, R. Merkel, J. Webber, R. Wieland, A. Ostmann, J. Wolf. 2003. 3D system integration technologies. In *Proceedings of the Materials Research Society Spring Meeting*, January, San Francisco, USA.

Wunderle B., J. Auersperg, V. Großer, E. Kaulfersch, O. Wittler, B. Michel. 2003. Modular parametric finite element modelling approach for reliability studies in electronic and MEMS packaging. In *Proceedings of the Symposium on Design, Test, Integration and Packaging of MEMS/MOEMS DTIP*, May 5-7, Cannes, France.

Site: **Karlsruhe Research Center**
                         Forschungszentrum Karlsruhe
                         P.O. Box 36 40
                         76021 Karlsruhe, Germany

Date Visited: April 4, 2005

WTEC Attendees: T. Hodgson (Report author), K. Ehmann, T. Kurfess,
                    K. Cooper, H. Ali

Hosts: Dr. Norbert Fabricius, Head of Nano- and Microsys-
                         tems Programs, Tel: +49 7247 82-8585,
                         Email: norbert.fabricius@nanomikro.fzk.de
                         Dr. Tilo Baumbach, Head of Institute for Synchrotron
                         Radiation, Tel: +49 7247 82-6820,
                         Email: tilo.baumbach@iss.fzk.de
                         Dr. Volker Saile, Institute for Microstructure Tech-
                         nology, Tel: +49 7247 82-2740,
                         Email: Volker.saile@imt.fzk.de
                         Dr. Ulrich Gengenbach, Institute for Applied Com-
                         puter Science, Tel: +49 7247 82-3769,
                         Email: ulrich.gengenbach@iai.fzk.de

BACKGROUND

The Karlsruhe Research Center (KRC) started in 1956, primarily for the development of nuclear technology. It has approximately 3,500 employees housed in 23 institutes and a budget of approximately €250 million. It is part of the Helmholtz-Gemeinschaft, which is more applied than the Max Plank Institute, but is more theoretical than the Fraunhofer Institute. Its mission is to develop the underlying science for high-potential applications. Its governmental support is approximately 90% from the Federal government and 10% from the state government. In varying degrees, the individual KRC Institutes receive funds from industrial partnerships ranging from 5–25% or more.

We were welcomed by the Director, who gave us an overview of the KRC. He commented that interdisciplinary work is a natural process but that you start with the disciplines first. From an educational standpoint, students (from Karlsruhe University) do not get into the laboratories for about three years, which is when the need for interdisciplinary work starts.

RESEARCH AND DEVELOPMENT ACTIVITIES

Activity #1: Institute for Synchrotron Radiation

Institute Head Dr. Tilo Baumbach hosted us. The synchrotron is the third largest in Germany. There are a number of facilities around the synchrotron that are used to tap the energy beam. Much of that is focused on X-ray lithography applications, which several of the facilities housed in cleanrooms. The facility is used to a great extent as a service facility to other elements of the KRC and to industry. Many of the users are involved in micromanufacturing applications.

Activity #2: Institute for Microstructure Technology

Institute Head Dr. Volker Saile hosted us. The institute has 110 employees. We were shown a number of applications. They have made gears for (very expensive) watches for Moser Watches. The gears are made of gold using X-ray lithography. The gears are made to 1 micron accuracy. They are 200–300 microns thick and are so smooth that they require no lubrication. They want now to automate both the manufacturing process and the quality assurance process to bring down the cost. Then they will transfer the technology to a consortium run by the watch industry.

Figure D.40. Layout of the synchrotron.

We were shown lenses for focusing X-rays. The X-rays are focused by a series of air refractions to 100–200 nanometers. This appears to be a very simple and very effective technology.

They are making molds for various manufacturers using X-ray lithography. The comment was made that the molds last for a long time, and thus repeat business is a long time coming.

They have developed spectrometer technology that is effective for a number of applications: Bilirubin-measurement (i.e., jaundice in newborn babies); determining the color of teeth; determining the color (and thus quality) of grain; and determining the color of diamonds. The technology works for the visible and infrared spectrum.

Figure D.41. Gold escapement wheel made on the synchrotron.

Finally, we were shown microoptical distance sensor which was made partially using X-ray lithography. In the next activity, we will see how the assembly of this device is being automated.

The comment was made that all but one of the Institute heads has joint appointments with Karlsruhe University. Thus there is an excellent relationship with the university. Dr. Saile noted that he had 15 PhD students working at the Institute.

Activity #3: Institute for Applied Computer Science

The head of the microassembly group, Dr. Ulrich Gengenbach, hosted us. We were shown an automated assembly system used for the assembly of the microoptical distance sensor noted above. It is essentially a small robot with X, Y, Z and Rot(Z) axis movement. Piezo-electric actuators power it. It has a series of end effectors that are stored next to it. The end effectors can be exchanged by the robot itself, depending on the task to be performed. Workpieces are mounted on standardised 2"-trays (the standard DIN 32561 covers the range from 1 to 8 inches). Generally, in these small dimensions picking up small parts is difficult, but placing pieces can be extremely challenging. They solve this problem by placing an auxiliary adhesive first to help in the placement. They later place other adhesives to permanently fix the part. The weak point of the system appears to be the limited force capabilities of the actuators, which is being mended by redesign of the Z-axis with stronger actuators. They also commented that with respect to retooling and manual reconfiguration they may have over-miniaturized the system.

Figure D.42. Application of spectrometer

In working with external partners of these types of projects, the fundamental philosophy is to get people to design so that accuracy is not an issue (i.e., design for assembly). Limiting factors in miniaturizing assembly equipment are electrical and fluid connectors. The connectors tend to be too large.

It was noted that in this area, as opposed to silicon based microsystems in hybrid microsystems, there are no design rules. Physical properties change in the microrange, and new solutions have to be found.

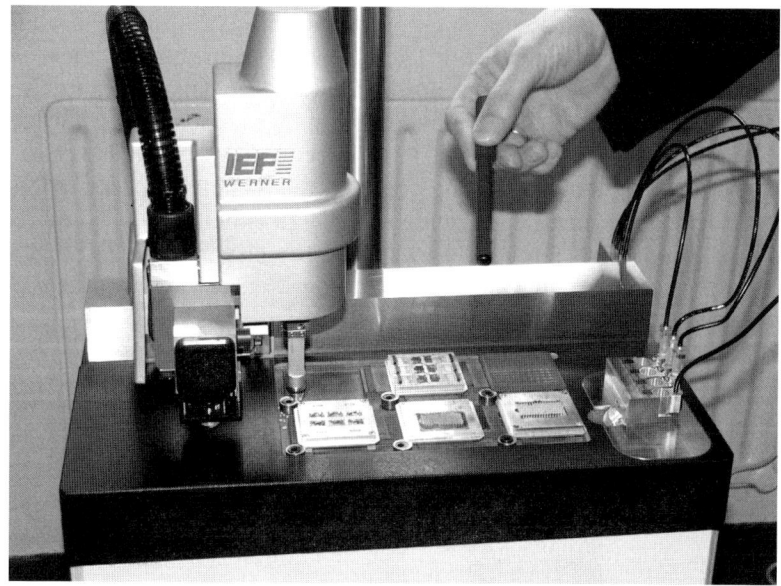

Figure D.43. Automated assembly system.

SUMMARY AND CONCLUSIONS

This was an extremely productive visit. There is very high quality work going on at the KRC. The KRC personnel were extremely open and would answer questions right up to the limits of what they were able to do without violating confidentiality requirements. They provided many interesting insights.

REFERENCES

Forschungszentrum Karlsruhe in der Helmholtz-Gemeinschaft, ANKA Annual Report, 2004.

Forschungszentrum Karlsruhe, Institute for Applied Computer Science, Micro-Assembly Group, CD containing group publications, 2004.

Site: **Klocke Nanotechnik**
                             Pascalstr. 17
                             D-52076 Aachen, Germany

WTEC Attendees: M. Culpepper (Report author), D. Bourell

Hosts: Dr. Volker Klocke, Chief Executive Officer,
                             Tel: +49-2408-950992-0,
                             Fax: +49-2408-950992-6,
                             Email: info@nanomotor.de

BACKGROUND

Klocke Nanotechnik is a privately owned company founded by Dr. Klocke in 1994. Klocke Nanotechnik employs a team of six scientists and engineers. There are several other technical (e.g., machinists) and service personnel (e.g., sales) who work on-site as contractors. The company has a network of sales agents in several countries: Japan, Singapore, Malaysia, Thailand, South Korea, and Taiwan. The company was solely founded and is sustained by revenue obtained via sales of their products. Klocke Nanotechnik is a member of the IVAM Microtechnology Association, a private and nonprofit group of over 130 companies and institutes that offer products related to microtechnology. The IVAM association is international, having members in eight European countries, Korea, Japan and the U.S. Klocke Nanotechnik serves the needs of micro- and nanoscale production/research in several fields such as scanning probe microscopy, microassembly, manipulation of carbon nanotubes, and life sciences.

RESEARCH AND DEVELOPMENT ACTIVITIES

Overview

Klocke Nanotechnik provides an array of meso- and macroscale modules which may be used to construct customized robotic systems for applications such as positioning, scanning, inspection, assembly, etc. Two main highlights of the visit were their optical position sensor technology and their Nanomotor actuators. The Nanomotor enables long-stroke (several mm) positioning with atomic resolution and 1% repeatability. The optical sensor technology was capable of providing nanometer-level position resolution over several mm range. The sensors are smaller than a dime and therefore are easily integrated into smaller-size machines for feedback control. Klocke Nanotechnik is focused upon providing their customers with highly customizable (in cost, performance, and capability) robotics platforms. The company makes components, sub-systems, and full production systems. The sub-systems may be "built-up" from the components and the full-production systems may be "built up" from the components and sub-systems. This modular approach enables customers to (1) buy only what is needed with a high level of customization; (2) perform development and proof-of-concept activities with sub-systems rather than buying/renting a full-production system; and (3) add or swap modules so that they may upgrade/modify their sub-systems to become production-ready equipment. All of the components which Klocke Nanotechnike sells may be controlled with the same electronics and software.

Component Level (Nanomotor, Optical Sensors, Grippers, and Tables)

The Nanomotor, shown in Fig. D.44, is a mesoscale actuator which is capable of several millimeters stroke with atomic resolution and 1% repeatability over the used stroke (without load change). The actuator operates using a piezoelectric actuator for fine positioning and for coarse positioning by inducing pulse waves. Speeds of up to 5 mm/s can be achieved. The actuators can be supplied with load capacities of 20 g. The Nanomotor is capable of operation in ultra-high vacuum, in low-temperature environments and within magnetic fields. The actuator may be integrated into either a robotic machine or used as a stand-alone actuator for positioning of end effectors and samples.

Klocke Nanotechnik utilizes a variety of end effectors to bridge the gap between their machines and meso-, micro-, and nanoscale parts. The grippers shown in Figure D.45 are obtained from partners in the IVAM. These grippers are designed to rapidly acquire, position and release parts for microassembly. The grippers are capable of nanometer-level resolution gripping motions, and they can control the applied gripping force so as to not damage small and fragile parts. The gripper shown in Figure D.45 has a gripping range of 0.6 mm and may be used to lift 2 grams. The gripper has been used to assemble optical components (e.g., fibers and lenses), gears and other microproducts.

We also observed a very interesting optical sensing technology, but unfortunately we were unable to procure an explanation of how this sensing technology worked. This technology is unique as it enables them to perform 2 nm resolution sensing of position over several mm with a sensor that is smaller than a dime. The small size of the sensor and its performance capabilities makes the sensor well-suited for use in smaller-sized equipment. Klocke Nanotechnik also makes backlash-free tables which may be used as platforms for positioning stages. This type of stage is readily available from several vendors and therefore was not a focus of our inquiry.

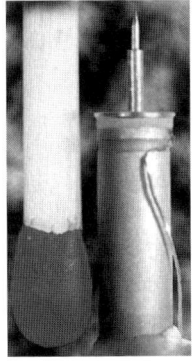

Figure D.44. Component-level—Nanomotor.

Figure D.45. Component-level—microgripper.

Subsystems

The preceding components may be combined with each other and structural components (e.g., mounts etc.) to form robotic sub-systems. These systems are particularly useful if one only has a minor positioning task, a specific research/experiment, or tasks of similar scope and complexity. There are many combinations of sub-systems which can be made from the components that are offered by Klocke Nanotechnik. Two examples are provided in Figures D.46 and D.47. Figure D.46 (left) shows a three-axis nanomanipulator which has been constructed from various structural components and four Nanomotors. The Nanomotor may be equipped with a variety of component-level end effectors, some of which are shown on the right side of Figure D.46. The end effectors include grippers, needles, and scribing tools.

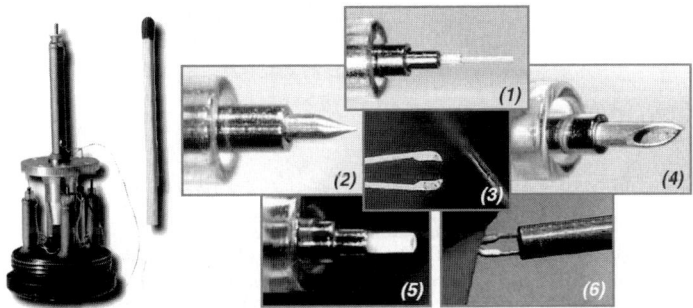

Figure D.46. Sub-system—nanomanipulator (left) and end effectors (right).

Figure D.47 shows a positioning stage which consists of several structural components, a Z-axis positioning device (top), and an X-Y positioning platform (bottom).

Figure D.47. Sub-system—micropositioning stage.

System Level

The preceding components and sub-systems may be combined with minor structural components (e.g., mounts) and larger-scale structural components (e.g., granite tables, granite structures, breadboard table tops, etc.) to form robotic positioning systems. The modular approach used by Klocke Nanotechnik enables one to customize the type of system that may be constructed. For example, Figure D.48 shows what one might call a medium-sized robotic system. This type of setup would be well-suited for medium-volume production runs, medium-volume experiments, and proof-of-concept development activities.

Figure D.48. Medium-sized stage which might be used for development/prototyping/process development.

For instance, this setup could be used to prove out a microassembly process before investing the resources in a full-production machine. In essence, one may use this set-up to develop one or more models of full production machines. The medium-size systems may then be upgraded/modified to full-scale production machines by adding or swapping components/sub-systems. Figure D.49 shows an example of a full-production machine that would result from such an upgrade.

Figure D.49. Large-scale stage for production.

Klocke Nanotechnik ensures that all of the components and sub-systems may be controlled with the same electronics and software, thereby ensuring compatibility between modules and simplifying the upgrade process.

SUMMARY AND CONCLUSIONS

Klocke Nanotechnik has an impressive array of modular components which may be used to construct a multiplicity of robotic systems for applications such as positioning, scanning, inspection, assembly, etc. This modular approach enables customers to (1) buy only what is needed with a high level of customization; (2) perform development and proof-of-concept activities with sub-systems rather than buying/renting a full-production machine; and (3) add or swap modules so that they may upgrade/modify their sub-systems to production-ready equipment. The optical metrology components and Nanomotors were two of the highlights of the visit. These components solve cost-performance problems which have been plaguing nanopositioning and precision positioning for several years.

REFERENCES

Klocke, V. and T. Gesang. 2002. Nanorobotics for micro production technology. In *Proceedings of the 2002 SPIE Photonics Fabrication Conference*, October 28, 2002-November 1, 2002, Brugge, Belgium.

Klocke Nanotechnik, Nanomotor, http://www.nanomotor.de (Accessed October 20, 2005).

Site: **Kugler GmbH**
Heiligenberger Strasse 100
D-88682 Salem, Germany
http://www.kugler-precision.com/

Date Visited: April 8, 2005

WTEC Attendees: T. Hodgson (Report author), K. Ehmann, T. Kurfess, K. Cooper, H. Ali

Hosts: Lothar Kugler, Tel: +49-7553-92 00-11,
Email: info@kugler-precision.com
Till Kugler, Tel: +49-7553-92 00-64,
Email: till.kugler@kugler-precision.com
Dr. Lehndorff, Tel: +49-7553-92 00-14,
Email: lehndorff@kugler-precision.com

BACKGROUND

Kugler GmbH is a family-owned company run by Lothar Kugler and his sons Till and Joerg. The company was founded in 1983. They have 50 employees; 20% are college graduates, three have PhDs. Their export rate runs 25–30% of sales, and R&D takes 15–20% of the budget. Kugler makes ultraprecision air-bearing or hydrostatic machines and multi-axis systems for micromachining. The machines can be equipped with mechanical spindles and lasers. Kugler also offers customized micromachining. Another big product field is reflective optical components infrared (IR) to visible light (VIS) (submicron accuracy), beam-bending components and systems for high-power CO_2 lasers (laser marking systems, cutting and welding heads, focusing heads, bending units, lightweight components). One of their major manufactured product lines are copper mirrors used for beam delivery systems of CO_2 lasers of up to ~40 kilowatts. The mirrors are water-cooled to maintain mirror geometry. Flatness accuracy is on the order of 0.15 micron, and roughness is ~2.0 nm. Regarding micromachining, in general Kugler does not manufacture end products. Rather, they manufacture the machines, sub-systems, molds, and components that enable mass micromanufacturing. They produce the building blocks of micromanufacturing. They noted that they have no real competitors in Germany at the ±0.1 micron level. Worldwide they do have competitors.

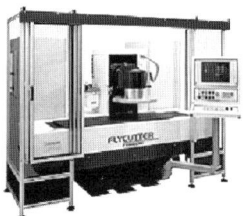

Figure D.50. Optical sample products

Figure D.51. Kugler Flycutter machine Type F1000 CNC

Figure D.52. Laser Hybrid Welding Head Type LK390H

RESEARCH AND DEVELOPMENT ACTIVITIES

Their machines are used primarily for optical application (e.g. optical surfaces on non-ferrous metals, beam bending mirrors, polygon wheels for industrial bar coding readers), but also for high-precision mechanical surfaces (e.g. reference surfaces for roughness calibration, pump housings, precision air-bearing surfaces, etc.). In answer to a question, they noted that you could not scale down conventional technology to achieve the accuracies they realized as the parameters change. The result is that there are no general rules that can be followed, and everything is empirical. For instance, it may be necessary to involve metrology equipment to produce such high-precision parts. Based on the results of the measurements, the process can be compensated to get the desired product. The limitation of accuracy is the crystal size of the copper used for the mirror (or whatever metal is being worked). The point is that better machines may not be the only answer to providing more precise products.

The fundamental structure of their machines is granite to provide the fundamental stability needed (e.g. thermal, vibrations). They use air or hydrostatic bearings for almost all that they do, in some cases preloaded for higher stiffness of the system. The Microgantry machine has integrated high-frequency machining spindles, cutter optics, and coordinate measuring machine (CMM) technology (see Figure D.53). It has a positioning and travel accuracy of 0.1 micron and rests on passive damping supports. Kugler has worked with a spindle manufacturer to develop an adaptive, thermal compensation scheme for spindles. It also offers a six degree-of-freedom (DOF) test probe with 0.01 micron resolution and one of the most compact tool-changing systems available, with 60 tools (~300 x 300 x 300 mm). The machine can perform both milling and laser machining (multi-functional). The connection between the linear motor driver and CNC system is performed by digital/optical fibre bus, to achieve real-time control loop. Tool tip sensing is accomplished with a contactless laser-measuring system. For the axis feedback, Heidenhain glass scales are used.

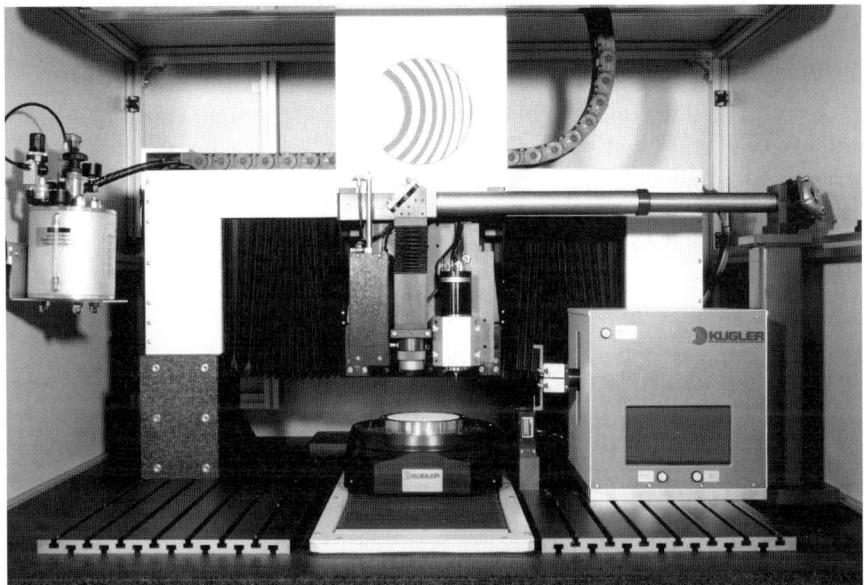

Figure D.53. Kugler Microgantry system.

They showed us two of their new machines that they are about to market. The two machines represent a significant departure from their historical product base. One is a very small microturn lathe with mechanical bearings (not much bigger than a jeweler's lathe). The MICROTURN MT is about 500 x 400 x 120 mm in size and weighs about 60 kg. It uses hand-scraped dovetail slides with a straightness of about 0.1 micron over the entire travel length. The drive is a lead screw; the spindle is made in cooperation with a partner. The travel is ±20 mm in each of the two axes. They feel that given their know-how in compensation, the replacement of their customary air bearings by mechanical components will not degrade accuracy, and submicron accuracy can still be achieved. It was not clear that they had yet achieved the accuracies that they hoped to obtain for this machine but were in the process of fixing some inadequacies. They commented that scaling down the size of machines also scales down the stability of the machine. However, the trend is to scale down the machines, so they were going to compete. The goal is to develop machines that are one-third the size, but still with the same accuracy.

The other machine was a scaled-down version of their larger machines, but with a fundamental granite structure. They seemed to feel that their superior CNC control and compensation system would give them an advantage in this new market (for both of the new machines).

We were shown a rotary table (on an air bearing with direct drive) that could rotate from 0.1 rotations/minute to 200 rotations/minute. Positioning accuracy was ±0.36 arc seconds of accuracy combined with an axial and radial runout of less than 0.1 micron. Spindle speeds on the high-speed machinery that we saw ran in the range of 160,000–200,000 rpm.

They have developed an outstanding metrology system using a scanning microscope and "Lab View" as a basis for measuring roughness. The graphical output from this system gives them an excellent tool for analysis and internal development. They appeared to believe that the system gave them an advantage in their development of systems.

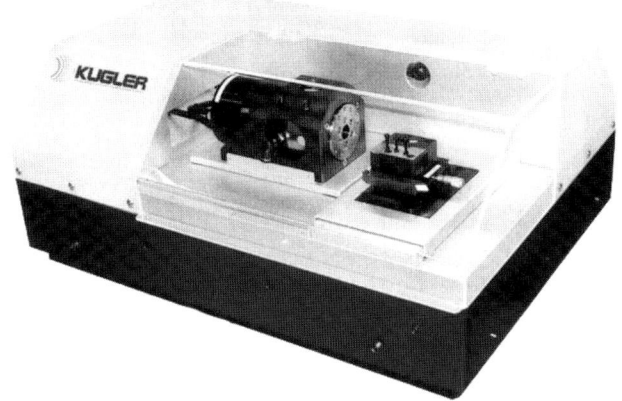

Figure D.54. Kugler's new MICROTURN MT.

When asked about the quality of the cutting tools they procured they noted that 30%-40% of the tools in a given batch might be insufficient for their use. They noted that this could not be blamed on the supplier in that at this level of accuracy they were pushing the envelope of capabilities and quality control is incredibly difficult.

Fixturing is an issue that is addressed on an *ad hoc* basis. Techniques used include vacuum chucking, melting into wax and freezing into ice cubes. Macroscale fixturing techniques cannot be scaled to the microdomain. Actually, all relevant parameters, namely feed, speed and depth of cut, need to be "reinvented."

Kugler micromachining technology today is slightly better than 1 micron (0.1 micron for a single axis), for the next generation they are shooting for ±0.1 micron (the accuracies they already obtain for their optics machining). The trend they envision is the development of small machines for small parts. Their goal is to develop a machine three times smaller than their current Microgantry GU. Ninety percent of the parts fit into ~50 mm cubes.

One of the more interesting comments made by Till Kugler was that it took a special type of worker to run high precision machines. The good ones are very neat in dress and are well-organized as that tends to carry over into their work. Without the "neatnik" personal qualities, they would not be able to get the best out of the machines.

SUMMARY AND CONCLUSIONS

Kugler GmbH is one of the world leaders in high-precision machining. They were very open in their discussions with us. They believe that their advanced bearing concepts and their controller system are superior and that it gives them a competitive advantage. They use virtually no roller bearing as that will not give them the accuracy they need to achieve. They have superior positioning technology. They have developed a number of enabling systems (in particular a metrology system) that enhance their ability to continue to deliver quality products.

Site: **Laser Zentrum Hannover e.V.**
                             Hollerithallee 8
                             D-30419 Hannover, Germany

WTEC Attendees: D. Bourell (Report author), M. Madou, M. Culpepper,
                             R. Rajurkar, J. Cao, S. Ankem, G. Hazelrigg

Hosts: Mr. Thorsten Temme, Production and Systems Devel-
                             opment Department, Head of Microtechnology
                             Group, Tel: +49 511-2788-319,
                             Fax: +49 511-2788-100, Email: tt@lzh.de
                             Dr. Boris Chichkov, Head of Nanotechnology De-
                             partment, Tel: +49 511-2788-316,
                             Fax: +49 511-2788-100, Email: ch@lzh.de

BACKGROUND

Laser Zentrum Hannover is strongly affiliated with the University of Hannover. Historically, there were three major divisions: the Institute for Materials Science, the Institute for Production Technology and Tooling Machines and the Institution for Quantum Physics. The mission of the Center is threefold: research, laser consulting and industrial training. The Center was founded in 1986. Currently, there are 215 staff members, including about 100 scientists and 30-40 technical staff. Income for the Center is €10.9 million. Funding comes from the State of Lower Saxony (10%), industry (30%) and other public funding (60%). The latter includes the European Commission and the German Ministry for Research and Education. The current facilities were completed in 1997 and include 5,700 m^2 of space with 28 labs. The Center spins off successfully developed technologies as newly formed companies. Fifteen such companies have been formed at the Center.

The Laser Zentrum Hannover has broad expertise in laser technology, including macro, micro and nano. Areas involving micro- and nanotechnology include thin films for optical coatings, laser development, production and systems technology (computer-aided design (CAD)/computer-aided manufacturing (CAM), process modeling, and micromachining), microrapid prototyping (including microstereolithography, two-proton polymerization and fused metal deposition (FDM)), glass processing, wire marking, laser welding, materials and process technology, nanotechnology, laser metrology (sensors and flow measurements) and biophotonics.

RESEARCH AND DEVELOPMENT ACTIVITIES

Nanotechnology Department

The department consists of three areas: top-down technologies, X-ray and extreme ultraviolet (EUV) and fine particles. Most of the effort was reported in the first area, including femtosecond laser micro- and nano-processing, two-photon polymerization (2PP) and nanoimprinting. Very striking is two photon polymerization, a rapid technique for creating free-form objects in a photopolymeric liquid like SU-8 photoresist. The laser interacts with the liquid only near the focal point where the photon density is high, creating radicals that initiate polymerization of the resin. Exquisite parts including statues and spiders demonstrate the versatility of the technique. Part dimensions are on the order of 20–100 μm. The polymer resin is an inorganic/organic hybrid polymer (ORMOCERS). Other parts include nano-sized 3D scaffold mesh structures with 220 nm lines on 450 nm centers. Lateral resolution of the 2PP process is 120 nm. Other application examples include spiders with 1 μm legs that are ~50 μm across and ~30 μm tall, free-standing LEGO hollow columns approximately 50 μm in diameter and 300–400 μm tall, three-dimensional tapered waveguides 2–10 μm in width, and nanoembossing. For the last process, molds were created with negative features, which, on embossing onto polymers, produce raised structures for optics applications. A design rule for 2PP is generally to build parts in such a way that the laser does not scan through cross-linked solid materials since there is a slight change in refractive index upon cross-linking.

Hole drilling is improved using femtosecond pulsed lasers. The quality of the hole includes considerations of cracking avoidance and clean, sharp hole features with no localized melting, burring or material build-up. It is possible to deposit gold nanobumps on glass. Using 10 nJ/pulse, 60 nm particles can be deposited. For patterning, 100 nm gold particles are arrayed to create photographic images microns in size.

A microelectromechanical systems (MEMS)-like technique has been developed at the Laser Zentrum Hannover called laser lithography. Using a mask and etch technique, parts can be produced.

Metrology for laser-generated parts is principally scanning electron microscopy for shape and size.

Tools, Processes, Applications and Trends in Laser Micromachining

Four areas are applications (medicine, electronics, automotive), processes (joining, drilling, ablation, stereolithography), materials, and lasers and machines (ultraviolet (UV)/vacuum ultraviolet (VUV), short pulsed, femtosecond).

Lasers available for projects at the Laser Zentrum Hannover vary from 1064 nm down to 157 nm. Femtosecond lasers are very short-pulsed, have extremely high-peak power density and are used for drilling high-quality holes. Since the pulse width is extremely short, the edges produced by ablation are extremely accurate and sharp. For example, a 70% silica optical glass (BK7) may be ablated as a square hole ~100 μm on a side and 10–20 μm deep. The hole is unacceptable when a 308 nm laser is used due to splintering. Good ablation occurred when a 193 nm laser was used.

Drilling was demonstrated by the application of injection nozzles for automotive applications. Stainless steel workpieces were laser drilled with 100–250 μm holes. Nano- and femtosecond lasers have been used to drill 60 μm holes and trepan a wide variety of materials: fused silica, titanium, nitinol, alumina and aluminum nitride. Holes tend to be tapered. Joining processes were exemplified by a lead-on pad surface mount application. Joints approximately 200 μm in size were made, producing environmentally friendly lead-free joints. Three-dimensional ablation was demonstrated in cutting of alumina carriers for airplane radar devices with eximer lasers. A second example of laser ablation was trimming of small vibrating systems (i.e., tuning forks) made from quartz and used for roll-over protection in automobiles. Trim spots were 10 μm deep and 80-100 μm wide.

A micomold technique called 3D Structuring has been developed at the Laser Zentrum Hannover. Silicon molds ~1 mm in length were made by MEMS etching. Replicas were made in sol-gel material to create a positive. The sol-gel parts were used as fixtures for mounting microlenses and fiber optics.

A discussion of trends in micromanufacturing was quite interesting. One was a trend from microprocessing to nanoprocessing. Associated is continued development of novel laser sources such as picosecond lasers. Related was reduction of the laser spot size. Finally, there is growing interest in functional surfaces spanning large areas. One application of the last trend was improved operating conditions by creating functional surfaces on turbine blades. A second area was wetability of polymers on surfaces. The wetting angle could be varied between 0–150° based on the functional surface morphology.

SUMMARY AND CONCLUSIONS

The Laser Zentrum Hannover reported on a diversity of microfabrication processes and research programs involving machining, laser cutting, stereolithography and ablation. Their specific area of laser expertise was in the lower energy laser processes, less than 300 watts.

REFERENCES

Presentation of Research at the Laser Zentrum Hannover. PowerPoint presentation (n.d.).

Serbin, J., A. Egbert, A. Ostendorf, B. N. Chichkov, R. Houbertz, G. Domann, J. Schulz, C. Cronauer, L. Frhlich, M. Popall. 2003. Femtosecond laser-induced two-photon polymerization of inorganic-organic hybrid materials for applications in photonics. *Optics Letters* 28:5, 301-303.

Serbin, J., A. Ovsianikov, B. Chichkov. 2004. Fabrication of woodpile structures by two-photon polymerization and investigation of their optical properties. *Optics Express* 12:21, 5221-5228.

Young, D., S. Sampath, B. Chichkov, D. B. Chrisey. 2005. The future of direct writing in electronics.
http://www.circuitree.com/CDA/ArticleInformation/coverstory/BNPCoverStoryItem/0,2 135,142466,00.html (Accessed October 20, 2005).

Site: **Institut für Mikrotechnik Mainz GmbH (IMM)**
                       Carl-Zeiss-Strasse18-20
                       D-55129 Mainz, Germany

WTEC Attendees: D. Bourell (Report author), M. Madou, M. Culpepper,
                    R. Rajurkar, J. Cao, S. Ankem, G. Hazelrigg

Hosts: Dr. Theodor Doll, Managing Director,
                          Tel: +49 6131/990-100, Fax: +49 6131/990-200,
                          Email: doll@imm-mainz.de
                       Josef Heun, General Manager, Tel: +49 6131/990-100,
                          Fax: +49 6131/990-200,
                          Email: heun@imm-mainz.de
                       Dr. Volker Hessel, Vice Director R&D, Chemical
                          Process Technology Department,
                          Tel: +49 6131/990-450, Fax: +49 6131/990-305,
                          Email: hessel@imm-mainz.de
                       Dr. Peter Detemple, Section Leader LIGA,
                          Tel: +49 6131/990-318, Fax: +49 6131/990-305,
                          Email: detemple@imm-mainz.de
                       Dr. Klaus Stefan Drese, Head of Fluidics and Simula-
                          tion Department, Tel: +49 6131/990-170,
                          Fax: +49 6131/990-305,
                          Email: drese@imm-mainz.de
                       Stefan Kunz, Head of Mechanical Fabrication,
                          Tel: +49 6131/990-185, Fax: +49 6131/990-205,
                          Email: kunz@imm-mainz.de

BACKGROUND

The Institut für Mikrotechnik Mainz GmbH (IMM) is a state-operated R&D facility. Founded in 1991, there are 105 employees, and 60-70% are technical. The company is a nonprofit organization similar to the Fraunhofers, with funding coming from the state rather than the Federal government. In Germany there are 182 such companies. Approximately 45% of the funding comes from Rheinland Palatinate with a roughly equal amount of funding coming from company research and development projects. Total annual funding is approximately €11 million. There are about 100 company projects each year. Most funding is national (34.8%), with 84% coming from Europe, 8% from Asia and 8% from the United States. Company leadership consists of a scientific board composed of persons from local universities and industry. There is a strong affiliation with the Johannes Gutenberg University of Mainz. Since its inception, the IMM has successfully spun off over 17 for-profit companies. The company is divided into four main areas: Micro- and Nanosystems, Fluidics and Simulation, Sales and Administration.

RESEARCH AND DEVELOPMENT ACTIVITIES

Overview

There were six main areas. The first was microelectron optics applications. Using mostly photo emission electron microscopy (PEEM), the application area is optopole alignment devices. The second area is non-silicon microchemical process technology. Applications are primarily molds at the microscale. The third area, micromixers and microreactors, emphasizes fluidics, biofluidics and simulation. Fourth is precision machining, which includes precision milling, turning and grinding as well as sink and wire EDM. Fifth is advanced silicon etching using the lithography, electroplating and molding (LIGA) approach. Finally, there is work in the area of optics. A significant effort is underway in plastic molds for embossing of mirrors and gratings. Overall, the materials in use are plastics including polycarbonate, stainless steel, aluminum, brass and silicon. The driver for selection is service properties rather than being particularly well-suited for micromanufacturing.

The *Chemical Process Technology Department* effort focuses on micromixers and microreactors. IMM is involved with catalog sales of these devices. In 2003, sales included Japan, the United States and Asia. Mixers and reactors are classified into three types (A, B and C), depending on the speed with which the mixing/reaction takes place. The micromixers and reactors have flow rates on the order of 0.01 m^3/hr, and guides are approximately 100 μm in size. The larger macromixers move 3,200 liters/hr at 0.7 bar. Another development was heat exchangers with microflow channels, 0.6 m^3/min at 60 bar. Most German automakers use microreactors for fuel processing. The areas are propane-steam processing and polymer exchange membrane (PEM) fuel cell gas control and pre-processing. One advantage of microprocessing is high-conversion yields. In one example, batch processing yielded about 83% crude product while microprocessing yields 95%. Specific examples of micromixers and micro-reactors are: creation of extra-bright dyes, reduced time for toluene direct fluorination (hazardous materials), and Kolbe-Schmitt synthesis with a yield ~440 times in time/space.

Fluidics and Simulation The modular lab-on-a-chip program combines several technologies to produce on-demand micromixers and microsensors. Modules contain blanks that may be quickly machined and installed, with

five days from concept to finished prototype. The main fluidics sections
are milled in a polymer sheet, cyclo olefin copolymer. Sheets are machined
and then stacked to create a sandwich structure. Channels are 0.8 mm to 4
mm, and the devices may include a number of components: a micropump,
channels, pipette fluid insertion points, flat membrane valves, turning
membrane (gate) valves made from Teflon. The smallest features are on
the order of 50 μm. The operation of the lab on a chip is effected using
LABVIEW software.

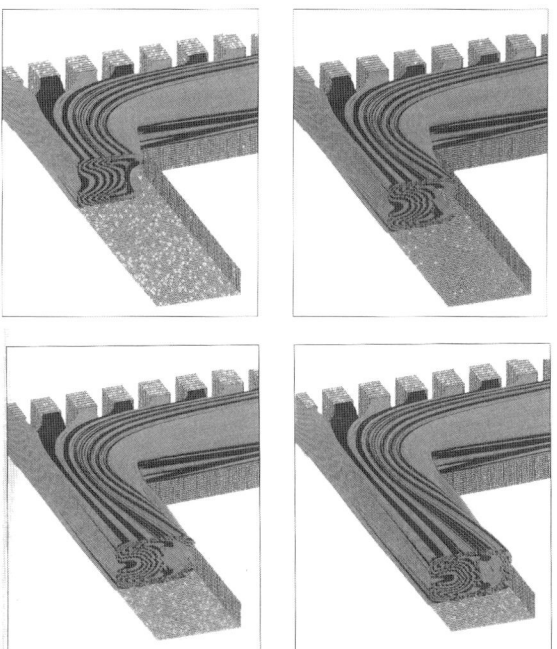

Figure D.55. Simulation of liquid flow in a microfluidics
channel.

An example of the lab on a chip concept was gel electrophoresis of an
amplified biomedical sample (DNA372 and primer). Separation was effected by selective attraction of fluid components to an electrode, such as 3
μm latex spheres.

Flow and mixing characteristics are modeled using ANSYS CVEX
software. The simulation software works well as a modeling tool down to
about 10 μm size for liquid flow. Below that, partial slip conditions must
be used, and then it becomes necessary to use molecular dynamics models.
Some advantages of the lab on a chip concept are low cost, a robust flexibility in chip design and function, reduced time to market when used for

prototypes, and the opportunity to concentrate on science rather than laboratory equipment.

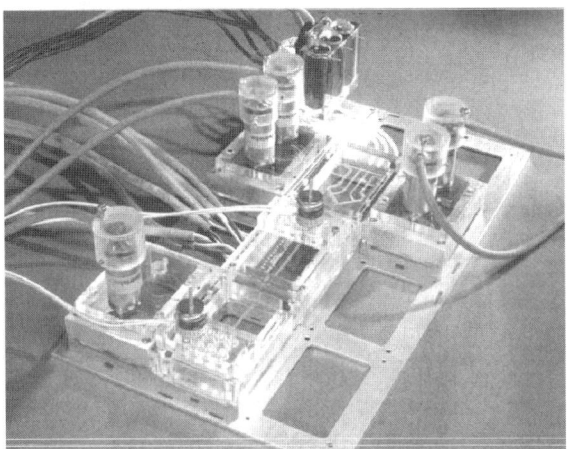

Figure D.56. Lab on a chip.

LIGA manufacturing and laser micromachining technologies are used for microstructuring technologies at the IMM. It is possible to build high-aspect ratio shapes, 500 µm tall with 85 µm wide features. The material was SU-8 photoresist. Electrode structures were also created in gold, platinum, aluminum and silicon nitride. Features as small as 1 µm thick and 10 µm wide were obtained. A second application was creation of platinum resistors spanning microbridge structures. Microfluidic structures may be produced using ASE™ (Advanced Silicon Etching). Channel widths as narrow as 40 µm and 300 µm deep may be produced in silicon as well as silver and copper. Silicon deep etching of membranes and cantilevers may be produced using ASE™ with 0.3–50 µm width and 0.1–10 µm height. A variety of materials are amenable to the process including silicon, silicon nitride, tungsten silicide, nickel and copper.

Two micron diameter hole arrays have been produced for isopore filter membranes using UV-lithography. LIGA techniques have been used to produce molds for plastic injection, molding a variety of shapes. Molds from nickel, steel and polyoxymethylene (POM) have been produced. An example application is a three-stage planetary gear ~50 µm in diameter.

Interdigital Electrode Structures (IDSs) are created using a lithographic approach. A mold is formed in silicon and is used to produce small polymer and nickel parts. Using oblique sputtering across grooved surfaces, it is possible to generate discrete sputter zones for sensors. Probes for brain insertion are prepared using a proprietary process. Wires are end finished to a 30–35 µm probe tip with a linear or helical sensor array further away from the tip.

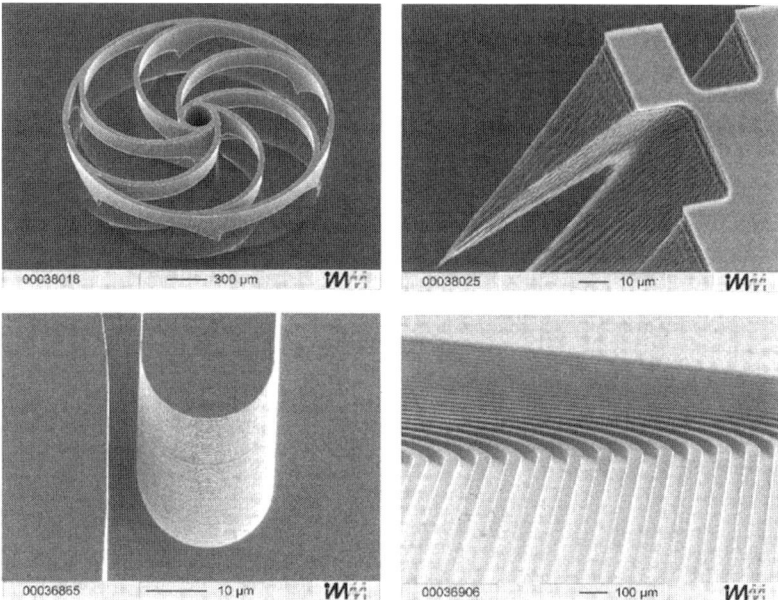

Figure D.57. Examples of advanced silicon etching.

There is an effort in precision machining and micro EDM. Parts are used in watches, medical, injection molding and stamping dies. Structural features as small as 0.5 µm are obtainable. Equipment includes two computer numerical control (CNC) machines with high-speed rotational spindles and 0.1 µm resolution, a Fehlman Picomax 54 2D CNC, and a Fehlman Picomax 60 3D CNC. They have several Swiss-made EDM machines, an AGIE HSS150F 0.025–0.25 mm diameter wire EDM, and an AGIE Vertex EDM with 0.02–0.2 mm diameter wire. Researchers at the IMM have used this approach to create ~80 µm diameter shafts with axial holes and 100 µm features. EDM machining time is about 20–30 hrs depending on the shape produced.

In collaboration with the University of Limerick, a 6 mm by 6 mm fan housing was created. Rotating parts have 25 µm clearance, and the fan blade is 15 µm at the thinnest section. The three-step process is to cut out a rotor blank 50 µm in size using wire EDM. Second, the part is precision machined using sink EDM. Last, a second sink EDM process is used to finish the blades.

Slit arrays 80 µm wide on about 200 µm centers have been produced in steel. LIGA microextrusion electrodes have been machined to create cross or hexagonal extrusion die openings. End mill tools 30 µm in diameter have been made using wire EDM in titanium boride, a conducting ceramic.

EDM turning of tungsten carbide is used to point the ends of wire. A nominally 100 µm-reduced section flares into a 150 µm sphere followed by a tip about 80 µm long and 10 µm in diameter at the end. Sink EDM tools with 80-300 µm features have been produced in tungsten carbide or copper. These tools are then used as sink EDM electrodes inserted in holes and run around the hole inner diameter to create internal non-circular cross section holes. EDM drilling of 50 µm holes in platinum and 25 µm holes in stainless steel foil has been demonstrated. Applications of micro EDM include superfocal mixer components, oval gear wheels and rib electrodes. Tolerance on the order of ±1 µm is achievable in wire EDM of stainless steel.

Ultraprecision machining includes use of diamond cutters for turning, milling and grinding of nonferrous metals and polymers. Less than 10 nm surface roughness and shape error less than 0.1 µm/100 mm are obtained. Used for optical applications, such as hot embossing tools for polycarbonate, lenses ~315 µm in diameter can be produced. A pyramidal surface feature in aluminum and brass is created using a rotational table.

Three precision machines are on-site. First is a two-axis NANOFORM 350 built by Precitech in the United States. A homemade fly cutter is also available, along with a milling machine developed by the Fraunhofer IPT in Aachen. Examples of CNC cutting applications include microchannel machining in poly methyl methacrylate (PMMA) using a 100 µm cutter as well as a 400 µm cutter to generate caterpillar mixers.

SUMMARY AND CONCLUSIONS

The Institut für Mikrotechnik Mainz has an impressive infrastructure for doing a wide variety of micromanufacturing processes. When asked about the market potential for micromanufacturing, General Manager Mr. Heun mentioned two areas: bioanalytical capabilities on a plastic chip and sensors. These highlight the "lab-on-a-chip" approach developed at the IMM.

The Institut für Mikrotechnik Mainz GmbH is an excellent example of a state-funded nonprofit research institution, spinning off almost 20 for-profit companies in the last five years. The research is very applied, comparable to work seen at the Fraunhofers. The main emphases are micro- and nanosystems, microfluidics, LIGA technology and precision machining. Equipment approached the state-of-the-art in machining and LIGA.

REFERENCES

Institut für Mikrotechnik Mainz GmbH: The Microengineers. Brochure (n.d.).
Institut für Mikrotechnik Mainz GmbH: Chemical Process Technology. Brochure (n.d.).

Institut für Mikrotechnik Mainz GmbH: Organic-Synthesis Bench-Scale Pilot Plant. Bro-
    chure (n.d.).
Institut für Mikrotechnik Mainz GmbH: Specialty Microstructured Heat Exchangers from
    the W to the kW Range. Brochure (n.d.).
Institut für Mikrotechnik Mainz GmbH: Fuel Processing. Brochure (n.d.).
Institut für Mikrotechnik Mainz GmbH: Micromixer. Brochure (n.d.).
Institut für Mikrotechnik Mainz GmbH: Chip-Based Lab. Brochure (n.d.).
Institut für Mikrotechnik Mainz GmbH: Bio-Fluidics. Brochure (n.d.).
Institut für Mikrotechnik Mainz GmbH: Computer Simulation. Brochure (n.d.).
Institut für Mikrotechnik Mainz GmbH: Lasers in Microtechnology. Brochure (n.d.).
Institut für Mikrotechnik Mainz GmbH: Advanced Silicon Etching. Brochure (n.d.).
Institut für Mikrotechnik Mainz GmbH: Electro Discharge Machining. Brochure (n.d.).
Löb, P, K. S. Drese, V. Hessel, S. Hardt, C. Hofmann, H. Löwe, R. Schenk, F. Schönfeld,
    F. Schönfeld, B. Werner. 2004. Steering of liquid mixing speed in interdigital micro
    mixers—from very fast to deliberately slow mixing. *Chemical Engineering and Tech-
    nology* 27:3, 340-345.
Schönfeld, F., V. Hessel, C. Hofmann. 2005. An optimised split-and-recombine micro
    mixer with uniform 'chaotic' mixing.' *Lab Chip* 4:1, 65-69.

Site: **Philips Center for Industrial Technology (CFT)**

Date Visited: April 8, 2005

WTEC Attendees: M. Madou (Report author), J. Cao, K. Rajurkar, S. Ankem, G. Hazelrigg

Hosts: Ir. Diederik van Lierop, Tel: +31 40 27 43407,
Fax: +31 40 27 42195,
Email: diederik.van.lierop@philips.com
Ir. Ronald Schneider, Mechatronics Designer, Mechatronics System Development,
Tel: +31 40 27 43150, Fax: +31 40 27 42195,
Email: Ronald.Scheider@philips.com
Eric van Grunsven, Project leader, Technologist,
Tel: +31 40 27 33410, Fax: +31 40 27 32850,
Email: e.c.e.van.grunsven@philips.com
Ir. Raymond Knaapen MTD, Mechatronics Designer,
Tel: +31 40 27 44177, Fax: +31 40 27 44625,
Email: Raymond.Knaapen@philips.com
Prof. Ir. Herman M.J.R. Soemers, Senior Mechatronics System, Tel: +31-40- 27 32887,
Fax: +31-40-27 33201,
Email: h.m.j.soemers@philips.com

BACKGROUND

Philips is a healthcare, technology, and lifestyle company that generates $30 billion in revenues and has over 161,500 employees. The Philips Center for Industrial Technology (CFT) at Eindhoven can trace its roots to 1914, when Philips established a physics research laboratory. Historically oriented toward the development of new products for Philips, CFT has recently expanded its scale and scope to encompass partnerships with industry. Today, the CFT High Tech Campus hosts other companies both independent firms and subsidiaries/spinoffs of Philips—and supports a business incubator program. Companies work jointly with Phillips and use the cleanroom and other research facilities currently available on the campus.

At the time of the site visit 15 companies were located on the campus, including ASML, Sun, IBM, ST, SiliconHive, MontaVista, Dalsa, and Polymer Vision. Approximately one company a month opens up an office on the campus.

RESEARCH AND DEVELOPMENT ACTIVITIES

Activities at Philips Applied Technologies

- *System in Package.* For its interconnect and packaging technology, Philips CFT is focusing on 3D assembly (e.g., cameras, lenses, microphones, solid-state lighting, and biosensors), wafer-scale packaging, and Ulthimo (described below).

- *Mechatronic Systems.* Invention, architecture, development, and design. The focus here is on stages and systems (e.g., magnetic levitation, sub-micron positioning, and the floating wafer reactor), precision techniques, and software solutions.

- *Smart Sensor Systems.* For ambient sensor technology, the focus here is on vision (e.g., smart cameras and devices for assembly verification), optics (e.g., solid-state lighting and optical metrology), and sensors (e.g., devices for auto body contouring).

- *Electromagnetics and Cooling.* Philips CFT is testing products for electromagnetic compliance. They conduct risk assessments and establish design rules. The work is performed in a state-of-the-art test house.

- *Packaging.* The goal here is to reduce large packaging volume for shipping. They are developing test packaging to prevent damage or loss.

- *Prototyping.* Philips CFT performs electronics prototyping and industrialization in a small production facility.

- *Industry Consultancy.* They are interested in fast market-driven innovation and ways to improve operational efficiency.

- *Environmental Services.* Lead-free soldering. Philips CFT provides business support for the implementation of sustainability.

Detailed Examples of Programs within these Activity Areas

Scanning Micromirror

Philips CFT is trying to unify its scattered microelectromechanical systems (MEMS) efforts. The goal here is to build a strong centralized MEMS basis around a scanning, tilting single-panel micromirror demonstrator project. The micromirror has a bandwidth of >1.5 kHz, offers two decoupled degrees of freedom (DOF), and uses a closed-loop control and electrostatic drive. The mirror is fabricated from AF45 borosilicate glass using a fine powder blasting technique developed by Philips. The diaphragm is polyimide, and the hub of the supporting pole is Ni. The sacrificial layer is fabricated from aluminum (Al), and almost all processes are uncommon.

Technical issues that are being addressed include squeeze film damping, insulation, and stress at the pole-diaphragm interface. A finished demonstrator is expected soon. Currently there are no defined applications for the scanning micromirror, although CFT foresees its use in head-up displays and projection devices.

Optical Structures

Engineers are developing optical structures that consist of a collection of optical features working in unison to achieve an optical function. The overall idea is to be able to machine very small features on very large substrates. Examples demonstrated to the visiting panelists included a very large Fresnel lens that could be used in large TVs and microlenses with a surface finish of 2 nm fabricated using electrolytic-in-process dressing (ELID) poly-grinding techniques. All the equipment needed for making these optical surfaces is made in-house.

During the site visit, CFT also demonstrated a linear Fresnel lathe equipped with a diamond tool, as well as a tool-grinding apparatus that had been developed in this center. They have also built a machine that can be used to make microlenses by indenting a concave shape in a material surface. ELID techniques are used, along with active control corrections to counteract waviness. The machine uses Lorentz actuators in flexible hinges and is capable of a correction rate of 0.6 μm over 250 mm.

Systems-in-Package (SIP)

SIP devices created at CFT are used for a wide variety of healthcare, lifestyle, and technology applications. Components include radio frequency (RF) modules, cell phone cameras, MEMS microphones, power transistors, and biosensors.

- Ulthimo technology is a very important and elegant new development that could be used in both ICs and MEMS. It uses a metal foil, consisting of a copper-aluminum-copper (Cu-Al-Cu) sandwich. A laminate is patterned, and the chip is integrated on the flexible laminate; the result is an ultra-thin laminate that is very useful for thermal and high-frequency processes. Double-level interconnects are feasible. The absence of wire bonding has proven much better for high-frequency applications and 3D packaging in general. In this process, Cu foil is folded and molded (or the reverse, say for a compass sensor with three sides). The technique is used for a camera module with five major components. Packaging by folding leaves spaces for other components (e.g., fiber optics or fluidics). Because passive and active components can be integrated, it is very much the ideal MEMS packaging.

- Leadframe packaging—chip size packaging-wafer level packaging (see also FGI). Power devices have a vertical geometry, making packaging requirements very different (for example, drain is on the back). A hole is drilled to the back of the chip and brought to the front. For a few holes a laser is preferred. For many holes reactive ion etching (RIE)—combined with wet etching from the back (via is 50 μm)—is better. Powder blasting is also used to address coils on the substrate.

The Microcontact Wave Printer

Improvements in microcontact printing (see Whitesides soft lithography) are achieved by applying a vacuum to the flexible mold. The image is then transferred from a polydimethylislioxane (PDMS) stamp to a glass backplate at a maximum contact pressure of 0.1 bar. This system enables printing without introducing deformations.

Wave printing solves the problems of roll alignment difficulties in roll printing and air inclusions in flat printing. In wave printing, valves are used to create a vacuum that evacuates the air gaps. This method enables low-cost resolutions to be obtained for dimensions to about 1 μm. Philips CFT envisions that applications for wave printing will include plastic electronics.

SUMMARY/CONCLUSIONS

Despite some transition problems, Philips CFT is still a trend-setting organization. They are attempting to reinvigorate their MEMS program by centralizing it and introducing a demonstrator problem (the single-tilting micromirror without Si). The Ulthimo 3D packaging effort is world-class and should be adapted in MEMS. Finally, microcontact wave printing increases the potential that soft lithography will eventually be feasible as a manufacturing option.

REFERENCES

Philips CFT. 2005. *Project spotlight: 47 solutions for as many questions.* Eindhoven: Philips.
Philips Research Facilities. http://www.research.philips.com/ (Accessed October 20, 2005).

Site: **Swiss Federal Institute of Technology – Zurich (ETHZ)**
Institute of Robotics and Intelligent Systems (IRIS)
ETH Zentrum, CLA H15.2
Tannenstrasse 3, CH-8092 Zurich, Switzerland
http://www.ethz.ch/

Date Visited: April 7, 2005

WTEC Attendees: K. Ehmann (Report author), H. Ali, J. Cao, K. Cooper, G. Hazelrigg, T. Hodgson, T. Kurfess, K. Rajurkar

Host(s):   Professor Dr. Bradley J. Nelson, Tel: +41-1-632-5529,
Fax: +41-1-632-1078,
Email: brad.nelson@iris.mavt.ethz.ch
Professor Dr. Christofer Hierold,
Tel: +41-1-632-3143, Fax: +41-1-632-1462,
Email: hierold@micro.mavt.ethz.ch

BACKGROUND

The Swiss Federal Institute of Technology (ETH) was founded in 1854 by the Swiss government and opened in Zurich in 1855. Today, it has two campuses (ETH-Zentrum and ETH-Hoenggenberg). ETHZ is one of two Federal Institutes of Technology, the other being EPFL (Lausanne) which began in 1969. It has 17 departments, 83 institutes and laboratories, 330 professorships, 840 lecturers (fulfilling teaching obligations and conducting research), a staff of more than 7,500 (25% women), 11,700 registered students, 1,250 yearly graduates, and has annual expenditures of $813 million. Five-hundred thirty students complete a doctoral thesis yearly, approximately half of whom are foreign. A total of 21 Nobel Prizes have been awarded to people who are or were professors at ETH. The department of Mechanical and Process Engineering (D-MAVT) has 10 Institutes with a total of 28 professorships, and about 850 undergraduate and 200 graduate students (2001). The Institute of Robotics and Intelligent Systems (IRIS), our host, headed by Professor Nelson, has a staff of 18. It is interesting to note that each professorship at ETH carries with it a number of guaranteed staff positions (7.5) and a guaranteed amount of annual funding (~$100,000).

RESEARCH AND DEVELOPMENT ACTIVITIES

Our visit to ETHZ lasted for about six hours. The activities in three areas were presented with an emphasis on IRIS.

Professorship for Micro- and Nanosystems (Professor Christofer Hierold)

The group consists of 14 members (seven Ph.D students) with an emphasis on microsystems defined as systems consisting of mechanical and non-mechanical parts. Typical examples of such systems include electrostatic bearings for microelectromechanical systems (MEMS), gyroscopes to facilitate free spinning (without support) disks, and microthermoelectric generators that harvest waste body heat to provide power for sensors and other systems. A major emphasis of the laboratory is on the characterization of materials for microstructures. In particular, applications and characteristics of polymers are investigated for microstructures because of their low cost and bio compatibility and degradability. As an example, they are developing a bulge testing method for the determination of Young's modulus, stresses, and other factors, in analogy to tests performed at the macroscale. Both static and dynamic tests are being developed with the primary aim of assessing their applicability in MEMS structures.

Further activities of the group center on carbon nanotube (CNT)-based technologies, in particular on: (1) the investigation of the possibility of utilizing the property of CNTs, which drastically change resistance with strain (gage factors between 600–1,000 possible), as force transducers in nanosystems; and (2) the investigation of the growth and alignment of CNTs in a controlled fashion for their efficient replication. An interesting concept being developed is the use of piezo-polymers as strain sensors to monitor bone growth based on the premise that as bone heals stresses are transferred to the inserts (e.g., plates and screws)

Institute of Mechanical Systems (IMES) (Professor Juerg Dual)

This Institute is among the largest in D-MAVT with six professorships. Three projects were presented.

1. *Ultrasonic micromanipulation*: The aim is to use ultrasound for the contactless manipulation of small particles (complementary to, for example, optical tweezers and dielectrophoresis). It is suitable for the manipulation of single or multiple particles (e.g., mechanical parts and biological cells), in the size range of between 10 mm and 100 mm. Piezo-actuators are used to excite a glass plate. This results in asymmetrical vibration and pressure in the fluid layer causing the particles to move to equilibrium positions (potential wells), as shown in Figure D.58.

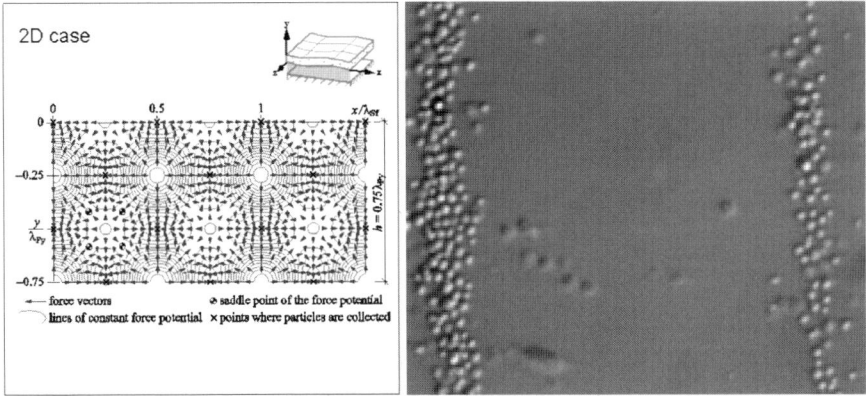

Figure D.58. Example of a simulated potential well and
aligned particles for a different well.

Current focus is on the minimization of the device in such a way that
in biological applications smaller clumps of cells are trapped in the
wells. Different plate movements are also being investigated to explore
the movability of whole clumps in a global sense.

2. *Nanosonics research*: The pump-probe method is investigated to
   measure the time it takes for a p-wave to travel through a thin layer,
   making multiple reflections. The rationale is that if the layer thickness
   is known then Young's modulus can be calculated and vice versa. The
   laser beam is split into two parts: pump—used for excitation (90%)
   and probe—used for measurement (10%). The relative time of arrival
   of the two beams at the surface of the specimen is altered by the variable path length. Pulse duration is 80 femtoseconds to be able to
   measure thin layers (100 nm range). Wavelength = 800 nm; infrared
   and spot size is 100 μm.

3. *Material characterization in the μm-range*: Presented by Udo Lang,
   PhD candidate. The goal is to determine the mechanical properties of
   materials used for the fabrication of micrometer-sized structures. The
   knowledge of design parameters, such as strength and elastic constants, allow the determination of the critical loading conditions for
   microstructures. Phenomena currently being investigated include the
   brittle failure of single crystal silicon structures, the size effects in thin
   copper foils, and the mechanical properties of polytronic materials.

**Institute of Robotics and Intelligent Systems (IRIS) (Professor Brad
Nelson)**

The research philosophy of IRIS is depicted in Figure D.59 showing the
convergence of solid state, molecular biology and physical chemistry into
nanobiotechnology that dictates the need for the development of intelligent

robotic devices for microassembly tasks at the nano- and microscales. The resulting activities at IRIS are driven by actual problems for which the necessary technologies are being sought and developed, rather than the other way around. The visit to IRIS was long and covered a broad range of activities through presentations and lab visits. Because of limited space only a sampling of projects can be highlighted here.

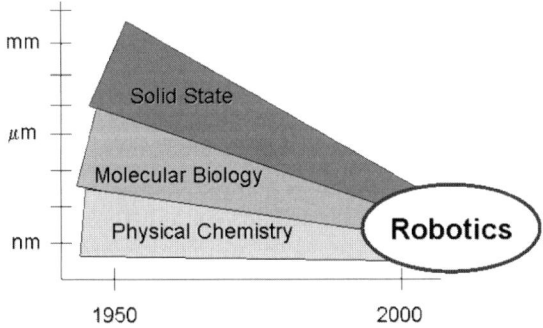

Figure D.59. Research philosophy of IRIS.

1. Biomicrorobotics. Focus areas include cell handling, force measurements, and cell membrane properties. An example of a MEMS-based multi-axis cellular force sensor is shown in Figure D.60a and its application to the measurement of forces exerted by a drosophila in flight is shown in Figure D.60b. The same sensor was used to measure the hardening of the egg cell membrane of a mouse after fertilization. A six degree-of-freedom (DOF) version of the sensor depicted in Figure D.60a is being patented. The sensor uses the capacitive principle and has a resolution of 10 nN.

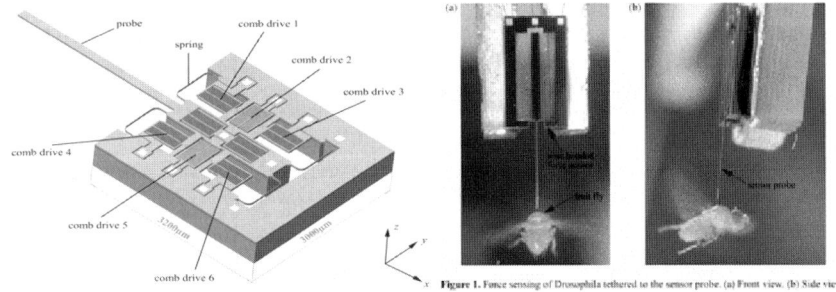

Figure D.60. (a) MEMS-based force sensor, and (b) its use
for drosophila flight measurement.

2. Robotic surgery. Minimally invasive surgery will require autonomous systems for both diagnostics and the procedure. The group is working

on the development of *in vivo* magnetically guided microrobots. One of the key problems is to deliver information and power to the device. The first prototype microassembled microrobot is about 800 μm long and 50 μm thick and is made of laser micromachined steel, electroplated nickel and silicone (Figure D.61). The device is being steered by magnetic coils that generate the necessary locomotion forces and torque. The application that drives these developments would be drug delivery into retinal arteries whose dimensions are on the order of 100 μm.

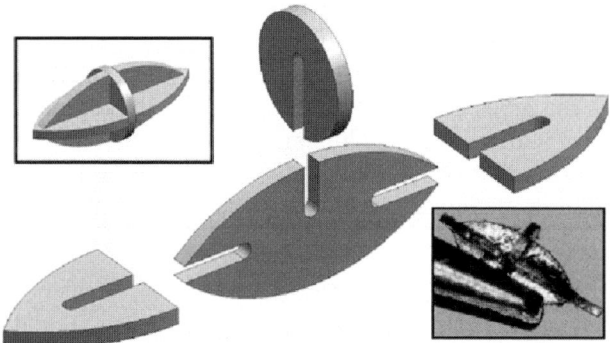

Figure D.61. *In vivo* magnetic microrobot.

3. CNT-based linear nanoservomotors. Figure D.62 depicts the idea. The intent is the development of a closed loop system for linear motion. So far it has been shown that translation is possible and the resulting forces were measured. Torque measurements and loop closure are problems under current research.

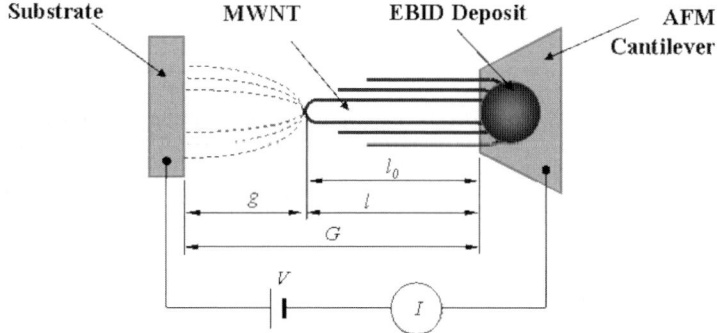

Figure D.62. Multi-walled nanotubes as linear motor.

4. Vision sensing. A CAD model base approach as opposed to stereovision is being developed to facilitate six DOF tracking for microas-

sembly operations. The advantage of the system is that not all the features of the object need to be visible, and a large separation of the microscope/vision axes is allowed for obstruction-free manipulation. An average precision amount of 1 μm and a relative accuracy of 2% have been achieved.

SUMMARY AND CONCLUSIONS

Three Institutes, with a wide-ranging program supporting a multitude of projects, were visited by the panel. The principal activities centered around nano- and microsystems addressing issues from material characterization and property evaluation for microstructures to nano- and microscale manipulation and assembly. The activities of IRIS were the most closely associated with the goals of our study. Of particular importance were the developments of microrobotic locomotion devices and the various technologies in support of microassembly operations (viz., vision tracking, force sensing, task modeling and microgripper development).

NOTES

ETHZ is changing their traditional diploma-based model for undergraduate education to a model more similar to the U.S. system based on the Bachelor of Science (BS)/Master of Science (MS) concept with the BS portion being three years in duration. This is influenced by the fact that Swiss elementary education is 13 years long.

Undergraduate students are trained in the use of microfabrication techniques.

ETHZ is working on intellectual property (IP) rules that break down as: one-third for university, one-third for private industry and one-third for the Institute.

REFERENCES

Eidgenössische Technische Hochschule Zürich. http://www.ethz.ch (Accessed October 20, 2005).

IRIS. http://www.iris.ethz.ch/ (Accessed October 20, 2005).

Micro and Nanosystems, Prof. Cristofer Hierold. http://www.micro.mavt.ethz.ch (Accessed October 20, 2005).

Site: **Technical University of Eindhoven**
                             Department of Mechanical Engineering
                             Den Dolech 2
                             P.O. Box 513, 5600 MB
                             Eindhoven, The Netherlands
                             Tel: 31-40-247-9111
                             http://www.tue.nl/en

Date of visit: April 8, 2005

WTEC Attendees: K. Rajurkar (Report author), S. Ankem, J. Cao, G.
                             Hazelrigg, M. Madou

Hosts: Dr. Marc Geers, Tu/e, 5600 MB Eindhoven, The
                             Netherlands, Tel: +31-40-247-5076,
                             Fax: +31-40-244-7355,
                             E-mail: M.G.D.Geers@tue.nl

BACKGROUND

The Technical University of Eindhoven (TU/e) is a research-driven, de-sign- and industry-oriented university of technology. Because of its focus on fundamental/strategic industry-relevant technological research, it holds a prominent position in the research fields of communication technology, materials technology, and catalysis. The TU/e is also strengthening the field of bioengineering. Besides research, the TU/e offers a Master's mechanical engineering degree program, in which there is one track with an emphasis on micro- and nanotechnology. It has state-of-the-art laboratories with equipment and instrumentation related to microenergetic, multi-scale and microscopy, and precision engineering. The education and research activities are focused on the manufacturing, processing and mechanics of sub-micron devices; on microcooling and microheating problems; and dynamics and control of particular microsystems or problems in systems engineering. Some of the current research activities (noted through presentations, posters, reports and publications collected during the site visit) are described below:

RESEARCH ACTIVITIES

The Mechanics of Materials program headed by Professor Marc Geers consists of sub-programs on structure-property and constitutive modeling, microscopic aspects of deformation and surface engineering, micromechanics of functional devices, multi-scale mechanics, damage and fracture in metals, and impact protection and injury biomechanics for automotive safety. The research group focuses on developing fundamental understanding of materials processing and engineering problems at different length scales based on the physics and mechanics of the underlying material microstructure. One of the areas of metal forming which is of main interest to this WTEC study is briefly discussed below (please refer to the report "Mechanics of Materials" prepared by Professor Geers' group for details on other research areas).

Professor Marc Geers leads a program that focuses on the fundamental understanding and modeling of various problems of metal forming stages on different scales. The group is working on developing structure-property relationships by designing and implementing single-scale models, scale transitions or embedded multiple scales in the mechanics of metals. Their science-based approach relates the microscopic level (discrete dislocation and dislocation structures, i.e. at nanometer scale) to the mesoscale level (single- and polycrystals, grain boundaries, i.e. at the micrometer scale) and the engineering macroscopic level (i.e. at the millimeter scale and above). The group has developed new multi-scale (microstructurally based) constitutive models, enhanced numerical descriptions of metals in metal-forming applications, and has established structure-property relations for metals. They are developing mesoscopic models for fatigue damage in metals and are studying miniaturization and forming of microparts.

The influence of grain sizes on the processing of thin metal sheets is being investigated. The approach is to reduce the sheet thickness at a constant grain size and then change the grain size at a constant sheet thickness. The material investigated was soft aluminum sheet, with thickness ranging from 0.17 to 2 mm. Grain sizes ranging from 0.016 to 600 mm^2 have been obtained. The yield strength as well as the maximum load decreases with a decreasing number of grains over the thickness. For grain sizes larger than the specimen thickness, the value of the yield strength increases with the grain size.

Another WTEC relevant activity in the mechanical engineering (ME) department of the TU/e is microinjection molding, which is part of the work in the group headed by Professor Dietzel, who also made a brief presentation of his group's activities. A first requirement for predicting the dimensional stability and accuracy of injection molded products is a correct modeling of the specific volume of those results from processing conditions. Currently available pVT constitutive models lack experimental validation and correctly obtained material parameters as input. The objective of this project is to investigate the influence of thermo-mechanical history on the resulting specific volume of semi-crystalline polymers. The influence of high cooling rates and mechanical deformation are especially of interest. A new dilatometer has been designed and built for experimental validation of current pVT constitutive equations. Also, the relationship between microstructure development and resulting specific volume will be investigated.

SUMMARY AND CONCLUSIONS

The mechanical engineering department of the Technical University of Eindhoven (TU/e) faculty works with the Netherlands Institute for Metals Research (NIMR) on many areas of manufacturing technologies (metal forming and welding), metals production (steel and aluminum), and lifetime properties (corrosion and surface and interface engineering). The visited laboratory at the ME department of the TU/e is funded by the University, Dutch National reseach programs and the Netherlands Institue of Metals Research (NIMR). The research effort involves theoretical and experimental work on micromechanical modeling of polymers, size effects in microforming, and FE models for multiscale second-order computational homogenization of multi-phase materials.

REFERENCE

Evers, L. P., W. A. M. Brekelmans, M. G. D. Geers. 2004. Non-local crystal plasticity model with intrinsic SSD and GND effects. *J Mech Phys Solids* 52, 2379-2401.

Evers, L. P., W.A.M. Brekelmans, M.G.D. Geers. 2004. Scale dependent crystal plasticity framework with dislocation density and grain boundary effects. *Int J Solids Structures* 41:18-19, 5209-5230.

Geers, M. G. D., Internal Evaluation Report, Mechanics of Materials, 2000-2004 (n.d.).

Geers, M. G. D., V. G. Kouznetsova, W. A. M. Brekelmans. 2003. Multi-scale first-order and second-order computational homogenization of microstructures towards continua. *Int Jnl Multiscale Comp Eng* 1:4, 371-386.

Kouznetsova, V. G., M. G. D. Geers, W. A. M. Brekelmans. 2004. Multi-scale second order computational homogenization of multi-phase materials: A nested finite element solution strategy. *Comput Methods Appl Mech Engrg* 193, 5225-5550.

Raulea, L. V., A. M. Goijaerts, L. E. Govaert, F. P. T. Baaijens. 2001. Size effects in the processing of thin metal sheets. *J Mat Proc Techn* 115, 44-48.

Technical University Eindhoven, Materials Technology Institute, http://www.mate.tue.nl (Accessed October 20, 2005).

van Dommelen, J. A. W., D. M. Parks, M. C. Boyce, W. A. M. Brekelmans, F. P. T. Baaijens. 2003. Micromechanical modeling of the elasto-viscoplastic behavior of semi-crystalline polymers. *J Mech Phys Solids* 51, 519-541.

Site: **Carl Zeiss Industrielle Messtechnik, GmbH**
                        73446 Oberkochen, Germany

Data Visited: April 4, 2005

WTEC Attendees: T. Kurfess (Report author), H. Ali, K. Cooper,
                        K. Ehmann, T. Hodgson

Hosts: Mr. Karl Seitz, Director New Technology,
                            Tel: +49-7364-20-4326, Fax: +49-7364-20-2101,
                            Email: seitz@zeiss.de
                        Mr. Roland Roth, New Technology Systems Integra-
                            tion, Tel: +49-7364-20-3940,
                            Fax: +49-7364-20-2101, Email: r.roth@zeiss.de
                        Mr. Mauricio de Campos Porath, New Technology,
                            Tel: +49-7364-20-8140, Fax: +49-7364-20-2101,
                            Email: m.deporath@zeiss.de

BACKGROUND

Carl Zeiss Industrial Measuring Technology (IMT) employs about 1,300 personnel globally. It is part of Carl Zeiss, Inc., which was founded in Jena, Germany in 1846. Currently, Carl Zeiss, Inc., has approximately 13,700 employees. The company generated revenues of approximately €2.1 billion last year.

RESEARCH AND DEVELOPMENT ACTIVITIES

The primary reason for the visit to Carl Zeiss IMT was to review their new small-scale coordinate measurement machine (CMM). Called the Zeiss F25, it is currently available for purchase. However, Zeiss continues to develop the capabilities of the F25. Figure 5.2 shows the F25. This report is broken into three separate sections. First, an overview of the F25 is presented, and then the major system innovations are discussed. Finally, the probing system used in the F25 is presented.

F25 Overview

The F25 is a three-axis prismatic coordinate measurement machine that is specifically designed to measure small-scale parts. The F25 employs two different sensors for metrology operations, a charged-couple device (CCD) optical sensor with a 13X objective for 2D sensing, and a tactile probe that can be used in either a point-to-point measurement mode or a scanning mode. Figure 5.4 shows the measuring head of the F25 with the tactile probe and the CCD sensor. A camera is also mounted on the F25 measurement head to provide visual feedback when using the tactile probe. The F25 is designed to use both the tactile sensor and the optical sensor in conjunction with each other. That is to say, that the geometric relationship between these two sensors is well known. Thus, measurements made with the tactile probe can be geometrically related to those made with the optical sensor. The measurement volume of the F25 is 100 mm × 100 mm × 100 mm when using both sensors in conjunction with each other. If a single sensor is used, the measurement volume is increased to 135 mm × 100 mm × 100 mm. The F25 is designed for a maximum part mass of 5 kg.

The F25 uses Heidenhain glass scales (2 μm physical grating steps with enhanced interpolation) for a resolution of 7.5 nm. The volumetric accuracy of the CCD sensor is approximately 500 nm. Currently, the repeatability of the CCD sensor is less than 200 nm (this depends on sub-pixel interpolation). The volumetric uncertainty of the tactile probe in point-to-point mode is 250 nm, with a repeatability of less than 50 nm and a resolution of 7.5 nm. The system capability in scanning mode has not been determined at the writing of this report. It has a maximum travel velocity of 20 mm/s along a single axis with a maximum 3D velocity of 35 mm/s. The maximum scanning speed with the tactile probe is 0.5 mm/s. The maximum acceleration of the F25 is 500 mm/s^2 per axis. Error mapping is used on the F25. All 21 errors (position, straightness, angular and squareness errors) are mapped for this system. Furthermore, dynamic error correction is also employed. Finally, the Zeiss personnel indicated that a new proprietary error correction technique was developed and employed on the F25.

F25 Technical Innovations

The F25 is an extremely well-designed machine. It employs state-of-the-art scales for metrology feedback. It is also enclosed in an environmentally controlled chamber. The temperature within the chamber is controlled to ±1°C. A major innovation of the F25 is the complete elimination of the Abbe offset in the X and Y directions (horizontal). The design is based on a PhD dissertation from the University of Eindhoven supported by Zeiss. The basic design, which is patented, allows for the entire metrology systems (glass scales) for the X- and Y-axes to move with the machine. The system is designed to have a Z-axis (vertical) Abbe error zero at the midpoint of the Z-axis stroke.

The F25 is also significantly more rigid than other metrology platforms, due mainly to a new generation of aerostatic linear air-bearing designs. While the stiffness of the machine is proprietary in nature, a significant amount of the stiffness increase was attributed to the improved stiffness of the components comprising the air bearing. The F25 employs a non-iron based linear drive enabling significantly increased acceleration.

Special fixturing was also developed for the F25 to hold two types of microcomponents, prismatic parts (see figure D.63) and rotationally symmetric parts (see Figure D.64).

Figure D.63. Fixture for prismatic parts.

Figure D.64. Fixture for rotationally symmetric parts.

The F25 is controlled by the Zeiss C99 controller, which is a controller used on a number of other Zeiss coordinate measurement machines. They have leveraged this controller as well as their CALYPSO software for geometric analysis (see Figure D.65). This is the standard metrology package that comes with Zeiss CMMs.

Probing System

Tactile probe is for full 3D measurement and is used in the same fashion as a standard CMM probe. Tactile probe is based on a small probe being mounted on a monolithic flexure, as shown in Figure 5.7. Figure 5.5 provides a perspective as to the size of the probe. The flexure is an etched silicone wafer that has a piezo-resistive membrane on it. When the probe contacts a part surface, it deflects, changing the resistivity of the piezo-resistive material. In a point-to-point mode, a specific change in resistivity generates a trigger indicating that the part has been contacted. In scanning mode, the servo drives of the machine keep a constant force on the probe, generating a constant deflection. Two sizes of probe are currently used. The larger probe employs a 0.300 mm diameter ruby sphere for its tip with a 0.200 mm shaft diameter. The smaller probe uses a 0.120 mm diameter spherical tip with a shaft diameter of 0.050 mm. The scanning force for the large probe is 0.5 mN, and the scanning force for the smaller probe is 0.250 mN. These forces generated a tip deflection of 1 μm for their respective tips. The entire probe tip length is 7 mm, enabling the F25 to measure a hole with a depth of 4 mm.

The optical CCD sensor is used for 2D measurement. It is a calibrated optical measurement system that relies on knowing that the scaling of the image is calibrated to size. It is employed in much the same manner as using an optical microscope for 2D metrology. The optical probe has several illumination options including a ring illumination that can be fully or partially activated (see Figure D.66), as well as back lighting capabilities.

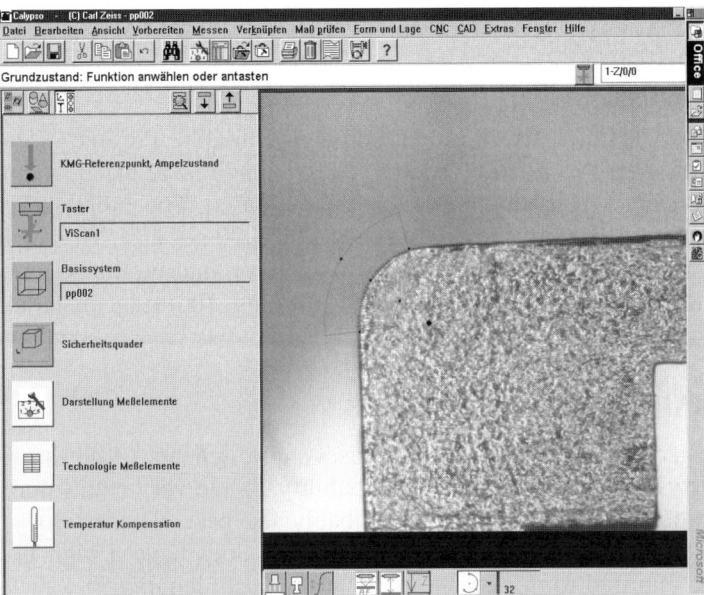

Figure D.65. Calypso software.

Figure D.66. Optical sensor with illumination source.

Other Pertinent Issues

This is an entirely new machine. Zeiss believe that there is a good market for such a machine. This particular machine required approximately $6 million to develop and prototype. It was the result of a joint project with the federal German government and Zeiss. The initial project was on the order of $4 million, with the government supplying approximately 30% of the initial project funds. Zeiss provided the extra $2 million required to complete the project.

Future Issues

The F25 is currently in production and available. Presently, lead times for ordering an F25 are 6-8 months. These should be reduced as more F25s are produced, and production levels are increased. The estimated cost of a well-equipped system is approximately $440,000 per unit. This includes an active vibration damping system, software for interpreting CAD models and simulating the inspection process, training, fixturing, installation and 12 styli.

SUMMARY AND CONCLUSIONS

From a metrology perspective, this system appears to be the first metrology system that is truly 3D in capability. In the very near future, it will be commercially available. It is probably the best 3D metrology system available for microcomponents. However, it does have a high price, and the time required to inspect the part can be fairly lengthy. It will serve the R&D community well but may not be able to meet the needs for process control in the production environment.

REFERENCES

Carl Zeiss IMT Corporation, http://www.zeiss.de/us/imt/home.nsf (Accessed October 20, 2005).

| Site: | **Zumtobel**
HochsterStrasse 8
A-6850 Dornbirn, Austria
http://www.z-werkzeugbau.com |

Date Visited: April 8, 2005

WTEC Attendees: K. Ehmann (report author), H. Ali, K. Cooper,
T. Hodgson

| Hosts: | Günter Konzilia, Tel: +43-55-72-3901223,
Fax: +43-55-72-390185,
Email: guenter.konzilia@z-werkzeugbau.com
Hans Schnutt, Tel: +43-55-72-390667,
Fax: +43-55-72-3909667,
Email: schnutt@z-werkzeugbau.com |

BACKGROUND

The Zumtobel Group (8,000 employees—€1.3 billion turnover) is a leading international concern in the lighting sector. The panel has visited Zumtobel's toolmaking division (Zumtobel-werkzeugbau). The toolmaking division employs 160 people and supports nearly 100 apprentices and eight trainers. The annual turnover is about €20 million, three-quarters of which is with companies not belonging to the Zumtobel Group. The company's core expertise is in mold-making (28%), automation and special purpose machines (30%), precision machining (17%), punching and forming (11%), prototyping (10%) and other (4%). In these areas the company has developed particularly strong expertise and capabilities in micromanufacturing, viz., micromachining, micromolding, and ultraprecision manufacturing in general. Zumtobel defines micromanufacturing in terms of part size (sub-millimeter to approximately 10 mm), feature size (few μm to about 1 mm), accuracy (0.5 μm to about 50 μm), and surface roughness (10 nm to 1 μm).

RESEARCH AND DEVELOPMENT ACTIVITIES

The visit to Zumtobel lasted for about 2 1/2 hours and included a detailed presentation of the company's technological capabilities and philosophy and a tour of the facilities.

Mr. Günter Konzilia briefed the panel on Zumtobel's approach of meeting customer requirements through close interaction and problem-solving from the inception of the initial design to realization. This process is frequently burdened by misunderstandings caused by terminology and lack of understanding of micromanufacturing possibilities and limitations by many of the customers. To overcome this, and to facilitate a rapid convergence to a solution, Zumtobel has developed charts that highlight process capabilities in terms of typical part attributes. Each entry has an associated price/time index/multiplier that expresses the degree of difficulty in realizing the particular part attributes and that is used to assess the cost of the final product. An excerpt from an example table is shown in Figure D.67.

An interesting analogy in approaching a design/manufacturing problem solution was stated by Mr. Konzilia by comparing the strength of the German oak to the resiliency/flexibility of the Chinese bamboo in a storm—the oak may break but the bamboo will survive. He has implied that in many cases the customary approach from the "strength" perspective might fail to yield an answer while an approach from a diametrically opposite "resilient/flexible" standpoint might lead to a solution. This latter philosophy is applicable in many engineering situations.

Production methode	Accuracy +/-	Radius R	Surface Ra	Factor for Production time
Die sinking EDM	0,01	0,04	1,6	3
Micro die sinking EDM	0,0025	0,015	0,2	20
Wire EDM D=0,1	0,006	0,07	0,8	2
Micro wire EDM D=0,02	0,002	0,02	0,1	15 - 20 (setup time)
Micro milling (D = 0,2mm	0,005	0,1	0,2	1
Micro milling (D = 0,1mm	0,002	0,05	0,1	6

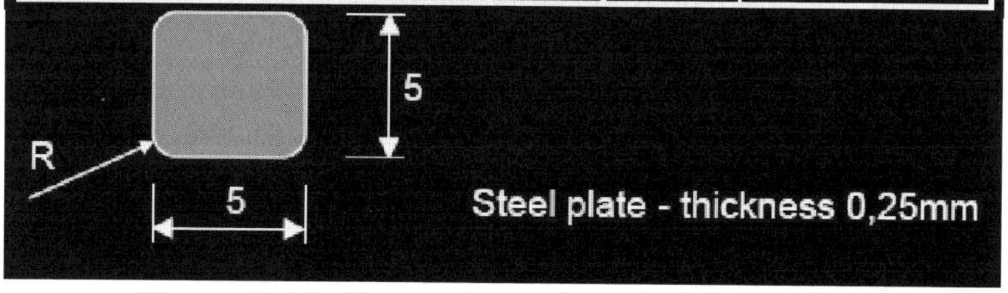

Figure D.67. Comparison of production methods.

Zumtobel has developed an impressive level of micromanufacturing capabilities. Some of the examples are illustrated in Figure D.68.

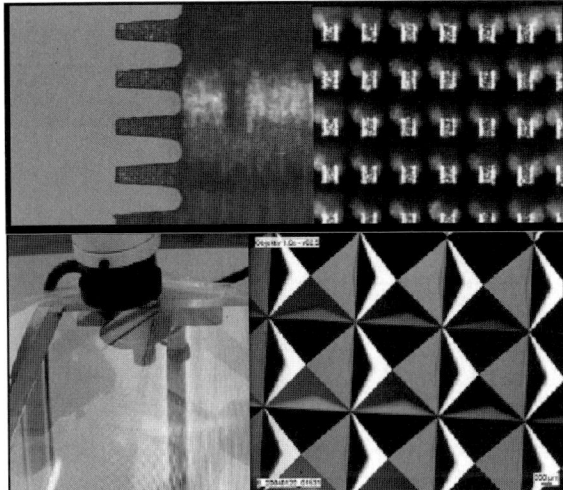

Figure D.68. Micromanufacturing examples: (a) microwire EDM—40 μm features, (b) diamond fly-cut features for molds.

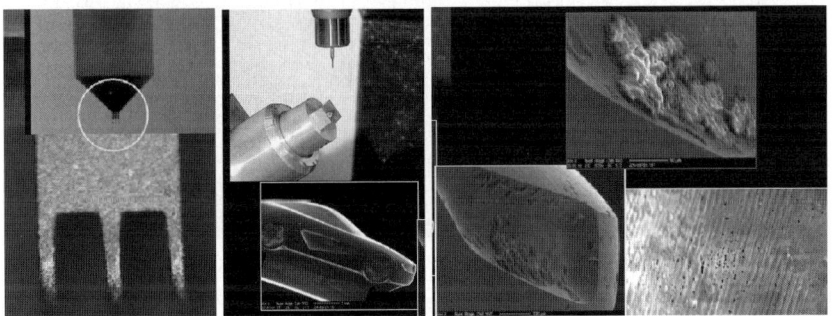

Figure D.69. Micromanufacturing examples: (a) sinking EDM electrode, (b) micromilling, (c) debris on micromilled surfaces.

Figure D.69b shows a particularly difficult-to-resolve issue in micromilling, namely cutting fluid remnants on the machined surfaces that are very difficult to remove at the scales in question.

Zumtobel's particular strength is in micromolding since they have resolved two central problems: the ability to manufacture precision micromolds and the ability to successfully execute the process on suitable microinjection molding machines (in their case the Battenfeld Microsystem 50). The advantage of the Battenfeld machine in comparison to conventional technology is strikingly illustrated in Figure D.70 by the dramatically reduced volume and size of the melt cushion. Zumtobel is working with Battenfeld on further improvements to this system. Examples of microinjection molded products are shown in Figure D.71.

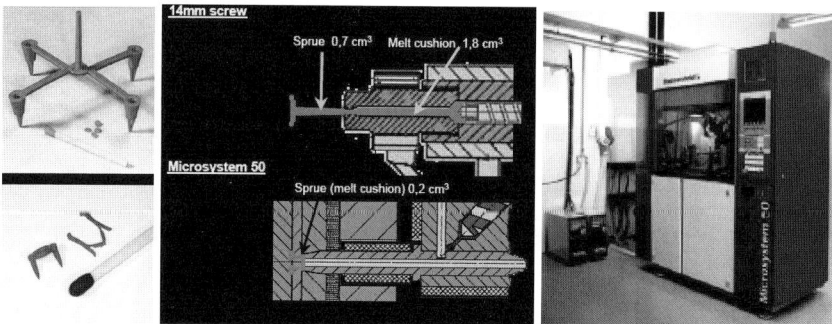

Figure D.70. Comparison of conventional vs. Zumtobel's technology: (a) comparison of sprue systems, (b) comparison of conventional (14 mm screw) and Battenfeld's injection mechanism, (c) Battenfeld Microsystem 50 injection molding machine.

Figure D.71. Microinjection molded components.

There are a number of additional technologies that Zumtobel employs in their business (e.g., lithography, electroplating and molding (LIGA), UV-LIGA, silicon technologies, electroplating, laser processing), but they are outsourced and realized through partnerships.

The tour of the facilities included conventional die and mold manufacturing and the micromanufacturing facility. Micromanufacturing processes discussed included micromilling operations on Kern machine tools, diamond fly-cutting and microinjection molding on the Battenfeld Microsystem 50 machine. A considerable number of micromolds were presented to the panel members.

SUMMARY AND CONCLUSIONS

The visit has concentrated on the discussion of Zumtobel's micromanufacturing technologies and on the tour of the corresponding facilities. The company is clearly in the leading group in their field of expertise (they have had contracts with NASA in the past). The panel gained a good overview but not much in terms of detailed technical insight into the technologies discussed and future plans and directions. Perhaps the most impressive part was the activities surrounding their microinjection molding capabilities.

REFERENCES

Zumtobel Staff Werkzeugbau. http://www.z-werkzeugbau.com (Accessed October 20, 2005).

APPENDIX E: GLOSSARY

2PP	Two-photon polymerization
ABS	Acrylonitrile butadiene styrene
AC	Alternating current
ACD	Artificial cellular devices
ACF	Anisotropic conductive film
ADME/Tox	Absorption, distribution, metabolism, and excretion/toxicology
AFM	Atomic force microscope
AGRIKOLA	*Arabidopsis* Genomic RNAi Knock-out Line Analysis (Europe)
AIST	National Institutes of Advanced Industrial Science and Technology (Japan)
ASE™	Advanced Silicon Etching (Germany)
APM	Asia Pacific Microsystems, Inc. (Taiwan)
ASIC	Application specific integrated circuit
ASME	American Society of Mechanical Engineers
IWF	Institute for Machine Tools and Factory Management (Germany) at the Berlin University of Technology
BGA	Ball grid array
BMBF	Federal Ministry of Education and Research (Germany)
BS	Bachelor of Science
BT	Biotechnology
CAD	Computer-aided design
CAM	Computer-aided manufacturing

CAX	Computer-aided application software
CBN	Cubic boron nitride
CCD	Charged-couple device
CDMA	Code-division multiple-access
CEBOT	Cellular Robotic System (Japan)
CFD	Computational fluid dynamics
CIRMM	Center for International Research on Micro Mechatronics (Japan) at the University of Tokyo
CMM	Coordinate measuring machine
CMP	Chemical mechanical polishing
CMOS	Complementary metal oxide semiconductor
CNC	Computer numerical control
CNTs	Carbon nanotubes
CPU	Central processing unit
CSP	Chip-scale packaging (also: chip-size package)
CTE	Coefficient of thermal expansion
CVD	Chemical vapour deposition
DARPA	Defense Advanced Research Projects Agency (U.S.)
Desktop Manufacturing Paradigm	Encompasses the creation of miniaturized unit or hybrid processes integrated with metrology, material handling and assembly to create 'microfactories' capable of producing microprecision products in a fully-automated manner at low cost
DES	Department of Engineering & Science (Taiwan) within the National Science Council
DFM	Dynamic force microscope
DFMS	Desktop flexible manufacturing systems

DFX	Design for X
DI	Direct input, direct injection or direct interface
DI water	De-ionized water
DNA	Deoxyribose nucleic acid
DOE	Department of Energy (U.S.)
DOF	Degrees of freedom
DRAMs	Dynamic random access memory semiconductors
DRIE	Deep reactive ion etching
DTF	Desktop factory or desktop fabrication
ECDM	Electrochemical discharge machining (Taiwan) at the National Taiwan University
ECM	Electro-chemical machining
EDM	Electro-discharge machining
EDX	Energy dispersive X-ray
EFAB	Electrochemical fabrication (U.S.) at MEMGen Corporation
ELID	Electrolytic-in-process dressing
EMC	Electromagnetic competency
EMF	Electromagnetic field
EPFL	École Polytechnique Fédérale de Lausanne (Switzerland)
EPSRC	Engineering and Physical Sciences Research Council (U.K.)
ERC	Engineering Research Center (U.S.) of the NSF
ESD	Electrostatic discharge
ET	Environmental technology
EU	European Union

FAB	Fast atom beam
FBAR	Film bulk acoustic resonators
FEA	Finite element analysis
FEM	Finite element methods
FE-TEM	Field emission transmission electron microscopy
FIB	Focused ion beam
FPGA	Field programmable gate arrays
FTS	Fast tool servo
GE	General Electric (U.S.)
GM	General Motors (U.S.)
GMO	Genetically modified organism
GPS	Global positioning system
HAM	Hole area modulation
HDTV	High-definition television
HeCd laser	Helium cadmium laser
HEPA	High efficiency particulate air
HP	Hewlett Packard
HRC	Hardness Rockwell C (also see RC)
HRTEM	High resolution transmission electron microscopy
ICs	Integrated circuits
ICP	Inductively coupled plasma
ID	Inside diameter
IDM	Integrated device manufacturers
IDSs	Interdigital Electrode Structures
IH process	Integrated harden polymer process

IIS	Institute of Industrial Science (Japan) at the University of Tokyo
ILT	Institute for Laser Technology (Germany) at RWTH-Aachen University
IMM	Institut für Mikrotechnik Mainz (Germany) GmbH
IMTS	International Manufacturing Technology Show
IP	Intellectual property
IPK	Institute for Production Systems and Design Technology (Germany) at Fraunhofer Institute
IPMC	Ionic polymer-metal composite
IPT	Institute for Production Technology (Germany) at RWTH-Aachen University
IR	Infrared
IT	Information technology
ITRC	Instrument Technology Research Center (Taiwan) of the National Applied Research Laboratories
ITRI	Industrial Technology Research Institute (Taiwan)
IZM	Institut Zuverlassigkeit und Mikointegration (Reliability and Micro-Integration) Fraunhofer Institute, Berlin
JV	Joint venture
KAIST	Korean Advanced Institute of Science and Technology (Korea)
KIMM	Korean Institute of Machinery and Metals (Korea)
KITECH	Korea Institute of Industrial Technology (Korea)
KOH	Heated potassium hydroxide
KOSEF	Korea Science and Engineering Foundation (Korea)

KRISS	Korea Research Institute of Standards and Science (Korea)
LASER	Light amplification by stimulated emission of radiation
LBM	Laser beam machining
LCD	Liquid crystal display
LED	Light-emitting diode
LIGA	X-ray Lithography, Galvanoformung and Abformung (Germany) X-ray lithography, electroplating and molding (English)
LSI	Large-scale integration
LZH	Laser Zentrum Hannover e.V. (Germany)
MATLAB	MATrix LABoratory
MB	Megabytes
MCL	Microcylindrical lenses
MCNC	Microelectronics Center of North Carolina (U.S.)
MD	Molecular dynamic
MEC	Mitsubishi Electric Corporation (Japan)
MEMS	Microelectromechanical systems
METI	Ministry of Economy, Trade and Industry (Japan)
MEXT	Ministry of Education, Culture, Sports, Science and Technology (Japan)
Micro SLA	Micro stereolithography
MIRDC	Metal Industries Research & Development Center (Taiwan)
MIRI	Metal Industries Research Institute (Taiwan)

MIRL	Mechanical Industry Research Laboratory (Taiwan) of ITRI
MIT	Massachusetts Institute of Technology (U.S.)
MMC	Metal matrix composites
MOCIE	Ministry of Commerce, Industry, and Energy (Korea)
MOST	Ministry of Science and Technology (Korea)
MPa	Megapascal
MPI	Max Planck Institutes (Germany)
MR	Magneto-rheological
MRAM	Magnetoresistive random access memory
MS	Master of Science degree
MSC	Micro spark coating
MSTC	Manufacturing Science and Technology Center (Japan)
MT-AMRI	Machine Tool Agile Manufacturing Research Institute (U.S.)
MUMPS	Multi-user MEMS processes
NA	Numerical aperture
NAIST	Nara Institute of Science and Technology (Japan)
NC	Numerical control
NCBI	National Center for Biotechnology Information (U.S.)
NCKU	National Cheng Kung University (Taiwan)
NEDO	New Energy and Industrial Technology Development Organization (Japan)
NEMS	Nanoelectromechanical systems

NIGMS	National Institute of General Medical Sciences (U.S.)
NIH	National Institutes of Health (U.S.)
NIL	Nano-imprint lithography
NIMR	Netherlands Institue of Metals Research (Netherlands)
NIST	National Institute of Standards and Technology (U.S.)
NLBMM	Non-lithography-based meso and microscale
NLI	National Laboratory Institute (Taiwan)
NSC	National Science Council (Taiwan)
NSF	National Science Foundation (U.S.)
NSRC	National Synchrotron Radiation Center (Taiwan)
NT	Nanotechnology
NTRC	Nanotechnology Research Center (Taiwan) at ITRI
NTU	National Taiwan University (Taiwan)
NURB	Non-uniform rational b-spline
OD	Outside diameter
ODM	Original design manufacturer
OEM	Original equipment manufacturer
OLED	Organic thin film transistors
ONR	Office of Naval Research (U.S.)
ORMOCERS	Inorganic/organic hybrid polymer
PBBA	Pentabromobenzyl acrylate
PBT	Polybutylene terephtalate
PC	Polycarbonate

PC	Personal computer
PDA	Personal digital assistant
PDMS	Polydimethylislioxane
PEEK	Polyetheretherketone
PEEM	Photo emission electron microscopy
PEM	Polymer exchange membrane
PFA	Perfluoroalkoxy
PI	Physik Instrumente (Germany)
PIM	Powder injection molding
PLCs	Programmable logic controllers
PMAC	Programmable motion and control
PMMA	Polymethylmethacrylate
POM	Polyoxymethylene
PSD	Position-sensitive detector
PVD	Physical vapor deposition
PZT	Lead zirconate titanate
RC	Rockwell C
RF	Radio frequency
RIE	Reactive ion etching
RISC	Reduced instruction set computing
RIKEN	Institute of Physical and Chemical Research (Japan)
RKEM	Reproducing Kernel Element Method (U.S.)
ROM	Read-only memory
RPL	Rapid Prototyping Laboratory (U.S.) at Stanford University

RPMD	Rapid Micro Product Development (Germany)
SAW	Surface acoustic wave
SCREAM	Single crystal reactive etching and metallization (U.S.) at Cornell University
SEM	Scanning electron microscopy
SiO	Silicon monoxide
SIP	System in packaging
SII	Seiko Instruments, Inc. (Japan)
SLA	Stereolithography
SOA	State of the art
SOI	Silicon on insulator
ST	Space technology
TaSiN	Tantalum silicon nitride
TEM	Tunneling electron microscope
TI	Texas Instruments (U.S.)
TPE	Thermo plastic elastomer
UCLA	University of California, Los Angeles (U. S.)
UG	Unigraphics
UIUC	University of Illinois at Urbana-Champaign (U.S.)
UV	Ultraviolet light
VC	Vanadium carbide
VCSEL	Vertical cavity surface emitting laser
VIS	Visible light
VLSI design	Very large scale integration design
VUV	Vacuum ultraviolet

WEDG	Wire electro-discharge grinding
WLI	White light interferometry
WMD	Weapons of mass destruction
ZEMI	Zentrum für Mikrosystemtechnik (Center for Micro Systems Technology)